PHILOSOPHY OF CHEMISTRY

LOUVAIN PHILOSOPHICAL STUDIES 15

PHILOSOPHY OF CHEMISTRY

BETWEEN THE MANIFEST AND THE SCIENTIFIC IMAGE

J. van Brakel

LEUVEN UNIVERSITY PRESS

Uitgegeven met de steun van de Universitaire Stichting van België en K.U. Leuven Commissie voor Publicaties.

First edition published: 2000
Reprint: 2013

© 2013 Leuven University Press / Presses Universitaires de Louvain / Universitaire Pers Leuven, Minderbroedersstraat 4, B-3000 Leuven/Louvain (Belgium)

All rights reserved. Except in those cases expressly determined by law, no part of this publication may be multiplied, saved in an automated data file or made public in any way whatsoever without the express prior written consent of the publishers.

ISBN 978 90 5867 063 2
D/2000/1869/58

Cover: Friedemann BVBA

preface

The chapters of this book cover different parts of the newly emerging discipline of philosophy of chemistry. The general philosophical issue addressed in all chapters is the relation between the manifest and scientific image, both of which claim to be the final arbiter of 'everything'. The terminology of manifest versus scientific imagery stems from Sellars. The former is the daily practice or common-sense-human-life-form, which concerns things like water, milk-lapping-cats and injustice-angry people, as well as 'people-in-the-world'. The scientific image is concerned with things like neurons, DNA, quarks, black holes, and the Schrödinger equation.

With respect to chemistry the question can be raised: where does it fit in - with the manifest or scientific image? Sellars and others probably would reply unhesitatingly: the scientific image. My intention is to raise doubts about that self-evidence. I argue that chemistry is *primarily* the science of *manifest* substances, whereas 'micro' or 'submicro' scientific talk, though important, useful, and insightful does not change what matters, namely the properties of *manifest* substances. These manifest substances, their properties and uses cannot be *reduced* to talk of molecules or solutions of the Schrödinger equation. The common assumption that chemistry can be reduced to physics is incorrect. My analysis of the notion of chemical substance supports the view that *if* the question of priority is raised, it is the manifest, macroscopic image that is prior to the scientific, microreductionistic image, not only in a methodological sense, but also in an epistemological and ontological sense. Similarly, within the scientific image it is not the case that chemical (or physical) microdescription will, in principle, always give a more complete and true description. If 'submicroscopic' quantum mechanics were to be wrong, it would not affect all (or any) 'microlevel' chemical knowledge of atoms and molecules (bonding, valence, structure, and so on). If molecular chemistry were to be wrong, it wouldn't disqualify all (or any) knowledge of, say, water - not at the 'macrolevel' (e.g. its' viscosity at 50 °C), nor at the pre- or proto-scientific manifest level (e.g. ice is frozen water).

The importance of fundamental (physical) theories in furthering understanding and prediction in chemistry and elsewhere is not denied; nor is there any rejection of the many important intertheoretic correlations between 'macro' and 'micro', and between 'macro' and 'manifest'. The question is whether the existence of these relations justifies the use of slogans like 'chemistry can be reduced to physics' - understood in the *in principle* sense that chemical concepts, laws, and explanations at the macrolevel are, notwithstanding their *practical* significance, strictly speaking, superfluous or epiphenomenal.

As there are relatively few publications on the philosophy of chemistry, I begin with a historiography of its initial neglect and recent emergence, situating its birth in 1994. Particular attention is given to Kant's influential view that chemistry is not a proper science and to the interest in the philosophy of chemistry in Eastern Europe in the period 1949-1984. I show that the neglect of the philosophy of chemistry can be ascribed to a number of related factors. Since Aristotle, the concept of *hyle* or 'stuff' has been suppressed in philosophy, in favour of Euclidean geometry and Newtonian mechanics as a universal model for scientific knowledge. The view of science as something that is mathematised, preferably axiomatised, and concerned with strict laws has dominated philosophy of science to the present. A related factor has been a bias towards the ontology of matter (what physics has to say about it), as distinct from the ontology of substances, i.e. different *kinds* of matter (and what chemistry might say about *that*). As a consequence of these biases views such as those of Ostwald and Duhem, who stressed the priority of thermodynamics and the basicness of notions such as pure substance (independent of atomic models), were marginalised. Traditional philosophy of science is interested in the foundations, structure, and change of mechanistic theories about microscopic and submicroscopic 'reality' (where theoretical physics merges increasingly with mathematics). It shows little interest in non-mechanistic theories concerned with macroscopic phenomena or with the capacity to make new things; nor, until recently, paid it much attention to how the experimental side of making things (substances, instruments) is interwoven with the other faces of natural science.

Chapter 2 introduces a number of general philosophical issues that figure in the background of the philosophy of chemistry. First I elaborate the tension between the manifest and scientific image and argue *against* the a priori arguments Sellars gives to support his view that the scientific image can claim priority over the manifest image. I then make the terminology manifest – macro – micro - submicro, already used in the paragraphs above, more precise, taking interfaces between solids, liquids, and gases as an example. As background to the discussion of the alleged reduction of chemistry to physics (and more generally the reduction of 'macro' to 'micro'), I review interdiscourse relations such as reduction, supervenience, and emergence. Interdiscourse relations are epistemic and/or ontological relations between different theories or disciplines. In an older, reduction-biased terminology, they are called bridge laws connecting different theoretical levels. To prepare the discussion of chemical substances and the question whether they can be said to have underlying 'essential' properties that determine asymmetrically their manifest properties, I review philosophical theories about natural kinds. I argue that we are confronted with a dilemma. Either we must favour a *plurality* of natural kinds, embedded in a variety of manifest, macro-, and micro-discourses. Or we must choose to eliminate them completely, at least all those kinds ordinary people might know of.

In chapter 3 I argue that the (chemical) notion of pure substance is methodologically, epistemologically, and ontologically *independent* of molecular chemistry and quantum mechanics and *dependent* on the manifest intuition of pure substance, which finds its scientifically more precise counterpart in the thermodynamics of substances. I start with moving the manifest/scientific debate to a concrete illus-

PREFACE

tration: manifest and scientific water. Then I discuss various approaches to elucidating or defining (chemical) pure substances and compounds. Although it is often assumed that a pure substance can be defined in terms of (atomic) composition and (molecular) structure, strictly macroscopic definitions are also possible. For example, a pure substance is one of which the macroproperties, such as temperature, density and electric conductivity, do not change during a phase conversion (as in boiling a liquid or melting a solid). On such a view chemistry is not primarily the science of molecules, but of substances. In the final section of this chapter I discuss the notorious polywater episode in physical chemistry. I show that in reaching the decision that polywater does not exist (in the intended sense), arguments were drawn from the manifest, macro-, micro-, and submicro-levels, without any one playing a more crucial role than another.

In chapter 4 I consider a form of essentialistic realism that became influential in the 1970s as one alternative to block the threat of incommensurability of successive scientific theories. Water is again the main example: Can we say that the *essence* of water is H_2O (assuming that 'water is H_2O' is true)? This is very much a 'philosopher's discussion' that proceeds with little knowledge of chemistry. I lay out the pitfalls of philosophical speculation about water not being H_2O on the planet Twin Earth. More substantially I argue that what is essential to being water is that it is the manifest or macroscopic natural kind or substance *water*. Its essence is *not* that it is H_2O and talk about essences is better dropped. However, *if* one insists on talking about essences, it turns out that *all* properties of natural kinds, including *all* its relations to other natural kinds, are, ceteris paribus, essential properties. This is so, first, because not only manifest or macroscopic, but microproperties too are dependent on the context. Second, it is not possible to distinguish between what is a counterfactual situation (a different possible world) and what another context. The appeal of the essentialist to possible world talk to flesh out the relation of the essence of natural kinds to the necessity of natural laws therefore fails.

In chapter 5 I start with a historical digression on the source of the idea that chemistry can be reduced to physics. Then I argue that macrochemistry (such as the thermodynamic theory of substances and compounds promoted in chapter 3) cannot be reduced to (microscopic) statistical mechanics and molecular chemistry cannot be reduced to quantum mechanics (via quantum chemistry). I show that the common assumption among physicists, philosophers of science, and many chemists that chemistry can be reduced in principle to physics derives from an extremely influential statement of Dirac. In contrast I show that all detailed investigations of the relation between chemical and physical theories show that it doesn't fit any standard model of reduction available in philosophy of science.

In chapters 6 and 7 on the ceteris paribus character of laws and models in chemistry and chemical engineering, I argue that *all* laws are ceteris paribus and that the relation between a model and what it allegedly represents is symmetrical: fitting the model is a matter of mutual attunement of both model and reality. Any law or theory abstracts from 'ordinary' contexts; hence it cannot be strictly *true* for real (actual) systems. The theory only describes the behaviour of systems in idealised contexts or model situations; strictly speaking it only applies to thought experiments. What applies to theories and laws in general applies to interdiscourse relations as well: they are ceteris paribus - not pieces of metaphysical glue. Moreover, the further down we go from manifest to macro to micro to submicro, the more idealisations are introduced, and the fewer the number of systems in the world that are, strictly speaking, described by the theoretical models. I elaborate on this issue via two case studies taken from applied science. The anomaly of two 'types' of capillary wetting of porous media illustrates a problem where the manifest and the macroscopic are intertwined, in a way quite analogous to the polywater case (both case studies stem from physical chemistry). Chapter 7 on the use of dimensionless numbers in chemical engineering, though embedded in the general theme of the status of models and interdiscourse relations, has in addition the purpose of drawing attention to lesser known themes in the philosophy of chemistry. After all, just as philosophy of science prefers theoretical physics over chemistry because it is more mathematical, and hence 'cleaner', philosophy of chemistry has a tendency to go for the more theoretical parts of chemistry and to neglect the messier aspects that deal with the large-scale production of chemicals. The case study in chemical engineering also illustrates that if one looks at the general features of the relation between theory and its application through models, there isn't much difference between the methodology of tailoring quantum mechanics to molecules in quantum chemistry and tailoring 'fundamental' phenomenological equations for heat and mass transfer through the model of dimensional analysis to design problems in chemical engineering.

If there are no strict interdiscourse relations (guaranteeing reduction or other metaphysical glue), then how do physics and chemistry fit into one world? In the concluding chapter I propose as a 'metaphysical model' the stance of anomalous monism. This is a theme discussed at great length in the philosophy of mind, but has not been applied to an 'easy' case of interdiscourse relations, viz. the relations between chemistry and physics. According to anomalous monism events cause one another not only independently of how they are described, but independently of how they are identified. It is not the case that there are first physical events and chemical events and then some physical events turn out to be identical with some chemical events. All events are *events*, no matter their description in chemical, mental, physical, moral or whatever terms. By assuming that each event that can be given a chemical description also has a physical description (although we can neither specify it, nor provide exact criteria of identification), some insight is gained in the autonomy of the chemical *and* the physical, while still keeping both in the same world. I conclude the chapter with a summary of a range of arguments against the priority of the scientific image that transcend specific issues raised in previous chapters.

A considerable part of this book consists of reworkings of earlier publications. This is indicated in notes at the appropriate places. Permission of publishers, where applicable, is gratefully acknowledged. In all cases the text has been extensively revised and expanded.

Some subsections may be inaccessible or less relevant to some readers, as I cover the whole range from 'pure' philosophy to (almost) 'pure' science. Some themes may be too philosophical for those readers primarily interested in chemistry (such as the Twin Earth story or the notion of supervenience). Some themes may be too technical for philosophers or even for chemists (such as the case studies on anomalous wetting behaviour and the use of dimensional analysis in chemical engineering). Although all are part of a cluster of arguments concerning the tension between the manifest and scientific image, each chapter stands quite independently. Cross-references between sections allow the reader to start almost anywhere in the book. Number codes between square brackets refer to sections (for example, '[6.1]' means 'see/compare section 6.1'). References such as 'Block' (followed by a number code) are used to direct the reader to text blocks, which function as 'long notes'. In citations that serve as illustrations, in most cases, I have left out ellipsis dots and adjusted punctuation for easy reading.

This is not a book on the foundations of chemistry. There will be many mistakes (from which I hope to learn). Its purpose is to open up a number of issues for discussion in the philosophy of chemistry.

contents

preface		v
contents		xi
list of text blocks		xiii
1 emergence of philosophy of chemistry		1
1.1	physical and chemical matter: some preliminaries	1
1.2	the heritage of Kant	7
1.3	from Hegel to Bachelard	13
1.4	'classical' philosophy of science and chemistry	17
1.5	philosophy of chemistry in Eastern Europe	22
1.6	the Chelintsev affair	27
1.7	philosophy of science opens up	34
1.8	birth of philosophy of chemistry proper	37
2 philosophical preliminaries		41
2.1	manifest and scientific image	41
2.2	surfaces	47
2.3	reduction	50
2.4	supervenience and emergence	53
2.5	natural kinds	58
2.6	causal theory of reference	66
3 chemical substances		71
3.1	the science of stuffs	71
3.2	manifest and scientific water	73
3.3	molecule structure and microreductionistic essences	78
3.4	a macroscopic definition of pure substance	82
3.5	polywater	87
4 essentialistic realism		99
4.1	Kripke and Putnam on water	99
4.2	molecules and atoms	106
4.3	possible worlds	110
4.4	proliferation of essences	116

5 the alleged reduction of chemistry — 119

5.1 the idea that chemistry can be reduced to physics — 119
5.2 thermodynamics and statistical mechanics — 123
5.3 quantum chemistry — 128
5.4 'ab initio' methods — 133
5.5 holism of quantum mechanics and shape of molecules — 142

6 ceteris paribus — 151

6.1 chemical laws and models — 151
6.2 ceteris paribus laws — 155
6.3 protochemistry — 160
6.4 the ubiquity of models and ceteris paribus conditions — 163

7 modelling in chemical engineering — 171

7.1 similarity considerations and dimensionless numbers — 171
7.2 dimensional analysis — 175
7.3 dimensional analysis in chemical engineering — 177
7.4 presuppositions of dimensional analysis — 180
7.5 the model of dimensional analysis — 186

8 conclusion — 191

8.1 how to fit it all together — 191
8.2 primacy of manifest over scientific image — 198

references — 203

index of names — 235

index of substances — 241

index of subjects — 243

list of text blocks

Block 1-1	The part of the world that physics leaves out	3
Block 1-2	Atoms and molecules	5
Block 1-3	Kant's growing absorption of the 'new' chemistry I	10
Block 1-4	Kant's growing absorption of the 'new' chemistry II	12
Block 1-5	Cassirer on the concept of 'atom'	15
Block 1-6	Some pre-1960 publications on philosophy of chemistry	19
Block 1-7	Chemical kinetics and chemical reactions	21
Block 1-8	DDR-journals publishing on philosophy of chemistry	23
Block 1-9	Isomer, tautomer, mesomer, resonance and aromaticity	28
Block 1-10	Views of western chemists on resonance structures	30
Block 1-11	Case studies involving chemistry	36
Block 1-12	Philosophy of chemistry at international conferences	38
Block 1-13	Post-1994 areas of interest in the philosophy of chemistry	39
Block 2-1	Manifest form(s) of life and congeners	43
Block 2-2	Tension between the manifest and scientific image	44
Block 2-3	Primacy of the manifest image or form(s) of life	47
Block 2-4	A few examples of surfaces or interfaces	48
Block 2-5	Examples of proposed reductions	51
Block 2-6	Variety of supervenience definitions	54
Block 2-7	Some examples of natural kinds	59
Block 2-8	A few proposed definitions of natural kinds	62
Block 2-9	Avoiding the issue of natural kinds	65
Block 2-10	Stereotype and essence of green and *lhenxa*	66
Block 2-11	Causal theory of reference cum essentialistic realism	67
Block 3-1	What is water?	76
Block 3-2	Eighteenth century dictionary definitions of water	77
Block 3-3	Preliminary characterisation of (chemical) substances	78
Block 3-4	Locke on the nominal and real essence of substances	80
Block 3-5	The 'given' which had to be interpreted	91
Block 3-6	Final word of Derjaguin	94
Block 3-7	First International Conference on Polywater	97
Block 4-1	Properties of water, heavy water and 'polywater'	102
Block 4-2	Alternative logical 'translations' of water is H_2O	105

Block 5-1	The spread of Dirac's views	120
Block 5-2	Unit operations in chemical engineering	124
Block 5-3	Churchland's elimination of temperature	125
Block 5-4	Relation thermodynamics and statistical mechanics	128
Block 5-5	Kopenhagen interpretation of quantum mechanics	129
Block 5-6	Feynman about quantum chemistry	130
Block 5-7	Chemical notions not derivable from quantum mechanics	132
Block 5-8	Text book authors on the choice of basis sets	135
Block 5-9	Views on orbitals and hybridisation	137
Block 5-10	Relation of quantum mechanics and chemistry	141
Block 5-11	EPR-correlations and reification of objects	145
Block 5-12	Different notions of (molecular) structure	146
Block 5-13	Bader's 'derivation' of atoms and functional groups	149
Block 6-1	Use of the term 'model' in chemistry	154
Block 6-2	Model types	156
Block 6-3	Ceteris paribus character of theoretical laws	158
Block 6-4	Protoscience	160
Block 6-5	Myth of the Given	162
Block 6-6	Capillary liquid transport in porous media	165
Block 6-7	List of variables and symbols used in chapters 6 and 7	167
Block 6-8	Shape factors	168
Block 7-1	Similarity criteria	172
Block 7-2	Dimensionless numbers	174
Block 7-3	Characteristic length parameters	180
Block 7-4	Number of fundamental or primary quantities or units	183
Block 7-5	Ceteris paribus conditions flow in a cylindrical pipe	184
Block 7-6	Ceteris paribus conditions Hagen-Poiseuille equation	185
Block 7-7	Variety of dimensionless numbers	186
Block 7-8	Model of models	187
Block 7-9	Problems with models	189
Block 8-1	Disorder of things	192
Block 8-2	Epistemic or pragmatic virtues	200

1 emergence of philosophy of chemistry

1.1 physical and chemical matter: some preliminaries[1]

In the history of western metaphysics, there is an extensive literature on concepts of substance and matter dominated by publications on Aristotle and Locke.[2] But many other philosophers, physicists and chemists, figure too: Averroës, Boyle, Descartes, Faraday, Gassendi, Leibniz, Mach, Newton, Ostwald, Priestley, Spinoza. However it is rare for these discussions to be related directly to chemistry. Typically, Düring's (1944) *Aristotle's Chemical Treatise: Meteorologica, Book IV* is a curiosity, in that it uses the word 'chemical' in the title. So too is Lewis (1996), whose Ph.D. thesis on Aristotle's *Meteorologica* is titled *Body, Matter and Mixture: The Metaphysical Foundations of Ancient Chemistry*.[3] Perhaps one could distinguish two lines of development in the interest in substance and matter in the western tradition: one a more philosophical, metaphysical line, with the key words 'substance' and 'essence' running from Thales to Aristotle via Locke to Kripke and Putnam and 'side-tracks' such as Mach and Ostwald. Second, a more scientific line concerned with the ontology of kinds of matter, with key words like 'element', 'compound/mixture', 'valence', and explanations of the properties of substances in terms of their composition and quantum chemistry. This line could also start with Thales and Aristotle and then move via Averroës, Gassendi and Boyle to Lavoisier, Dalton, Kekulé, Lewis, London and Pauling.

A slightly different way of putting this is to suggest that there are two separate issues: first, the ontology of *matter in general*, to be dealt with in relation to micro- and astrophysics; second, the ontology of *particular kinds of matter*, i.e. chemical kinds (van Brakel 1991c). Before 1900 the distinction was common. Lavoisier's (1789) view was that chemistry is the quantitative science of the macroproperties of substances and their transformations. He supported an atomic view of matter as an empirical hypothesis,[4] not as a metaphysical a priori and he did not use atoms and molecules in the explanations of observed chemical phe-

---

[1] Chapter 1 covers the same themes as van Brakel (1999a) though in much greater detail and drawing out more conclusions.
[2] On Aristotle's notion of substance see Witt (1989a); on Locke see Ayers (1991) and (Block 3-4).
[3] For Aristotle's notions of element and substance from a philosophy of chemistry perspective see G. Böhme (1980), Schummer (1996a:98-122 and *passim*). Metaphysicians since Aristotle define a substance as an object that does not depend for its existence upon any other object (Lowe 1998:10).
[4] According to Duhem, Lavoisier and his school led chemists to adopt the idea of a simple body as 'one that we have not been able to decompose, a body which has resisted every means of analysis in laboratories' (quoted from Duhem 1974:128; originally in Duhem 1902). Lavoisier himself said: 'Nous ne pouvons donc pas assurer que ce que nous regardons comme simple aujourd'hui le soit en effet: tout ce que nous pouvons dire, c'est que telle substance est le terme actuel auquel arrive l'analyse chimique, & qu'elle ne peut plus se subdiviser au-delà dans l'état actuel de nos connaissances' (Lavoisier 1789:I,194). *Cf.* Kant's anti-atomism (Malzkorn 1998): 'aber wir haben keine Materie, die sich mit Gewissheit nicht theilen läst' (AA28:664 – cf. note 21).

nomena (Abbri 1996). Lavoisier's view is reflected by Kekulé in his textbook of 1859.[5] The chemist Williamson wrote in the *Chemical Gazette* of 1851:[6]

> We are all agreed that chemistry is concerned with the material process of the transformations and changes which matter undergoes, and that the study of the properties of matter in themselves, as long as they undergo no change, belongs to physics.

Similarly Mendeleev aimed both at the unification of the many natural sciences and the autonomy of chemistry: atoms are the ultimate constituents of nature; chemistry studies their properties *qua* difference from each other, and physics their properties *qua* similarity (Block 1-1):[7]

> Chemistry, I am convinced, must occupy a place among the natural sciences side by side with mechanics; for mechanics treats of matter as a system of ponderable points having scarcely any individuality and only standing in a certain state of mobile equilibrium. For chemistry, matter is an entire world of life, with an infinite variety of individuality both in the elements and in their combinations.

The philosopher-scientist Peirce made the distinction between matter and particular kinds of matter in 1861 (Tursman 1989), as did the chemist-philosopher Ostwald (1907:5) and, following him, Cassirer (1923:204) in his philosophy of the exact natural sciences.

Discussions in both the history of philosophy and the philosophy of science have been concerned mainly with the first issue, the ontology of matter in general. For example, Dieks and de Regt (1998:46) say:

> It is part and parcel of the modern natural sciences to assume, at least as a working hypothesis that the description of matter given by fundamental physics is in principle complete.

This statement may sound plausible, but it is not obvious what 'matter' refers to; nor is it evident that 'matter' is a concept in theoretical physics. Presumably, matter has something to do with what physicists call 'elementary particles'. Even if we accept the identification of matter with the weird entities high-energy physics studies, there remains the vagueness of 'in principle complete'. Why wouldn't there be ever more 'particles' when we dig deeper and deeper, so that description is never complete? And would a complete description of 'matter' entail a complete description of all *types* of matter, such as water, jade, and silicic acid sol?

Related to the concept of matter is that of atomism (Block 1-2). Though discussed in numerous publications, Paneth (1962:1) correctly noted:

> The physical aspect (divisibility, mutual attraction and repulsion, and so on) has been discussed much more than the chemical (qualitative characteristics, valency, etc.).

---

[5] Kekulé (1859:3): 'Die Chemie ist also die Lehre von den stofflichen Metamorphosen der Materie. Ihr wesentlicher Gegenstand ist nicht die existierende Substanz, sondern vielmehr ihre Vergangenheit und ihre Zukunft.'
[6] Quoted in Benfey (1963).
[7] Mendeleev (1869), quoted in Kultgen (1958:177). See also Kedrov (1969) and Mendeleev (1889).

Block 1-1   The part of the world that physics leaves out.[9]

> MENDELEEV
> Kant said that there are in the world 'two things which never cease to call for the admiration and reverence of man: the moral law within ourselves, and the stellar sky above us.' But when we turn our thoughts towards the nature of the elements and the periodic law, we must add a third subject, namely, 'the nature of the elementary individuals which we discover everywhere around us.'
>
> PAULING
> Physicists in general tend to restrict themselves to the small part of the physical world with which they deal, and to leave out of their studies all such features as the structure and properties of substances in relation to their chemical composition, and the reactions that change one substance into another. Chemists, however – and biologists also – are interested in different kinds of matter, not only in the form of substances but also in their more complicated aspects, the objects that are built out of substances.
>
> POLANYI
> Chemistry answers questions regarding the interaction of more or less stable substances, and these questions cannot be answered without the experience of these substances and of the practical conditions in which they are to be handled. While quantum mechanics can explain in principle all chemical reactions it cannot replace, even in principle, our knowledge of chemistry. We may acknowledge this as the incipient separation of two forms of existence.

Presumably, the common identification of chemical and physical atomism is based on the assumption that the final ontology of everything (including chemistry) is whatever physics says it is. According to the received view, Boyle and Dalton tailored atomism to the needs of chemistry. However, contrary to the received view, perhaps the notion of chemical compound (pure substance) should *not* be seen as part of a continuous line of development from antique atomism via Boyle, but as emerging from the practice of sixteenth and seventeenth-century metallurgists and pharmacists.[8]

Although Boyle has been described as 'the father of chemistry',[10] it is wrong to stress the connection between Boyle's chemistry and his corpuscular philosophy. That the seventeenth century 'corpuscular philosophy' was inconsistent with the development of the *chemical* notions of element and compound was already emphasised by Kuhn (1952). Boyle's theory doesn't include the notion of *chemical* compound. It only covers compounds of corpuscles, without taking into account the relation to chemical synthesis and analysis, the combination and separation of substances, and concepts like conservation, reversibility, and homogeneity (and the empirical regularities concerning these features). A corpuscular theory only

---

[8] This paragraph and the next draw heavily on Klein (1994a, 1994b). See Siegfried and Dobbs (1968) for the influence of Swedish mineralogists on these developments.
[9] Polanyi (1958:394), Mendeleev (1889:168), Pauling (1950).
[10] Cf. the title of Pilkington (1959). Alexander (1985) rather incorrectly ascribes to Boyle the attempt 'to develop a genuinely scientific chemistry'. Boyle encouraged the wrong interpretation because he himself advertised his work as mediating between 'both chymists and Corpuscularians' and showing that 'many chymical experiments may be happily explicated by corpuscularian notions' (quoted in Klein 1994b:63). Whitaker (1996:63-66) more correctly describes Dalton as the 'founder of (chemical) atomic theory'.

conserves the number of particles, whereas a chemical theory conserves, in some way, substance-specificity and the related notion of pure substance. There is therefore reason to distinguish physical and chemical atomism. Dalton can perhaps be seen as the modern originator of chemical atomism, though not Boyle. Dalton can be taken as following the Democritean tradition, but his atoms are of distinct types, whose instances differ in size, weight, and mutual attraction (affinity).

Perhaps, the first occasion when the concept of chemical compound was presented to a scientific audience was in 1718 when Geoffroy presented his *Table des différent rapports* to the Royal Academy of Science in Paris. According to Geoffroy there is no difference between the synthesis of chemical artifacts based on chemical affinity and the resynthesis of natural bodies - a difference that had governed both theory and practice until then. The concepts Geoffroy was using and developing could no longer be explained by a theory of natural philosophy, but emerged from the practice of chemical operations and experiments. Geoffroy's *Table* was the prototype for a whole range of tables of affinity that were developed on the model of Newton's gravitational theory. But the original *Table* of Geoffroy was not modelled on Newton's theory, but on a systematisation of the practical knowledge of metallurgists and pharmacists.

The nineteenth century debate about the existence of atoms is well known. Mach (1883:589) famously said:

> The atomic theory plays a part in physics similar to that of certain auxiliary concepts in mathematics; it is a mathematical *model* for facilitating the mental reproduction of facts.

Until the 1870s Kekulé rejected the idea that chemical formulae represent 'real' molecules. Until 1886, the atomic theory was not taught in French secondary schools. Boltzmann describes his molecular gas theory as nothing but a conventional analogy. Poincaré said 'we' accept the atomic theory because 'we' are familiar with the game of billiards.

The last well known chemist to give up his resistance to 19th century atomism was Ostwald. As late as 1901 Ostwald argued vehemently in his lectures against the 'dead track' of molecular structure (Ostwald 1902). In his Faraday Lecture of 1904 he said:

> Chemical dynamics [i.e. Ostwald's approach], has, therefore, made the atomic hypothesis unnecessary for this purpose and has put the theory of stoichiometrical laws on more secure ground than that furnished by a mere hypothesis.

Instead of atomic mass Ostwald uses 'relative combining weight', e.g. by saying: 'it is possible to ascribe to each element a certain relative combining weight in such a way that every combination between the elements can be expressed by these weights or their multiples'. Gay-Lussac's gas law too is expressed without appeal to the hypothesis of Avogadro (Ostwald 1907). In the preface to the latter work he contemplates:

> Three years ago, on the occasion of the Faraday lecture, I made an attempt to arouse the interest of chemists in these matters, but the result

Block 1-2   Atoms and molecules.[11]

> Across centuries 'atomism' can, for example, mean: atoms are the four elements water, fire, earth, and air, or the three principles Mercury, Sulphur, and Salt, or Lavoisier's 33 simple substances, or atoms, or molecules, or 'electrons, protons, neutrons, positrons, mesons and neutrinos', or quarks, or the referents of a hidden variable interpretation of quantum mechanics. Hence 'atom' is a regulative, formal concept with permanent characteristics, though its *content* may completely change (Block 1-5).
> The notion of molecule came much later. Pre-Avogadro chemists tended to use the terms 'atom' and 'molecule' interchangeably. The word 'molecule' was introduced as a neologism by Gassendi in 1637. The idea of a molecule, as a category distinct from 'atom', 'body', and 'substance' was perhaps first formulated by Beeckmans in 1620. For Beeckmans *homogenea physica* were substance-characteristic particles ('substantial individuals') consisting of a number of atoms of different form and size, organised in a structure - a concept already 'predicting' isomerism.
> Proust and Richter tried to establish empirically the distinction between compounds and mixtures (law of definite proportions), but it was only with the acceptance of Dalton's atomic theory that a theoretical distinction between compounds and mixtures could be drawn (law of multiple proportions). According to Dalton there exist 'simple elementary particles' (atoms), which may combine into 'compound particles' (molecules).

> was not very encouraging. I know from personal experience that patient and continued labour can accomplish wonders even when the case seems hopeless. One must wait for the right time, and I am convinced that the time for this matter has arrived.

But by the end of the book (ibid:326), when addressing metamerism, polymerism, and the role of valence in structure theory,[12] he acknowledges that when compared with 'structure theory' his 'purely empirical' account in terms of differences in energy content

> predicts, however, nothing whatever about the chemical reactions which are to be expected, and is therefore not applicable as an aid to building up a system. Energy in the sense in which the word is used here is expressible by a mere number; it has no further properties, and is therefore not of value in expressing the qualitative differences belonging to chemical reactions.

He further acknowledges 'the introduction of a new factor, one involving differences in the spatial arrangement of elements, and in a few cases even this assumption appears insufficient' - though its status for Ostwald doesn't rise above that of 'a very important aid' (ibid:329). When the results of the experiments of Perrin on Brownian motion became available, Ostwald in 1908 finally surrendered:[13]

---

[11] See on atomism for example Pyle (1997), Pullman (1998). For an early account of *chemical* atomism see Ritchie (1945). On the early history of the notion of molecule Kubbinga (1984, 1988).
[12] The valence of elements is the expression of a definite property, that belongs to them independently of their chemical affinity.
[13] W. Ostwald, *Grundriss der physikalischen Chemie* (1908). Quoted in the translation of Nye (1972:151), where more details on Ostwald's 'conversion' can be found. According to Ruthenberg

> I have satisfied myself that we arrived a short time ago at the possession of experimental proof for the discrete or particulate nature of matter - proof which the atomic hypothesis has vainly sought for a hundred years, even a thousand years. Even the cautious scientists [is entitled] to speak of an experimental proof for the atomistic constitution of space-filled matter.

It has been suggested that Ostwald was forced to accept the atomic hypothesis because the issue of isomeric substances forced it upon him. Isomeric substances can be defined as substances having the same composition but different energy content,[14] but as this says nothing about which chemical reactions to expect, Ostwald was allegedly forced to admit to a kind of 'grainedness' in the physico-chemical world, on pain of having 'to admit alternative structures of *nothing in particular*!'.[15] Though this may sound utterly plausible, the appeal to the self-evident nonsensicality of 'structures of *nothing in particular*' breaks down under scrutiny, as later developments such as debates about the reality of resonance structures or orbitals show.[16]

In his anti-atomism, philosopher-chemist Ostwald aimed, first, to distinguish between what is given in experience and what is postulated by the mind: nothing *compels* us to affirm that mercury oxide 'contains' mercury and oxygen - a form of empiricism akin to that of Mach. Second, he aimed to show that energy is the most general concept of the physical sciences. The latter implies that thermodynamics is the most basic physical science. For different reasons, Duhem, like Ostwald, claimed that thermodynamics was the most basic science. In a series of articles, Needham has shown that Duhem developed what he regarded as an essentially Aristotelian view of chemistry, as an alternative to the corpuscular view.[17] He based his understanding of phenomenological thermodynamics on the Aristotelian idea of 'two contradictory opinions of the nature of a mixt' (Duhem 1902:15). It is against this background that Duhem's attack on the indefiniteness of the atomic hypotheses has to be judged; in his view these hypotheses are otiose, appealing to *a priori* considerations. For example he did not accept that van 't Hoff's stereochemical representations should be understood as a fully fledged geometric picture of something real. Molecular structures do not have to be understood in terms of atoms, but can equally well be understood in terms of the chemical properties of the compounds.[18]

Developments since the late 19th century will be discussed more fully later on. Here are some pointers. The appearance of isotopes meant that chemical elements, as ordered in the Periodic Table, were not considered simple anymore (as Dalton assumed). Rather, they came to have a complex internal structure. One

---

and Psarros (1994) the Czech chemist Wald resisted the atomic theory until the end of his life (in 1930).
[14] Cf. Block 1-9.
[15] Bradley (1955), italics and exclamation mark in original.
[16] See [1.6] and Block 5-9.
[17] Needham (1996a, 1996b, 1996c).
[18] Needham (1998:52-9). Nevertheless, according to Needham, it would be incorrect to describe Duhem as an anti-realist or instrumentalist.

element of their taxonomy covers a number of isotopes, each having different macroproperties (such as density or boiling point). Furthermore, atoms can disintegrate and fuse and the idea of a chemical atom as an indivisible unit of a type of substance had to be given up. With the disappearance of the German energeticists led by Ostwald and the discovery of radioactivity (and in particular the experiments of Rutherford around 1900), the atom changed hands from the (German) chemists to the (English) physicists. One of the main actors[19] was Maxwell. Typically, neither in his address on molecules to the British Association for the Advancement of Science in 1873, nor in the article on Atoms in the 9th edition of the *Encyclopaedia Britannica*, was Dalton mentioned (Gavroglu and Simoes 1994, Gavroglu 1996).

In the past century there has been a shift from defining substances in terms only of composition (of chemical elements), to the situation where the structure of molecules was considered a more fundamental property. At the same time, in a development stretching from Kekulé to Pauling, the idea of structure as a static architectonics of atoms, and of bonds as the defining structure, was slowly undermined. In the 'resonance structures' of a benzene molecule, the 'real' microsituation is a kind of mixture of a number of (logically) possible fixed arrangements of nuclei (of atoms) and electrons. The subsequent development of quantum mechanical accounts of the organisation of electrons in a molecule has led to a situation in which it is not too far-fetched to suggest that with the advance of quantum mechanics, atomism finally dug its own grave [5.5]. Hence, in a way, there is now support from modern physics for the views of Mach and Ostwald who, at the beginning of the century, opposed the reality of atoms as material objects.[20]

1.2 the heritage of Kant[21]

Kant's relation to chemistry is ambiguous and contested. Did he make detailed contributions to the chemistry of his time (Carrier 1990) or was he a dilettante (Heinig 1975)? Did he revise his transcendental system under the influence of the work of Lavoisier or is his view that chemistry is not a proper science, his final judgement? The latter is the widely known view which has had a large impact on both scientists and philosophers. In later life Kant may have changed his views,[22] but this is scarcely known: research on Kant's *Transition* or *Übergang* - his thinking after he had written the three Critiques - only came off the ground in the last decade.

---

[19] In the sense of Latour (1987, 1999), for whom also non-human actants are actors.
[20] Cf. Donnan (1933:12) in his Ostwald Memorial Lecture delivered before *The Chemical Society* (London) on January 27th, 1933: 'Although [Ostwald's] conceptual scheme was too simple, it must be admitted that in certain respects his ideas were nearer to the theories of modern physics than those of his contemporaries.' In the terminology of quantum chemistry, one might say that the quantum mechanical Hamiltonian of a 'molecule' is a structural description in terms of energy.
[21] All references to Kant are to the *Akademie Ausgabe* (*Kant's gesammelte Schriften*), published between 1900 and 2000, referred to as 'AA', followed by volume and page number. Page numbers of the *Critique of Pure Reason* are given in the well-known A/B format. English translations of texts from the *Opus Postumum* are from Kant (1993).
[22] Friedman (1992:264-90), Vasconi (1996), Dussort (1956).

Though Kant had made brief comments about science in his *Kritik der reinen Vernunft* and in the *Prolegomena*, his most worked out views can be found in the preface of his *Metaphysische Anfangsgründe der Naturwissenschaft* (1786).[23] The title already indicates that science is only possible because of certain metaphysical foundations. The use of mathematics introduces the pure part in science and at the highest level of abstraction there is the metaphysical *a priori*. A natural science is 'proper' to the extent that mathematics is applied within it. Sciences like chemistry and psychology are rational (because they use logical reasoning), though not *proper* sciences, because they miss the basis of the synthetic *a priori*. For Kant the difference between science that is, and is not proper, is in modern terms roughly the difference between pure and applied science, where for Kant 'purity' is guaranteed by metaphysics and mathematics. The distinction runs parallel to that between primary and secondary qualities and that between hard and soft (or 'special') sciences.[24]

Though in 1786 Kant considered Stahl (founder of the phlogiston theory), Galilei, and Toricelli paradigmatic practitioners of the scientific method, he is here thinking of empirical science, not of 'proper' science. Chemistry as a *practical* science is more like moral science than like physics, as when Kant says that 'a procedure which resembles chemistry', is a procedure needed in the analysis of moral common sense - what Körner (1991) has referred to as 'the quasi-chemical method'.[25] Hence, an empirical science such as chemistry uses a rational method of inquiry, but it is not a *proper* science; it is an *uneigentliche Wissenschaft*.[26] In the crucial passage Kant not only says that chemistry and psychology do not count as proper sciences because they use no mathematics, but that this requirement would be difficult *ever* to fulfil (Block 1-4).

There are some ambiguities if one compares this with older texts of Kant, in particular those referring to psychology. In the nineteenth century this led to extensive debates among interpreters. Some scholars even went so far as to say that the often-quoted passage is an oversight, inconsistent with the rest of Kant's writings (Drews 1894:259). It was also suggested that psychology might score better as a 'proper science' than chemistry, because Kant works with the distinction of 'bodily nature' (*res extensa*) and 'thinking nature' (*res cogitans*). In his lectures on metaphysics Kant described psychology as 'metaphysical empirical science of people'[27] and used the expression *mathesis intensorum*, suggesting that even if not now, psychology nonetheless *could* become a proper science.[28]

At the end of his life Kant came to realise that something was missing in his

---

[23] See Watkins (1998) for the argumentative structure of the *Metaphysical Foundations of Natural Science*.
[24] On such view, which is prevalent in contemporary analytic philosophy, there is physics and there are the special sciences (Fodor 1974).
[25] Kant, *Kritik der praktischen Vernunft* (1787): 'ein der Chemie ähnliches Verfahren' (AA5:163; see also AA5:92).
[26] For a recent discussion in English see Nayak and Sotnak (1995), though little is added to what had been discussed in the German literature of the past century.
[27] 'Nachricht von der Einrichtung Vorlesungen Winterhalbjahren von 1765-66' (AA 2:316): 'metaphysische Erfahrungswissenschaft vom Menschen'.
[28] Though it should be noted that these are quotations from Kant's 'pre-critical' period.

system. In the years 1796-1803 he was working on a draft of a work he had entitled *Transition from the Metaphysical Foundations of Natural Science to Physics*.[29] It is usually referred to as the *Opus postumum* - although among the hundreds pages of text there is at best a draft of a first chapter; the rest is 'working notes'. Kant had become convinced that a new a priori science must be added to his 1786 *Metaphysische Anfangsgründe der Naturwissenschaft*; without this new a priori science the 'pure doctrine of nature' remains incomplete. He says:[30]

> This treatise is directed towards filling what is still a gap in the pure doctrine of nature and in general in the system from a priori principles - and thus towards accomplishing completely my metaphysical task.

Kant here comes back on his famous statement in the *Critique of Pure Judgement* (1790) where he says in the Preface: 'hereby I bring my entire critical undertaking to a close' (AA5:170).

An important reason for the gap is that if the synthetic a priori has to guarantee the possibility of all experience, a lot is missing if one is stuck with physics (i.e. Newtonian mechanics). As he says in his physics lectures of 1785, the *Danziger Physik* (AA29:97-8):

> Chemistry has raised itself to greater perfection in recent times; it also rightfully deserves the claim to the entire doctrine of nature: for only the fewest appearances of nature can be explained mathematically - only the smallest part of the occurrences of nature can be mathematically demonstrated. Thus, e.g., it can, to be sure, be explained according to mathematical propositions when snow falls to the earth; but why vapours transform into drops or are able to dissolve - here mathematics yields no elucidation, but this must be explained from universal empirical laws of chemistry, and philosophy always belongs to chemistry, for it is a matter for the philosopher: to discover the universal laws of the action of matter through experience and to derive everything therefrom systematically.

Friedman (1992) shows that by 1785 Kant has become aware of the new discoveries in pneumatic chemistry; between 1785 and 1790 he has assimilated the developments in the science of heat; between 1790 and 1795 he has completed the conversion to Lavoisier's system of chemistry (Block 1-3 and Block 1-4).[31]

---

[29] *Übergang von den Metaphysischen Anfangsgründen der Naturwissenschaft zur Physik*. For exegesis see contributions in Forum für Philosophie Bad Homburg (1991), Hoppe (1969), Schulze (1994), Lequan (2000) - though there is by far no consensus how to interpret the *Übergang*. Förster (1991) comments that in the Kant literature 'herrscht darüber weitgehend Ratlosigkeit'. In the main text I follow Friedman (1992).
[30] See AA21:626; cf. AA21:640 and Kant's letter to Kiesewetter (AA12:257-8), where he also speaks of 'a gap in the system of critical philosophy'. Rothbart and Scherer (1997) discuss the relevance of Kant's *Critique of Judgement* for the scientific investigation of matter at length, refer to Friedman (1992) and mention the requirement that Kant needs 'a "transition", filling the gap on *a priori* grounds between the metaphysical foundations of nature, a pure product of though, and nature', but they do not seem to have picked up the relevance of Kant's *Übergang* for the status of chemistry.
[31] Vasconi (1996:157) suggests 'it is reasonable to assume that Kant was already familiar with the new discoveries as early as 1793', referring to a letter from Erhard (AA11:408), which refers to Girtanner's *Anfangsgründe der antiphlogistischen Chemie*.

Block 1-3   Kant's growing absorption of the 'new' chemistry I (for quotations in German see Block 1-4).[32]

1785 *Danziger Physik*
Physics is based on a priori principles and, in particular, on mathematics. Chemistry is based on a posteriori principles.

1786 *Metaphysical Foundations of Natural Science*
Chemistry can be no more than a systematic art or experimental study, but never a proper science, because chemical phenomena don't lend themselves to mathematical treatment which would connect them to the a priori.

1787 first *Critique*, second edition
Kant adds a physico-chemical example (state transition from fluidity to solidity) to the transcendental deduction (B163); but he still mentions Stahl's theory of the calcination of metals (Bxiii).

1789-90 Nachlass, fragment R. 66
first mention of Lavoisier

1790? *Opus postumum*
Kant states the caloric theory of gases; connects the problem of solidity with the concept of latent heat; and associates the science of chemistry with Lavoisier's doctrine of the composition of water.

1792
Kant receives a copy of Girtanner's textbook of anti-phlogistic chemistry (published that year).

1792-93 lectures on metaphysics (*Metaphysik Dohna*)
Kant refers to the compositeness of water.

1795a letter to Soemmerring
Kant juxtaposes the composition of water with Lavoisier's caloric theory of gases.

1797 Preface to the First Part of *The Metaphysics of Morals*
'there is only one chemistry (that according to Lavoisier)'

1798a *Opus postumum*
Caloric, as basis of the moving forces and primitive material, is contrasted with the elements of Lavoisier's systematic chemistry - conceived as derivative materials.

1798b *Anthropology From a Pragmatic Point of View*
Kant mentions Archimedes, Newton, and Lavoisier as representatives of great scientists.

---

[32] Sources in Block 1-3 and Block 1-4 are from Kant, *Akademie Ausgabe*, 29:97 (1785), 4:462,471 (1786), 14:489 (1789/90), 21:417,424 (1790), 11:408 (1792), 28:664 (1792/3), 12:33 (1795a), 21:453 (1795b), 6: 207 (1797), 21:605 (1798a), 7:326 (1798b). Other references to chemistry can be found, amongst others, in fragment 61 (AA 14.470): 'Chemie ist bloß physisch' and in the *Opus postumum*: 'Die ganze Chemie gehort zur Physik - in der Topik aber ist vom Übergange zu ihr die Rede' (AA21:288); 'Die Chemie ist ein Theil der Physik aber nicht ein bloßer Übergang von der Metaph. zur Physik. - Dieser enthält blos die Bedingungen der Möglichkeit Erfahrungen anzustellen' (AA21:316).

That Kant followed the debate over Lavoisier's system with ever-increasing interest in the years 1792-5 seems now well documented. Friedman refers to the passage in the 1797 Preface to the First Part of *The Metaphysics of Morals* where Kant has Lavoisier 'represent' chemistry (instead of Stahl). Förster (1993) has argued, on the basis of Kant's correspondence with Kiesewetter, that the idea of the *Transition* project goes back at least to 1790. Moreover around 1795 Kant develops his own version of the caloric theory of the states of aggregation. These developments could form the basis of a new kind of physical chemistry around 1798 (though physical chemistry in the modern sense had to wait another century).

But all this is the result of recent scholarship on Kant's later work. For a long time it was assumed that Kant could not have done any serious philosophy in his old age. Typically, Adickes, charged with editing the *Akademie Ausgabe* says in 1924 that in the 1790s Kant could not have been following the chemical literature seriously.[33] Kant's *Opus postumum* appeared in German (in the Academy edition) after more than a century of problems delaying its publication. And what was published is generally considered a mess, mere texts in random order without editing.[34] The first abridged English edition appeared in 1993.

The general point of Kant's 'chemical conversion' was perhaps first made, albeit much more briefly, by Dussort (1956) who argued that Kant remained a supporter of the phlogiston theory for a long time, though this changed in the *Übergang*. Lavoisier's work, in which the use of a balance played a crucial role, made possible the application of mathematics to chemistry. And Dussort points to similarities between the views of Lavoisier and those of Kant, in particular the assumption crucial to Kant, that sensation (experience) must be based on the movement of matter.[35] Moreover, the notion of caloric occurs on almost every page of the *Opus postumum*. Although Lavoisier didn't invent the term 'caloric', he certainly made it respectable. Dussort suggests that 'caloric' in the *Opus postumum* starts to take on the role of the transcendental object. The (a priori) properties of the *Wärmestoff* (or *Aether*) should determine the sensible properties of matter.[36]

---

[33] 'Auf jeden Fall kann keine Rede davon sein, daß Kant in den 90er Jahren hinsichtlich der chemische Literatur auch nur einigermaßen auf dem laufenden gewesen wäre' (Adickes 1924:I,63f).

[34] 'Die Bände XXI und XXII tragen den prätentiösen Titel *Opus postumum*, aber sie bieten keine Ausgabe der Vorarbeiten Kants zu seinem geplanten Werk, sondern drucken die zufällig zusammengeratene Papiermasse aus dem Besitz der Familie Krause ab. Die Herausgeber erzeugen durch die blinde Wiedergabe der Notizen den Eindruck, der alternde Philosoph habe nicht mehr zwischen einer Ätherdeduktion und seinen Rotweinflaschen unterscheiden können' (Brandt 1991:8).

[35] For Lavoisier's views on this see in particular his *Réflexions sur le phlogistique* of 1783 (*Œvres*, vol. II, p. 623). Duhem (1899:214) quotes Lavoisier as saying: 'la science des affinités est à la chimie ordinaire ce que la géométrie transcendante est à la géométrie élémentaire'.

[36] AA21:223,563,576,579,600; AA22:160,161,551. On the issue of using *Aether* and *Wärmestoff* often interchangeable see Schulze (1994:134-141). Carrier (1991:223) suggests that Kant's use of the words *Wärmestoff* and *Aether* in the *Opus postumum* is three-fold; it has a transcendental, a chemical, and a cosmological function.

Block 1-4 Kant's growing absorption of the 'new' chemistry II – for paraphrase in English see Block 1-3.

1785 *Danziger Physik*
Die Mathematik reicht gar nicht zu, den chemischen Erfolg zu erklären oder man hat noch keinen einzigen chemischen Versuch mathematisch erklären können; daher ließ man die Chemie aus der Naturlehre aus, weil sie keine Prinzipien a priori hat.
Die Chemie hat sich in neueren Zeiten zur großen Vollkommenheit emporgehoben; sie verdient auch mit allem Recht den Anspruch auf die gesamte Naturlehre: denn nur die wenigsten Erscheinungen der Natur lassen sich mathematisch erklären - nur der kleinste Teil der Naturbegebenheiten kann mathematisch erwiesen werden.

1786 *Metaphysische Anfangsgründe der Naturwissenschaft*
So lange also noch für die chemischen Wirkungen der Materien auf einander kein Begriff ausgefunden wird, der sich konstruieren läßt, d.i. kein Gesetz der Annäherung oder Entfernung der Teile angeben läßt, nach welchem etwa in Proportion ihrer Dichtigkeiten u.d.g. ihre Bewegungen samt ihren Folgen sich im Raume *a priori* anschaulich machen und darstellen lassen (eine Forderung, die schwerlich jemals erfüllt werden wird), so kann Chemie nichts mehr als systematische Kunst oder Experimentallehre, niemals aber eigentliche Wissenschaft werden, weil die Prinzipien derselben bloß empirisch und keine Darstellung *a priori* in der Anschauung erlauben, folglich die Grundsätze chemischer Erscheinungen ihrer Möglichkeit nach nicht im mindesten begreiflich machen, weil sie der Anwendung der Mathematik unfähig sind.

1789-90 Nachlaß, Fragment R. 66
Nach Lavoisier, wenn etwas (nach Stahl) dephlogistirt wird, so kommt etwas hinzu (reine Luft); wird es phlogisticirt, so wird etwas (reine Luft) weggenommen.

1792-93 *Metaphysik Dohna*
Ist Wasser Element? Nein denn es läßt sich noch auflösen, es besteht aus Lebensluft und brennbarer Luft, und wir nennen etwas was keine Spezies enthält elementarisch.

1795a enclosure of letter to S. T. Soemmerring
Das reine, bis vor kurzem noch für chemisches Element gehaltene, gemeine Wasser wird jetzt durch pneumatische Versuche in zwei verschiedene Luftarten geschieden.

1795b *Opus postumum*
Was ist Chemie? Die Wissenschaft der inneren Kräfte der Materie.
Auflösung ist entweder quantitativ wenn die Materie in gleichartige oder qualitativ wenn sie in ihre ungleichartige (spezifisch verschiedene) Materien geteilt wird.

1797 *Metaphysik der Sitten*, Vorrede
So sagt der Moralist mit Recht: es gibt nur eine Tugend und Lehre derselben, d.i. ein einziges System, das alle Tugendpflichten durch ein Prinzip verbindet; der Chymist: es gibt nur eine Chemie (die nach Lavoisier) ...

1798b *Der Anthropologie - zweiter Teil*
Welche Masse von Kenntnissen, welche Erfindung neuer Methoden, würde nun schon vorrätig da liegen, wenn ein Archimed, ein Newton, oder Lavoisier mit seinem Fleiß und Talent ohne Verminderung der Lebenskraft von der Natur mit einem Jahrhunderte durch fortdauernden Alter wäre begünstigt worden?

However, as indicated, these interpretations of Kant's later work have had as yet no impact. But his earlier views on science have been extremely influential, far beyond the narrow circle of Kant specialists. The chemist Meyer (1889:101) in an address delivered to The Association of German Naturalists and Physicians,[37] later published in the *Journal of the American Chemical Society*, mentions in passing that Kant's view on chemistry was referred to in *Deutsche Rundschau* of November 1889. In the English speaking world, Kant's view that chemistry is not a genuine science is 'exemplified by its place in William Paley's *Natural Theology*, the mandatory textbook read by every Cambridge gentleman throughout the nineteenth century' (Nye 1993:5). Paneth (1962:7-8) argues that Kant's definition results in an extremely narrow and inappropriate conception of science and declares 'chemistry, too, [is] a true science, even in those branches where it contains little or no mathematics'.[38] But such occasional opposition from chemists to Kant's views has never caught on.[39]

For the 'received' Kant a necessary requirement of 'proper' science is its tie to metaphysics and mathematics. Though the metaphysics has been dropped, the idea that 'proper' science is something that uses mathematics has been prevalent to the present day. To put it crudely, the logical positivists, following Hume, threw out metaphysics, replacing it by logic to line it up side by side with mathematics as the 'metaphysical foundation' of all 'proper' science.

1.3   from Hegel to Bachelard

Since the eighteenth century interaction between philosophy and chemistry can be found among those philosophers who, in their general philosophy, were influenced by knowledge of chemistry. This includes the philosophy of nature of the German philosophers Kant, Hegel, and Schelling (and to a lesser extent Fichte and Schopenhauer),[40] as well as such French philosophers of science as Duhem, Bachelard and Meyerson. Apart from being influenced by Kant, one thing the nineteenth century German *Naturphilosophen* did was pick up the notion of affinity or valency from chemistry making it the basis of a very general notion of chemism.[41] Chemism played an important role in Hegel's philosophy of nature. For him chemism stands between mechanism and teleology. A 'chemical object' (not restricted to 'ordinary' chemical objects) is intriguingly, an independent totality that is (nevertheless) defined in terms of its relation to other things.[42] Chemical objects in this sense share a family connection that separates them off

---

[37] The Gesellschaft deutscher Naturforscher und Ärzte.
[38] Many German chemists seem to have wrestled with Kant's predicament of chemistry.
[39] Heinig (1975) dismisses Kant's often quoted view that chemistry is not a real science as nothing special at the time Kant held this view, referring to a dissertation defended in Königsberg in 1789 by J.B. Richter entitled *De usu mathesos in chemia*. He also notes that already in 1741 Lomonossov wrote about 'the elements of mathematical chemistry'.
[40] For references to the older literature on this theme see von Engelhardt (1986).
[41] See for the history of the central chemical notion of affinity Levere (1971), Stengers (1995), Bensaude and Stengers (1993). The term 'chemism' is rarely used these days; an example is the title of chapter 2 in Bunge (1979).
[42] See Burbidge (1996), Snelders (1993), von Engelhardt (1993) and references given there.

from mechanical and teleological objects. Hegel's logic of chemism is not simply generalised from the empirical evidence of chemistry, but develops its own theoretical perspective, which can then be applied more generally in a number of spheres.

Von Engelhardt (1984, 1993) has stressed how important Hegel's philosophy of nature is. Its neglect is due to the incorrect ascription to Hegel (by Hegel scholars) of a contempt for both empirical work and modern research.[43] Only recently has the crucial influence of chemistry on Hegel's philosophy has been brought to the fore. For example (Burbidge 1993:615-6):

> Hegel has had to rework his philosophy of nature in the light of the results of empirical chemistry. His first attempt to force it into a tripartite scheme did not do justice to the significant differences between amalgamation, oxidation and acidification, and completely ignored the use of chemistry in refining. It suggests that in 1817 Hegel revised the syllogistic structure of the logic of chemism in the light of empirical chemistry.

It might be argued that Hegel considers chemistry to be a 'proper' science, and the philosophy of chemistry be regarded as a crucial part of Hegel's philosophy of nature.

Schelling developed his philosophy of nature in response to Hegel. In Book II of *Ideas for a Philosophy of Nature*, Schelling gives a detailed account of the chemical properties of bodies and chemical processes on the basis of a dynamic account of matter. According to Schelling 'attractive and repulsive forces constitute the *essence* of matter itself' (1988 [1797]:165). This German tradition of philosophy of nature also had an effect on practising chemists with philosophical interests. Mittasch (1948), a well-known chemist in the history of catalysis, wrote widely on the philosophy of nature, on Schopenhauer and chemistry, on Nietzsche and chemistry (including hundreds of pages of unpublished manuscripts on Nietzsche's philosophy of nature), on catalysis and determinism, and on the concept of causality of Robert Mayer. Without exception this had no impact on twentieth century philosophy of science, which was completely separated from 'metaphysical' philosophy of nature. The interest of German chemists in *Naturphilosophie* however has remained to the present day.[44]

Quite different from the tradition of German *Naturphilosophie*, Peirce too believed that the notion of valency as developed by Frankland, Mendeleev and others was one of the most important ideas in the history of science.[45] He used the con-

---

[43] For a detailed account of Hegel's familiarity with chemistry see Ruschig (1997), who shows that Hegel is thoroughly familiar with the chemical debates of his time and takes a positivist position similar to that of Duhem towards the atomism of Dalton. Hegel agrees with Dalton's *experimental* results but argues that they are couched in the worst kind of metaphysics ('in die schlechteste Form einer atomistischen Metaphysik eingehüllt') and in the section on Measure ('Das Maass') in his *Wissenschaft der Logik*,he gives a non-atomistic understanding of the law of multiple proportions. For Hegel the chemical object is determined by its relation to other objects via chemical reactions and the quantitative laws that govern them (Ruschig 1987, 1997).
[44] For example Müller and Hörz (1996), Müller (1998), Niedersen (1994).
[45] Note that the idea of valency as a number is in accord with the conception of chemistry in which structure is taken to be fundamental. Alternatively, the idea of valency can be seen as the combining power of atoms as a mental construction. That is, either one can put structure first and makes atoms

Block 1-5   Cassirer on the concept of 'atom'.[47]

> To the first naïve consideration, the atom appears as a fixed substantial kernel, from which different properties can be successively distinguished and separated out, while, conversely, from the standpoint of the critique of knowledge, precisely those 'properties' and their mutual relations form the real empirical data, for which the concept of the atom is created.
> The atom functions as the conceived unitary system of coordinates, in which we conceive all assertions concerning the various groups of chemical properties arranged.
> The concept, while obeying the facts, at the same time gains intellectual dominance over the facts.
> The objective existence of the different types of atoms is presupposed; it is only necessary to discover their *properties*, and to define them more exactly.

cept in what he called phanerochemistry, which, in going beyond chemistry and being applicable whenever there is a 'connection-of-two' has similarities to Hegel's chemism, though it is closely intertwined with the idiosyncracies of Peirce's three categories and notion of the semiotic (Tursman 1989).

Mill thought chemistry was less successful than physics, because of the problem of causal interaction. Two forces in mechanics 'combine', each retaining its original capacity or tendency. The law of vector addition gives a precise expression of this idea. In chemistry an acid and a base neutralise each other - each destroying the chemical power of the other.

There are few philosophers in the twentieth century, writing about natural science who as much as mention chemistry and then merely as a practically orientated attachment to physics. An exception is Cassirer (1923:203) who said:

> The conceptual construction of exact natural science is incomplete on the logical side as long as it does not take into consideration the fundamental concepts of chemistry.

His account of exact natural science, and in particular of chemistry, show similarities not only to the 'world-making' view of Goodman (who acknowledges Cassirer's influence), but also to Quine. According to the latter, ontologically speaking, theoretical terms are 'merely' posits in the (scientific) web of belief.[46] Drawing on Duhem (1902) and Ostwald (1904) among others, Cassirer argues that the concept of a (chemical) atom is (since Dalton's law of multiple proportions) a relational, regulative ideal concept, 'a mere relative resting point - a cross-section, which thought makes in the continuous flow of process' (Block 1-5).

Bachelard, initially a chemistry teacher at a provincial French college, later became a philosopher of science with a strong influence in France (notably on Foucault). However, in the few English publications on Bachelard's philosophy of science, chemistry is never mentioned.[48] Typically, in Tiles' (1985) book on

---

and bonds subsidiary concepts and valency only a number, or one derives structure from the 'combining power' of atoms with valency bonds.
[46] Quine (1994:183): 'what matters for any objects, concrete or abstract, is not what they are but what they contribute to our overall theory of the world as neutral nodes in its logical structure.'
[47] Quoted from Cassirer (1923:207-10,219-20).
[48] For example Gutting (1987), Tiles (1987), Tijiattas (1991).

Bachelard's philosophy of science, the two publications with 'chemistry' in their title (Bachelard 1932, 1971) aren't mentioned. Bachelard's historical approach is not always appreciated. As one of the reviewers of a reprint of Bachelard's *Le pluralisme cohérent de la chimie moderne* (1932) said: It is nothing but a historical story translated in philosophical terminology.[49] Theobald, who published rather extensively on the philosophy of chemistry [1.4], summarised the little that Bachelard had *specifically* said on chemistry as follows:[50]

- 'The thought of the chemist' oscillates between pluralism on the one hand and the reduction of this pluralism on the other.
- Each chemical substance refers to all the others - knowing about a chemical substance includes knowing how it is 'located' among other substances and how it behaves in all chemical reactions in which it can take part. That is to say, chemistry has a pluralistic ontology; an ontology of relations, more than of 'substances', let alone atoms, leading on to the importance of chemical synthesis. Schummer has stressed Bachelard's observation that the purity of substances should not be taken as a 'natural-given', but as the result of human operations.[51]
- Continuity is opposed: 'incompatible' sets of concepts and principles can be applied to phenomena.[52] For example, Bachelard (1934:175) offers the intriguing comment that experiments on the stretching of colloidal gels allow 'the physicist to act upon the *chemical nature* of substances' (emphasis in original).

Further, one may find in Bachelard's work the unworked out suggestion that, compared with (mathematical) physics, the philosophical neglect of chemistry parallels modern philosophical preference for form at the expense of substance. Bachelard puts this in a way that won't win him supporters in the prevailing tradition of philosophy of science.[53] For example, he speaks of: 'chemistry's unconscious dreamwork' and 'water is, in other words, a universal glue'. Or, less poetically:[54]

> The hypotheses of 'naive chemistry' which stems from the work of *homo faber* are at least as important psychologically as the ideas of 'natural geometry'. Both the remotest reverie and the harshest toil [should] be reintegrated into the psychology of *homo faber*.

Not all history reviewed in the past two sections is equally important to present concerns. One issue stands out as relevant to contemporary philosophy of chemistry - as well as contributing to its neglect:

---

[49] 'Cette philosophie de la chimie est en fait une stylisation de l'histoire, simple traduction de l'événement historique en style philosophique' (Bensaude 1974). For a critique of Bachelard's approach to the history of chemistry see Stengers (1995).
[50] Theobald (1982); see also Vinti (1996).
[51] Schummer (1996a:180,182,226); cf. Bachelard (1934:79).
[52] Cf. note 76.
[53] Or from French chemists for that matter: Delhez (1974).
[54] Quotations from Bachelard (1942:142-8), as translated by McAllester Jones (1991).

- Since Aristotle, the concept of *hyle* or 'stuff' has been suppressed in philosophy (of science), in favour of Euclidean geometry and Newtonian mechanics as a universal model for scientific knowledge, as is illustrated by Kant's influential view that chemistry is not a proper science.
- Specifically, views such as those of Ostwald and Duhem, stressing the priority of thermodynamics and basicness of notions such as pure substance (independent of atomic models or interpretations), were marginalised.

1.4 'classical' philosophy of science and chemistry

Until about 1960, philosophy of science, dominated by logical empiricism, mainly consisted of philosophy of physics, with marginal attention to philosophy of mathematics, biology, psychology, history, and social science.[55] It is instructive to consult the content pages of *The Journal of Unified Science (Erkenntnis)*, the 'house journal' of the *Vienna Circle* - home of the logical positivists - or the titles of the contributions to the *International Congress for the Unity of Science* from 1931 to 1940. Nowhere is there any indication that chemistry might be part of science, despite numerous references to logic, mathematics, physics, biology, psychology, some references to sociology, semantics, language, astronomy, medical psychology, economics, history; and even - on occasion - to values and aesthetics. No word however on chemistry. Similarly, in the 18 parts of the *Foundations of the Unity of Science: Toward an International Encyclopedia of Unified Science*, published between 1938 and 1970, the only reference to chemistry is found in Kuhn's (1962) *The Structure of Scientific Revolutions*, together with a few pages on chemical bonding in Frank's (1946) contribution on the foundation of physics. In the two main philosophy of science journals, *The British Journal for the Philosophy of Science* (since 1949) and *Philosophy of Science* (since 1933) there have been very few articles on the philosophy of chemistry (Block 1-6).

While there has been more interest in the *history* of chemistry, this area too, compared with physics, was relatively neglected – though substantial compared with the virtual non-existence of the philosophy of chemistry.[56] The history of chemistry is relevant to the philosophy of chemistry in a number of ways. For example, studies asking when chemistry 'became' a science will have to use a philosophy of science to establish whether the practice of chemistry meets the criterion of being a science.[57] Similarly if the name of Lavoisier is associated with the first quantitative chemical law, then this is simultaneously a statement in the philosophy of science. Finally, any historical episode and its later renderings that bear on the relation of, or distinction between physics and chemistry, is relevant;

---

[55] Canonical texts are Nagel (1961) and Hempel (1965); cf. Salmon (1999).
[56] See Brush (1978) for a rather old, but still pertinent assessment of why there has been relatively little interest in the history of chemistry. For the substantial pre-1994 interest in the history of chemistry see the bibliography at the Hyle website (http://www.uni-karlsruhe.de/~philosophie/hyle.html).
[57] See for some recent discussions on the 'emerging' science of chemistry the special issue of *Science in Context*, vol. 9, No. 3 (1996).

if only because of the constant rewriting of a history of science that privileges physics.[58]

I suggest the main reasons for the absence of interest in chemistry in traditional philosophy of science were as follows.[59]

Pre-1960 interest was almost exclusively in *theoretical* science. Thus experimental and applied physics (at least 90% of a physics department?) was as much neglected as chemistry. That chemistry is not a theoretical science in the sense of theoretical physics tends to be supported by chemists themselves who, in the words of the chemist-philosopher Polanyi (1958:156), have always been wary of theoretical 'speculation' unsupported by detailed experimental observations - a recurrent theme in public lectures of well-known chemists. Physicists put the same point more pejoratively; Dingle (1949a) for example, echoing Kant:[60]

> The truth is that chemistry indeed has no place in the strict scientific scheme. The part played by chemistry in the growth of science has been a pragmatical, heuristic one.

Placing the emphasis differently, physics and chemistry were lumped together as *exact* natural sciences with emphasis on the study of their *logical* structure. This meant that the interest was in laws in the sense of mathematical equations stating relations between magnitudes and theories that were axiomatisable, at least in principle. In that sense there are few laws and theories in chemistry. If anything, philosophers took this one step further. For example, Hartmann (1948), again echoing Kant says: 'all of chemistry that is lawlike is pure physics'. Mulckhuyse (1961) is an early attempt to do philosophy of chemistry that meets the formalistic standards of traditional philosophy of science. He presents an axiomatisation in first order logic of the 'classical' (van 't Hoff type) theory of chemical structure, though it had no further impact.[61]

Further, in the natural sciences from 1500 to 1900, there were more 'big' physical than chemical theories. Since the end of the nineteenth century philosophical interest had primarily been in the idea of unified theories (and hence unified science, the most fundamental theories, and concomitant ideas about reduction). There have been no theoretical developments in chemistry that could compete on these terms with physics. Typically, Causey devotes less than four pages to chemistry in his book *Unity of Science* (1977: 51-4). In so far as chemistry is mentioned in the philosophy of science, it is taken for granted that it can be reduced to physics [5.1].

---

[58] For historical studies bearing on the history of the distinction of physics and chemistry see for example Nye (1993), Bensaude and Stengers (1993); see also Hoffmann (1995).
[59] For reasons more 'internal' to chemistry that might explain the 'neglect' of the philosophy of chemistry see Bensaude-Vincent, (1994), Eisvogel (1996), Han (1996), Janich (1994b), Ruthenberg (1996), Psarros (1996a), Scerri and McIntyre (1997), Villani (1996) and brief observations in Bunge (1982), Janich (1992:173, 1994a:86f), Liegener and Del Re (1987b), Plesch (1999), Ströker (1967), Del Re (1996), Vancik (1999).
[60] Dingle (1949) adds: 'Reluctant as I am, and as a loyal physicist should be, to say anything good of chemistry, I cannot deny that, quite apart from its necessity for the amenities of life, it has been indispensable in making possible the rapid progress of physics.'
[61] For a critique of the limited scope of his project see Schummer (1996a:248f,255).

Block 1-6 Pre-1960 publications in *The British Journal for the Philosophy of Science* and *Philosophy of Science* on the philosophy of chemistry.

Malisoff	1941	on chemistry and 'emergence without mystifications'
Pirie	1952	concept of pure substance
Feibleman	1954	chemistry mentioned in the context of von Bertalanffy's theory of general systems theory
Bradley	1955	attempts an operational interpretation (in the sense of Bridgman and following the lead of Ostwald) of classical chemistry (Dalton, Gay-Lussac, Avogadro, Cannizzaro)
Ellis	1957	uses Gay-Lussac's law of combining volumes as an example to compare process and non-process explanations
Kent	1958	on scientific naming in which Lavoisier and Faraday play a central role
Kultgen	1958	philosophical conceptions in Mendeleev's *Principles of Chemistry*
Caldin	1959	on chemical theories

Mainstream philosophy of science simply regarded chemistry as part of physics and an unimportant part at that. Chemical examples were used, for example in discussing the status of dispositional properties. However, such examples had nothing to do with anything that might be considered *typically* chemical (as distinct from physical). There was only one 'early' subject that had some specific connection to chemistry (and to biology and psychology) and that concerned the idea of emergence.[62] This is the view of Broad (1925:58-9):[63]

> There need not be any peculiar *component* which is present in all things that behave in a certain way. The characteristic behaviour of the whole *could* not, even in theory, be deduced from the most complete knowledge of the behaviour of its components taken separately or in other combinations, and of their proportions and arrangements of this whole.

In the 1960s a few publications specifically on the philosophy of chemistry appeared. It was suggested that 'theory' means something different to a chemist than to a (theoretical) physicist; what comes under a chemical law or theory is not 'the same', but a 'similar' or 'analogous' thing. Caldin (1959) notes that often experiments in chemistry are not meant to test a model or theory, but are attempts to make it more precise.[64] This was echoed by Theobald (1976):

> Theories are rarely highly controlled by observation in chemistry, since [they] are generally rationalising constructions covering vast arrays of experimental data, rather than precise mathematical formulations vulnerable to a single quantitative misfortune.

---

[62] See Hempel (1965:259-64); Nagel (1961:366-80).
[63] Broad (1925:58-9); emphasis in original. See further [2.4].
[64] Caldin (1961) is a short introduction to the philosophy of science using chemistry as example; as to the philosophy of chemistry it doesn't add much to Caldin (1959).

This applies in particular to the refinement of molecular structure according to Theobald. And in a comment on Caldin (1959), MacDonald (1960) wrote:

> It would appear that the chemist regards theories - or perhaps better *his* theories (!) - as far less sacrosanct [than the physicist], and perhaps in extreme cases is prepared to modify them *continually* as each bit of new experimental evidence comes in.

Though chemists may sometimes test theoretical models in critical cases, this is not characteristic of chemistry: 'the thesis that theories are tested by attempts to falsify them is not supported' (ibid). If anything, chemistry supports Lakatos' methodology of research programmes.

Further, Churchman and Buchanan (1969), as well as Theobald (1976), found that the Hempel-Oppenheim scheme of deductive-nomological explanation (Hempel 1965) is not applicable to examples from chemistry. This however may be more to do with the highly abstract characteristics of the Hempel-Oppenheim scheme and the sort of criticisms that can be levelled at it from a more pragmatic stance (van Fraassen 1980), than to any difference between physics and chemistry.

A major issue in the philosophy of science has always been the discussion about scientific realism, but it is difficult to find references to chemistry in these debates. Gascoigne (1961) argued that molecules, atoms, electrons, and bonds score very differently when the question of realism is raised: Molecules (and crystals seem to refer to concrete things which can, in principle, exist by themselves. Atoms, ions, and electrons are essentially parts of things and normally, at least not in ordinary terrestrial contexts, do not exist by themselves. Bonds refer to abstract geometrical entities. Molecules and crystals are not, in fact, built or formed from atoms; chemical substances are not made from atoms, but from other chemical substances. However, though clearly relevant, chemistry never figured in the debate on scientific realism until the 1990s.

The question of differences between physics and chemistry with respect to subject matter and method has often been addressed, but it is typical for the neglect of the philosophy of chemistry that these tend to be isolated observations. At a text book level the difference has been ascribed to that between energy and matter, to that between all matter and individual kinds of matter, and to that between the study of microscopic and macroscopic phenomena. Early substantial contributions all focus on the historical development of chemistry, in particular its roots in alchemy.[65]

Although chemical *synthesis* is one of the most characteristic features that distinguishes chemistry from other sciences, philosophers have paid virtually no attention to it since the nineteenth century interest in chemism. Lévy (1979a, 1979b) has argued that chemical kinetics (Block 1-7) is a 'purely' chemical theory

---

[65] For example, in a book written from a phenomenological (Husserlian) perspective, it is argued that chemistry cannot restrict itself purely to mathematical-functional relationships (Ströker 1967:8): 'Doch handelt es sich bei ihnen [die Chemie] um solche [Vorgänge], die, anders als die rein physikalisch betrachteten, den Charakter *stofflicher* Veränderung haben. Die Chemie kann auf die Frage, *was* die Stoffe und ihre Qualitäten *sind* nicht verzichten; sie bildet gerade ihre leitende Problemstellung.' Cf. Block 6-4.

Block 1-7   Chemical kinetics and chemical reactions.[66]

> Because chemistry is *essentially* about transformations, the concept of time is a necessary ingredient of any theory that is truly chemical. It 'enters' chemistry via the second law of thermodynamics, the concept of molecular diffusion, and reaction kinetics proper. Although the first two could be taken as 'physical', reaction kinetics is essential to chemistry. The second law of thermodynamics can predict from a present state but has nothing to say about the time required to attain a future one.
> Chemical reactions can be defined as a change of number, kind or mass of pure materials, while number, kind and mass of elementary materials remain constant. Chemical reactions play at least three different roles: they provide an inventory of possible chemical changes; they yield information on the structure of molecules; they contribute to the 'emergence' of new compounds.
> Foundational discussions on chemical kinetics would have to elucidate concepts like reactivity, steric interaction, stability, equilibrium, (chemical) bond, bond strength, electron orbitals, solvation, valency, transition state, catalyst.

that cannot even be reduced to the 'physics-infected' parts of physical chemistry (thermodynamics, quantum chemistry). There are no physical principles that help the chemist to determine the order of a reaction, the details of its mechanism or the role of catalysts. But until recently these issues have hardly been touched

A closely related characteristic of chemistry is the 'language of chemistry' or chemical sign system, developed since the 1860s. This is not merely (or not at all) a pictorial or iconic system, but a sophisticated carrier of information about material properties, in particular dynamic properties like reactivities and reaction mechanisms. Organic chemistry has been described as the largest single body of knowledge; knowledge that is primarily expressed with lines and groupings of symbols subject to special iconic conventions. The chemical sign system can be seen as naming the nodes in the network of chemical space [3.1] and providing information about the structure of chemical space (i.e. chemically possible substances). The complex system of naming used in chemistry is 'exact', but not mathematical.[67] But again, until the 1990s, these issues were only sporadically touched upon.

Most of the publications on the philosophy of chemistry in this early period seem to be by chemists who developed an interest in the philosophy of science, not by 'professional' philosophers of science, not even if they were familiar with chemistry. This applies most particularly to the more substantial publications. As Bunge (1985:219f) noted:

> Given the popularity and prestige of chemistry, it is strange that the corresponding philosophy hardly exists. The publications in this field are only a handful Not even the distinguished philosophers of science Meyerson, Broad and Bachelard, who started out as chemists, made any

---

[66] Benfey (1963), Schummer (1998b), Lévy (1979b).
[67] Early publications on the language of chemistry (usually not tied to philosophy of science concerns) include Hoffmann and Laszlo (1991), Laszlo (1993), and Mounin (1981) reviewing Mestrallet (1980) - a very detailed semiological study.

significant contributions to the philosophy of chemistry: they preferred to write about other sciences.

Bunge's list of examples is not completely convincing. Polanyi (1958) could be added to the list. On the other hand, Broad started his study in Cambridge with physics, chemistry, botany, and mineralogy; but after two years he decided his future lay in philosophy. There isn't one reference to chemistry in the Broad volume in the *Library of Living Philosophers* (Schilpp 1959), although he used sodium chloride as an example in his writings on 'emergence'. Meyerson (1991), in his historical account of explanation in the sciences, discusses the resistance to Lavoisier's theory at some length (ibid:546-563); but in his long chapters on Hegel and Schelling there is no reference to chemistry, though there are occasional references elsewhere, for example (Meyerson 1930:31):

> The existence of the silver-element is only a hypothesis which is obtained after many deductions; and pure silver, like the mathematical lever, the ideal gas, or the perfect crystal are abstractions created by a theory.

Even though there appeared an occasional paper on the philosophy of chemistry in philosophy of science journals (Block 1-6), it had no further impact. Typically, in 1962 the *British Journal for the Philosophy of Science* published in two instalments a translation of Paneth's 1931 paper on the epistemological status of the chemical concept of element. In this paper Paneth argues that[68]

> I have referred throughout to 'basic substance' [*Grundstoff*] whenever the indestructible substance present in compounds and simple substances was to be denoted and to 'simple substance' [*einfacher Stoff*] whenever that form of occurrence was meant in which an isolated basic substance uncombined with any other appears to our senses.

There were only two extremely brief responses to Paneth's paper, disputing historical details concerning Locke and Lavoisier by Bradley (1962, 1963). Neither the republication of Paneth's article, nor the earlier exchange between Caldin (1959) and MacDonald (1960), in the same journal had any further impact.

1.5 philosophy of chemistry in Eastern Europe

In the period 1949 to 1986 a number of primarily German and Russian publications on the philosophy of chemistry appeared in Eastern Europe.[69] This included a range of books devoted solely to the subject.[70] In the former DDR (German Democratic Republic) twenty-four dissertations (Ph.D. or higher degree) were written on the philosophy of chemistry (Schummer 1996b). Although about a

---

[68] Paneth (1962:150), italics removed. Ruthenberg (1997) has pointed out the Kantian echo in Paneth's view: the simple substances are the *Phenomena* and the basic substances are the *Noumena*. See van der Vet (1979a) on the Paneth-Fajans debate and chemical identity.
[69] Many contributions from Rumania were by E. Bellu, some of which were published in English or French (Bellu 1973, 1979).
[70] Including Budreiko (1970), Garkovenko (1970), Ionidi (1958), Kedrov (1969), Laitko and Sprung (1970), Shakhparanov (1957, 1962), Simon, Niedersen and Kertscher (1982), Zhdanov (1960).

Block 1-8   Journals in the DDR which published on the philosophy of chemistry in the period 1962-1983.

Deutsche Zeitschrift für Philosophie
Rostocker Philosophische Manuskripte
Wissenschaft und Fortschritt
Wissenschaftliche Zeitschrift Ernst-Moritz-Arndt-Universität Greifswald
Wissenschaftliche Zeitschrift Friedrich-Schiller-Universität Jena
Wissenschaftliche Zeitschrift Humboldt-Universität Berlin
Wissenschaftliche Zeitschrift Pädagogischen Hochschule 'Wolfgang Ratke' Köthen
Wissenschaftliche Zeitschrift Technischen Hochschule für Chemie 'Carl Schorlemmer'
Chemiker Zeitung/Chemische Apparatur
Chemie in der Schule

third were on historical, economic, or educational subjects, most concerned the philosophy of chemistry proper. Publications on the history and philosophy of chemistry appeared in many journals (Block 1-8).[71]

In the *Deutsche Zeitschrift für Philosophie*, between 1962 and 1983, thirteen articles appeared on the philosophy of chemistry. The number appearing in the Russian journal *Voprosy filosofii* [*Problems of Philosophy*] was about the same.[72] The most influential writers in the USSR who published on the philosophy of chemistry were the philosophers of science Kedrov and Kuznetsov and the chemist Zhdanov. Kedrov published numerous books and articles in the philosophy of natural science. One of his earlier publications was a thorough-going article on Dalton which was translated into English (Kedrov 1949). He was among the high-ranking Russian philosophers and gave one of the plenary lectures at the *XV World Congress on Philosophy* (1973, Varna, Bulgaria). Zhdanov published widely, in philosophy and chemical journals, on esthetics and in the *Pravda*. Kuznetsov published less, but often about the philosophy of chemistry.[73] In addition there was explicit attention in philosophy journals to the chemical industry and 'the use of chemistry in the construction of communism'.[74]

Of the three hundred articles on the philosophy of the natural and social sciences that appeared in volumes 1-25 (1953-1977) of the *Deutsche Zeitschrift für Philosophie*, only a small number were on the philosophy of chemistry. Laitko (1996), who gives an insider's account of the socio-historical factors that stimulated interest in the philosophy of chemistry in the DDR, stresses that it wasn't much, compared with the interest in physics and biology. This would seem to be true if one notes that, apart from a few references to the history of chemistry, there is no philosophy of chemistry in Woodward and Cohen (1991), a volume

---

[71] The major contributors were [number in square brackets is number of (co-authored) publications listed in Schummer 1996b]: H.-J. Bittrich [6], K. Buttker [6], W. Fleischer [7], G. Fuchs [14], H. Laitko [9] U. Niedersen [14], R. Simon [23], W.-D. Sprung [19], I. Strube [7], F. Welsch [12].
[72] Not including many articles on philosophical problems of molecular biology and biochemistry such as Karpinskaya (1966), Semjonov (1959), Sisakjan (1959), Zbarsky (1963).
[73] See for example Kedrov (1956, 1962, 1965, 1969), Kuznetsov (1963, 1964), Kuznetsov and Pechenkin (1972), Kuznetsov and Shamin (1993), Zhdanov (1960, 1963, 1965, 1972).
[74] Editorial (1964), Garkovenko (1964).

containing a large number of papers from a German-American Summer Institute on science studies organised in the DDR in 1988. Nonetheless the number of philosophy of chemistry articles published in the *Deutsche Zeitschrift für Philosophie* was larger than the number of articles appearing in the same period in *all* English language philosophy journals. The same is true when comparing Russian and English language publications.

Moreover course books were written on 'philosophical problems of chemistry' and used in teacher training and technical colleges. Such books were also prescriptions for how to talk about chemistry from the perspective of dialectical materialism. But they included the latest developments of discussions in philosophical literature. A typical format for such a book would be to have chapters on philosophy and natural science, materialism and dialectics in chemistry, epistemological problems in chemistry, chemistry and society, philosophical views of particular chemists (e.g. Ostwald and Mittasch).[75]

One reason for the interest in chemistry in dialectical materialism goes back to Hegel, who 'used' chemistry to illustrate the dialectic of quantity and quality; in the authorised USSR wording (Bochénski 1963:17):[76]

> The law of the transition of quantitative changes into qualitative is a law according to which small, at first insignificant, quantitative changes, having reached a certain point, break (*narusajut*) the measure of the object and (thereby) evoke fundamental (*korennye*) qualitative changes. As a consequence, objects change, the old quality disappears and a new quality comes to be.

Chemistry was seen as a prototypical domain that illustrates this principle. For the present purpose dialectical materialism can be taken as the view that everything that exists consists of matter-energy; this matter-energy develops in accordance with universal laws; knowledge is the result of a complex interaction between human(s) and their 'external' world (but both are part of the same material world). It would not be difficult to transcribe a contemporary view in analytic philosophy, which combined non-reductive materialism with a 'dynamic', 'interactive' cognition theory, into dialectical materialism (and have it accepted for publication in a journal in Eastern Europe in the 1960s or 1970s).

Hegel's point of the dialectics of quantity and quality had been taken up by Engels (1940) in his *Dialectics of Nature*. This book consists of unfinished notes. Kedrov made a chrestomathic[77] reconstruction of how the book might have looked if Engels had finished it (Engels 1979). Kedrov used extensive texts from Engels' *Nachlass* and added parts from Engels' other publications and from the Marx-Engels correspondence, as well as brief connecting texts to unify the whole. Engels had been strongly influenced in his view on the natural sciences by Carl

---

[75] Simon, Niedersen and Kertscher (1982).
[76] See for the application of the law to chemistry, Kedrov (1962) and Kuznetsov (1963, 1964). Bellu (1979) used Bachelard's (1949) in support of the contradictory unity of conservation and transformation.
[77] In this context 'chrestomathic' means presenting a selection of passages of an author in a coherent whole.

Schorlemmer's book on the history of organic chemistry. Schorlemmer was the first Marxist natural scientist and friend of Marx and Engels. This explains the significant presence of chemistry in the *Dialectics of Nature* - even more so in the extended 'Kedrov'-version (Engels 1979) than in the 'Haldane'-version (Engels 1940). This 'historical' aspect of the connection between dialectical materialism and chemistry was stressed by Kedrov in numerous publications.

In applying the conceptual scheme of dialectical materialism to the natural sciences, a number of issues directly relevant to the philosophy of chemistry emerged. For example, in Kedrov's (1962) influential account, any type of change (or 'movement' in Engels' terminology), must correspond to a particular type of matter (Engels 1940:35):

> Motion in the most general sense, conceived as the mode of existence, the inherent attribute of matter, comprehends all changes and processes occurring in the universe, from mere change of place right to thinking.

Kedrov distinguished the following 'levels' of change and matter: a) nuclear physical; b) electrical (as in atoms), c) chemical (within a molecule), d) molecular-physical (as in liquids), e) geological (as in minerals). From level c) onwards there is also a line towards biological forms of change. The essence of each qualitatively new form of 'movement' is to be found in the interaction between elements at a lower level.[78] In general, any form of strict reductionism is opposed; physics and chemistry are 'both separate and together'. This is the principle of subordination and objectivity.[79] The irreducibly macroscopic character of typical chemical phenomena was stressed; e.g. reaction velocity only makes macroscopic sense. That chemical laws had a higher 'specificity, complexity, and individuality' was also used to argue against reductionism. Kedrov's views were discussed in many German publications, which often criticised his failure to go into chemical detail. In one typical example he incorrectly assumed all chemical change involves molecules.[80] In defence of this 'mistake' of Kedrov, Garkovenko (1963) criticised attempts to expand the notion of molecule to include ions, free radicals, macromolecules and crystals. The diversity should be recognised as qualitatively *different* 'forms of movement'.

In the DDR much was written on the relation between physics and chemistry.[81] Richter and Laitko (1962) argued that chemistry and molecular physics should be considered as two branches of macrophysics. Derivatively, there was considerable interest in the subject matter of chemistry. In the terminology of Engels and Kedrov the question was: what is specific to the *chemical* form of movement? In brief the conclusion reached was that the subject of chemistry is a set of laws (Richter and Laitko 1962, Laitko 1967), governing 'chemical forms of movement' (Rosenthal 1982) and the transformation of chemical matter.[82] Each

---

[78] See on the chemical form of the motion of matter also Dobrotin and Barzakovsky (1963) and Garkovenko (1963); and on the periodic law Faustov (1955).
[79] See on these issues: Niedersen (1983), Poller (1966), Richter and Laitko (1962), Simon (1975).
[80] Here, the material carriers of chemical change are assumed to be atoms, molecules, radicals and ions (both of atoms and atom groups).
[81] For example Laitko (1965), Laitko and Sprung (1970), Rosenthal (1982).
[82] 'Continuity is characteristic of chemical substances as well as discontinuity' (Kuznetsov 1963).

chemical transformation corresponds to a chemical law. It was argued that 'transformation of matter' includes polymorphous transformations and radioactive decomposition, leading to an extensive discussion of the precise definition of chemistry. It was pointed out that Engels' definition - 'the science of qualitative changes in bodies, which take place in conformity with change in their quantitative composition' - is better than many text book definitions, but it excludes isomeric transformations. Hence it was proposed that 'chemistry is the science of the qualitative changes of bodies that occur under the influence of changes in their quantitative composition and structure' (Zhdanov 1965).

Many publications appeared drawing negative conclusions concerning the reduction of chemistry to quantum mechanics (via quantum chemistry).[83] For example Laitko (1967) argues, against the reductionism of Kitaygorodsky (1966), that chemistry can be 'brought back' (is *zurückbar*), but not reduced to quantum mechanics. He expresses agreement with Shakhparanov (1962:32) who emphasises that the 'higher' notion of individualisation makes chemistry more specific and more concrete (whereas quantum mechanics is more general and more abstract). The connection between the valence-structure method and classical chemical formulae is discussed, as too is the underlying issue of 'model': do quantum chemists use (approximate) models of quantum physical systems or do they use chemical theories? And it is argued that the chemical and quantum mechanical structure of molecules should be distinguished. Because quantum mechanics in its standard Kopenhagen interpretation was seen as undermining materialistic principles, chemistry became the 'first science' to deal with the material properties of the world. If quantum mechanics is to chemistry as statistics is to economics (Kedrov 1962, Vihalemm 1974), the concept of molecular structure would vie for priority as the most fundamental concept of chemistry, though having no grounding in quantum mechanics [5.5].

In this literature *both* the idea that chemistry is qualitatively different from physics in an absolute sense *and* the idea that chemistry can be reduced to physics via quantum chemistry, was regarded as a fetish. Although phrased in dialectical jargon, the view would not seem very different from that of supervenience in analytic philosophy. For example Poller (1966) argued that the relation of quantum mechanics and chemistry should not be seen either as the thesis of absorbing chemistry in quantum mechanics, nor as a metaphysical separation of the two. The relation has to be understood dialectically as two things that can be unified and in contrast at the same time.[84] The arguments are often similar to issues

---

[83] See, amongst others, Haberditzl and Laitko (1967), Kitaygorodsky (1966), Poller (1966); Niedersen (1983), Fuchs (1964, 1965).
[84] 'Dem Verhältnis von Quantenmechanik und Chemie wird man jedoch weder durch die These von der Absorption der Chemie durch die Quantenmechanik noch durch eine metaphysische Trennung und eklektische Nebeneinanderstellung beider gerecht [tun]. Vielmehr muß dieses Verhältnis im dialektische Sinne als eine Einheit beider Seiten, die ihren Gegensatz nicht ausschließt, erfaßt werden. Auch daraus folgt, daß Quantenmechanik und Chemie nicht identisch sein können, denn "Bewegung ist nicht bloß Ortsveränderung", sondern "sie ist auf den übermechanischen Gebieten auch Qualitätsänderung".' Quotations inside the citation are from Marx/Engels, *Werke*, Bd. 20, p. 517.

raised in 'western' literature on the relation of chemistry to quantum mechanics Block 5-10. For example Laitko (1967) says:

> The quantum chemical methods of approximation are of such a kind that when introducing the simplifications, chemical considerations play an essential role, considerations which are of course supported by experimentation and the traditional chemical ways of thinking.

A general reason for the interest in the philosophy of chemistry was a strong concern not to separate science, education, and philosophy. For example in the DDR, similar articles on the philosophy of chemistry could appear in *Chemie in der Schule* and in the *Deutsche Zeitschrift für Philosophie*. Although at times the writing is dominated by ideologically-correct qualifiers, as when different views on the Brønsted theory of acids are formulated in terms of 'reactionary bourgeois theories', 'Maoist accounts' and the 'correct dialectic' approach (Simon 1975), this is epiphenomenal to the substantial issues being discussed. In a few cases the principles of dialectical materialism were a true stimulus to tackle philosophical issues; in most cases however the obligatory jargon was simply tacked on.[85] For example Poller (1966:334) writes: 'Both [redox] reactions must proceed simultaneously (Principle of the unity of the 'struggle' of contradictions!)'.[86] But this sentence stands out amidst a set of sophisticated arguments against the suggestion that quantum chemistry could, in principle, give an exact account of *what* is going on in a redox reaction. Similarly, Poller expresses the view that knowing *what* happens is not the same as knowing *how* it happens (with reference to Marx and Engels), an argument not much different from that of some recent critiques of reductionism (such as Dupré 1993).

1.6 the Chelintsev affair

An extra stimulus for the interest that developed in Eastern Europe in the philosophy of chemistry may have been an article in *Voprosy filosofii* in 1949 by Tatevskii and Shakhparanov. They argued that 'the physical theory of resonance is erroneous and the philosophical setting of its authors and propagandists is Machistic'[87] and 'hostile to the Marxist view'.[88] This publication was itself triggered by a book by the chemist G.V. Chelintsev, a professor of chemical warfare at the Voroshilov Military Academy, who had attacked Pauling's resonance theory of aromatic organic compounds and proposed an anti-resonance benzene theory.[89]

---

[85] For example, in the preface of a book on the history of spectral analysis, Kedrov (1956) writes: 'Am Beispiel der Entwicklungsgeschichte der Spektroanalyse wird versucht, mit Hilfe der marxistisch-dialektischer Logik allgemeine Wege zur Erkenntnis der Wahrheit durch den Menschen zu verfolgen'.
[86] 'Beide [redox] Reaktionen müssen gleichzeitig ablaufen (Prinzip von der Einheit und vom „Kampf" der Gegensätze!)' (scare quotes around 'Kampf' and exclamation mark in original).
[87] 'Machistic': referring to Lenin's criticism of Mach in his chief philosophical work *Materialism and Empirio-Criticism* (1st Russian edition 1909).
[88] For extensive quotations in English and a critique of the mathematical physics of Tatevskii and Shakhparanov (1949) see Wheland (1955:613-5).
[89] See for a detailed, though rambling, description of the following events (Graham 1971:ch.7) and further Hunsberger (1954), Vermeeren (1986) and Stork (1963).

Block 1-9   The notions isomer, tautomer, mesomer, resonance and aromaticity.

> Isomers are compounds composed of the same elements in the same proportions, but different in properties because of differences in structure. Enantiomers are species containing equal amounts of two optical stereoisomers. Diastereomers are stereoisomers that are not mirror images.
> Tautomers are isomers that change into one another rapidly and are usually in equilibrium with one another. The structure of isomers and tautomers is 'real' in a way that the 'resonance' (Pauling) or 'mesomer' (Ingold) structure of benzene and other (pseudo-)aromatic compounds is not.
> The aromatic character of a compound is usually defined as possessing delocalised electrons, typically associated with benzenoid aromatics. If aromaticity is held to be a purely electronic property (i.e. of cyclic electronic delocalisation), aromaticity may also occur in cyclic molecules containing nitrogen, boron, or even sulphur or phosphorus atoms.

Apparently Chelintsev had Lysenkoistic aspirations - though unsuccessfully as we shall see.

Brush (1999) has given a detailed account of the subsequent models of the structure of benzene from 1865 to 1945. During the first quarter of the twentieth century many benzene structures were discussed, but experiments failed to show which was correct. Then the new quantum mechanics offered two possible approaches: molecular orbitals and valence bonds.[90] The latter approach, promoted by Pauling's notion of resonance and Ingold's notion of mesomerism (Block 1-9), became the more popular one, at least for some time. Brush concludes that 'by 1945 chemists accepted a modernised version of Kekulé's oscillation theory of benzene, formulated in the language of Pauling's resonance theory, which brought with it the prestige of quantum mechanics.'

Pauling had introduced the term 'resonance' and his theory was extended and completed by Wheland. In response to the article of Tatevskii and Shakhparanov, Wheland (1955:613) dismissed their criticism: 'these authors have been misled by the carelessly worded expositions of the theory, which give an erroneous impression of its physical meaning, and which are unfortunately all too common'. But note that Wheland found the criticism serious enough to warrant rebuttal.

The theory of resonance can be understood as a synthesis of classical structural ideas and quantum mechanical concepts. This presents a possible problem for materialism in the following sense. How could something ('resonance') that has no material base in a particular molecular structure be the cause of anything? Quotations of Wheland and Pauling (Block 1-10) were reiterated over and over again in discussions among chemists in Moscow that reached their height in 1951.[91] The

---

[90] Valence bond theory considers a structure to be formed by bringing together complete atoms and then allowing them to interact to form bonds. Chemical bonding is seen as a result of the superposition of two orbitals of two different atoms in which either each orbital contains one electron or one orbital is empty and the second contains an electron pair. In molecular orbital theory nuclei (or nuclei and inner filled shells) are brought together and then electrons are placed in molecular orbitals. In the latter case there are no discrete chemical bonds; rather a set of orbitals is generated which allows electrons to roam over many nuclei (perhaps an entire molecule).
[91] Usually quoting the Russian translation (1948) of Wheland's 1944 *The Theory of Resonance and Its Application to Organic Chemistry* and the Russian translation of the second edition of Pauling's

issue fell on fertile ground, not so much because of the principles of dialectical materialism, but because there had been priority disputes since 1863 about the originators of the theory of chemical structure (Butlerov, Couper and Kekulé), in particular between German and Russian historians. Butlerov was probably the first to use the term 'chemical structure'[92] and to give a clear definition of tautomerism (though he didn't use the term). He also suggested that the correct structural formula can be obtained by studying different methods of synthesis of a compound and that one should consider the relation between atoms not directly connected by valence bonds. Nevertheless his name is rarely mentioned in Western publications. In English and German publications the name of Kekulé dominates; Couper is mentioned briefly and Butlerov often not at all – an obvious bias. On the other hand it is equally wrong to claim that 'Butlerov is the true creator of the theory of chemical structure', as Arbuzov (1952) does in an otherwise nuanced review of the developments. Probably it is fair to say that Butlerov, Couper, and Kekulé between them made all the important contributions. Because they knew each others' work and met one another on their journeys, a ranking in terms of priority, makes no sense.

In 1950 and 1951 two conferences were held at the Institute of Organic Chemistry of the USSR Academy of Sciences in Moscow. The report of the first meeting (Kursanov et al 1950) stresses the importance of Butlerov and the incompatibility of the 'idealistic' resonance theory with Butlerov's concept of structure. Although the mesomer concept is considered useful as a mathematical tool, it is argued that:

> The basic assertion of the resonance theory, that 'resonance' in some way can determine the properties of molecules, is devoid of meaning.

But significantly the report rejected Chelintsev's positive proposals for an alternative:

> A number of Soviet chemists have repeatedly called G.V. Chelintsev 's attention to these and many other unavoidable contradictions between the basic facts of physics and chemistry on the one hand and his [Chelintsev's] postulates with their consequences on the other. It is significant that even the author himself does not apply his 'new structural theory' in his writings.

The second conference confirming the decisions of the first, rejected Chelintsev's views even more explicitly, argued that the theory of resonance leads to agnosticism and replaced it by a theory of mutual influences (more akin to what had become known as the 'molecular orbit' method). The final report criticised

---

The Nature of the Chemical Bond.
[92] When presenting a paper on September 19, 1861 at the 36th congress of German naturalists and physicians in Speyer, entitled 'Einiges über die chemische Struktur der Körper'. For some quotes in English from the published Russian version of Butlerov's paper see Arbuzov (1952). Butlerov worked his ideas into a textbook on organic chemistry published in Russian in 1864 and translated into German in 1868.

PHILOSOPHY OF CHEMISTRY

Block 1-10   Views of western chemists on resonance structures.[94]

> COULSON 1939
> After all everyone knows that resonance is an excellent and useful approximation, which enables quite a large number of deductions and predictions to be made; but it would be a pity if anyone ever came to believe that resonance was real, and that therefore without appealing to its phraseology and technique, we could not interpret many of the facts of theoretical chemistry.
>
> PAULING 1940
> Regarding the element of arbitrariness in the concept of resonance we might say that the molecule cannot be satisfactorily represented by any single-valence bond structure, and abandon the effort to correlate its structure and properties with those of other molecules. The properties of the molecule are essentially those expected for an average of the two valence-bond structures, except for the stabilising effect of the resonance energy.
>
> PAULING 1954
> In the description of the theory of resonance in chemistry there has been a perhaps unnecessarily strong emphasis on its arbitrary character. The convenience and usefulness of the concept of resonance in the discussion of chemical problems are so great as to make the disadvantage of the element of arbitrariness of little significance.
>
> WHELAND 1955
> The individual structures which contribute to the state of a resonance hybrid are merely intellectual constructions, and hence they do not correspond to any actual molecules. We see that resonance is a man-made concept in a more fundamental sense than most other physical theories. It does not correspond to any intrinsic property of the molecule itself, but instead it is only a mathematical device, deliberately invented by the physicist or chemist for his own convenience.

Soviet philosophers, chemists, and physicists, each slightly differently, but fundamentally for not noting this infiltration of idealistic tendencies.[93]

The conclusions of these reports were quite different from those of the 1948 meeting at the Academy of Agricultural Sciences where Lysenkoism was the outcome. The greatness of the 'Russian scientific genius Aleksandr Mikhailovich Butlerov' was heralded; the term 'resonance' could not be used - its 'idealist' associations were to be shunned; chemists who had used the term 'resonance' in their writings were required to acknowledge their mistake. For example, Syrkin and Dyatkina (the previous 'champions' of resonance theory in the USSR) were forced to admit their errors, analysing five sources of their mistakes, concluding (1952:978):

> We shall in our future work make every effort towards the rectification of our mistakes. Our future task will be to contribute to the development of knowledge of the structure of molecules on the lines of the teachings

---

[93] Of the second report only a stenographed version was published and I've drawn on Graham's (1971:307-312) summary of its major parts. Probably it still formed the basis of a 1954 report of the USSR Academy of Science that was also published in German (AN SSSR 1954).
[94] Wheland (1955:612), Pauling (1940:11, 1945:127f), C.A. Coulson, letter to Urey, 20 July 1939, quoted in Simoes and Gavroglu (1999:378n50); Pauling (1964:433-4) - his 1954 Nobel Lecture.

of dialectical materialism and of the ideas of Butlerov, Mendeleev, Markovnikov, and other coryphaei of Russian science.

However at both conferences, Chelintsev's 'new structural theory', was decisively, and with little discussion, rejected as worthless. At the level of chemical research the difference between 'bourgeois' resonance theory and the continuation of the 'great tradition of Butlerov' amounted to little more than substituting 'mutual influences' for 'resonance'. Chelintsev made another attempt to gain influence when an edited collection of Butlerov's works appeared in 1951 (Graham 1971:312), though it had no impact whatsoever.

When Stalin died in 1953 the rhetoric disappeared. The last 'official' commitment to a 'purely Butlerovean' theory of structure for organic chemistry (incorporating new developments in quantum mechanics) appeared in 1954.[95] Two issues remained. The first was the old priority dispute, which will probably remain unresolved forever, though according to Rocke (1981:51):[96]

> The principal developers of those ideas had reached a satisfactory settlement by 1868 [including Butlerov]. It has been historians who have since muddled this interpretation.

Butlerov and Kekulé were personal friends and as far as is known had no quarrel about priorities. The second issue is the philosophical problem of what sense might be made out of saying that something, which does not exist (viz. the different resonance structures of benzene), can be used to explain all kinds of things. This criticism of resonance theory was most fully presented in a book by Shakhparanov (1957) and still present in Zhdanov (1960), but fizzled out (Budreiko 1970), partly because interests in the philosophy of chemistry became more determined by issues surrounding the influential work of Kedrov.

The translation of Shakhparanov's short book in German was not well received by philosophers of chemistry in the DDR, primarily because Shakhparanov's chemical expertise was unconvincing and his arguments were considered sloppy in the following sense:

- He confused the use of visual models with being committed to idealism.
- He failed to realise that resonance structures illustrate the dialectical character of nature, a point already made in 1939 by Haldane, the translator of Engels' *Dialectics of Nature*.[97]
- He failed to realise that even 'ordinary' double and triple bonds as depicted in structural formulae of molecules are not always 'pointing to' identical material entities, which illustrated his lack of chemical expertise.[98]

---

[95] That is the report mentioned in the previous note (AN SSSR 1954).
[96] For non-biased views on the Butlerov-Kekulé controversy from the Russian and the German side respectively, see Kuznetsov and Shamin (1993) and Stork (1963).
[97] Haldane (1938:91f) refers to Pauling's resonance theory as 'a beautiful example of dialectical thinking, of the refusal to admit that two alternatives which are put before you are necessarily quite exclusive'.
[98] As Pauling (1964:434) put it in his 1954 Nobel Lecture: 'The description of the propane molecule as involving carbon-carbon single bonds and carbon-hydrogen single bonds is arbitrary; the concepts themselves are idealisations, in the same way as the concept of Kekulé structures that are described as contributing to the normal state of the benzene molecule.'

In the USSR too, Shakhparanov's book marked the end rather than the beginning of a discussion. In a review of it in *Voprosy filosofii*, the situation was well summed up by Abramova, Garkovenko and Solonov (1963):[99]

> The resonance theory as a method for quantum mechanical calculations of molecules has been surpassed by better methods and it is doubtful whether it is necessary to repeat again old and not very well supported criticisms of Machism and agnosticism at the defenders of resonance theory. The relation of scientific abstraction to objective reality is not as simple as the critics of resonance theory have often tried to present it.

That this example of the politicisation of science had more to do with nationalism than with dialectical materialism is confirmed by the fact that in the DDR, the general issue of the significance of the development of quantum chemistry was often discussed, though the Russian 'hype' about Butlerov had little impact - after all, Kekulé was German. Stork (1963) completely rejects the Russian accusation that the resonance theory developed by Pauling, Wheland, and Ingold fails both empirically and philosophically. He explicitly criticises 'the Russians' for not mentioning important western authors like Ingold, at the same time placing much weight on '(older) Russian chemists'. He suggests they give the wrong impression that the state of the art in structural chemistry is due mainly to Russian chemists, whereas 'western' chemists (including Germans) have produced little else but a false resonance theory (Stork 1963:616). In retrospect Eastern German chemists considered the earlier attack on resonance theory as 'exaggerated' (Gey and Parthey 1984:283).

The issue in Eastern Europe became even more complicated when Pauling, the main architect of modern resonance theory, started to publish on 'life and death in the atomic era' (Pauling 1964). After the 1951 meeting and report from Moscow, the *Chemical and Engineering News* (September 10, 1951) carried the headline 'Soviets Blast Pauling/ Repudiate Resonance Theory' and as late as 1965 Russian chemistry text books had to avoid the term 'resonance'. But in November 1961 Pauling gave a lecture entitled 'The theory of resonance in chemistry' at the Institute of Organic Chemistry of the USSR Academy of Science in Moscow to an audience of about twelve hundred people. The lecture was translated and published in a Russian journal (Pauling 1962), after which Pauling became the most admired Western chemist in the Soviet bloc. Laitko (1967) for example starts his overview of philosophical questions concerning chemistry with a reference to Pauling:

> Important chemists like Lines Pauling have become conscious of their humanistic responsibility and raise their voice against the misuse of science for aggressive, antihuman purposes. In the socialist world stability and military security contributes considerably to increasing welfare and better public health.

---

[99] See also Kuznetsov and Pechenkin (1972).

Although there were no longer constraints on talk about 'resonance', the philosophical discussion about the status of resonance theory continued and moved to the more general theme of the relation of chemistry to quantum mechanics. Whereas chemistry moved on (to the molecular-orbital method and further developments of quantum chemistry), the philosophical questions surrounding bonding theory were there to stay. Vermeeren (1986) points out that *because* of the language barrier and the political overtones of the exchange, the scientific and philosophical importance of the problem was and still is underestimated. A pragmatic chemist might be happy to say that a chemical bond is 'a figment of the imagination' and in the same breath that 'few would say that there is no such thing as a chemical bond' (Sutcliffe 1996:654). This however won't satisfy philosophers of science.

The general thrust of my account on the resonance issue is different from Laitko (1996). Though it would go too far to discuss the disagreement in detail, let me just mention two points. Laitko was himself an important contributor to discussions in the DDR,[100] and seems vulnerable to what one might call the Lysenko fallacy. Second, possibly related to the first point, he seems too sensitive to any suggestion that the nature of chemical bonds cannot be explained exclusively by quantum mechanics, a sensitivity common in the DDR when such statements appeared in Russian publications.[101] For example Laitko (1965:334n14) regrets that 'some marxist authors' have unjustly criticised mesomerism and resonance theories for being 'idealistic' (referring to Shakhparanov 1962:100-2). Instead he argues that 'marxist philosophy' should further the permeation of chemistry by quantum mechanical methods,[102] a view quite different from the views expressed in Laitko (1967) referred to above.

With the decline of dialectical materialism as a political force, discussions about the philosophy of chemistry in Eastern Europe disappeared, though some of the authors involved are still active in the philosophy of chemistry.[103] In conclusion the following may be said concerning this rather extraordinary period of interest in Eastern Europe in the philosophy of chemistry.

- Very general metaphysical background assumptions constrain priorities of research of philosophers of science. Philosophy of science in the tradition of logical empiricism was interested in what Kant called proper science, i.e. exact (mathematised, if possible axiomatised) science. Under dialectical materialism in Eastern Europe there was more room for the philosophy of the science of different kinds of matter.

---

[100] Laitko (1965, 1967); Laitko and Sprung (1970); and many more; *cf.* Schummer (1996b).
[101] Cf. Heber (1964) and Laitko (1965:334n16).
[102] 'Leider wird von einigen marxistischen Autoren bis in die jüngste Zeit die Meinung vertreten, die Mesomerie- und Resonanzvorstellungen selbst (nicht wohl bestimmte Arten ihrer philosophischen Interpretation) seien ihrer Grundlage nach idealistisch. Die marxistische Philosophie sollte die progressive Tendenz der Durchdringung der Chemie mit quantenmechanischen Anschauungen und Verfahren mit allen Mitteln fördern.'
[103] R. Vihalemm from Estonia, referred to above (Vihalemm 1974) presented papers at the 10th International Congress of Logic, Methodology and Philosophy of Science held in Florence (1995) and at a symposium on the philosophy of chemistry in Ilkley, U.K. (1997); Laitko and Niedersen published extensively in the DDR period and also, amongst others Laitko (1996) and Niedersen (1994). See also the reference to Kuznetsov and Shamin (1993) in note 73.

- Many of the ideas put forward by the best philosophers of chemistry in the USSR and the DDR, phrased in the terminology of Engels, Marx, and Lenin, are surprisingly similar to those that arose as part of the naturalistic turn in Anglo-American philosophy after the end of the cold war when philosophical discussions about supervenience and emergence started to *re*-emerge.
- The debates about the theory of resonance show that nationalistic concerns were of much greater importance than the dogma's of dialectical materialism. The concerns of the debates and power machinations in Moscow promoting Butlerov had no followers in the DDR - though it is no doubt correct that Butlerov's importance had been suppressed in German and English text books that gave reviews of the history of organic chemistry.
- The Chelintsev affair shows that even when political dogma's merge with the power hunger of mediocre scientists, extreme situations like the Lysenko affair do not necessarily arise. The more common situation is that either the political dogma's are tagged on to the scientific results without influencing scientific work or, in more foundational or philosophical issues, the dogma's may actually stimulate the scientist-philosopher to explore new avenues, without being bound by canonical reading of those dogma's.

1.7 philosophy of science opens up

Since the 1970s, though many different strands of philosophy of science have emerged which generated more interest in case studies from chemistry, until the late 1990s, this rarely led to addressing anything *specific* to chemistry.

Since the 1960s a steady interest arose in historical case studies, both within philosophy of science 'proper' and in the sociology (or anthropology) of science. This was a reaction against the static 'structural' approach in the philosophy of science. As a result, philosophical reflection on chemistry increased roughly in proportion to the number of 'big' theories in chemistry. For example in Kuhn (1962), Boyle, Dalton, Lavoisier, and Priestley figure prominently. In a work, entitled *Method and Appraisal in the Physical Sciences* (note that *physical sciences* includes chemistry), there are five big case studies: atomism versus thermodynamics, Young versus Newton, oxygen versus phlogiston, Einstein versus Lorentz, rejection of Avogadro's hypothesis (Howson 1976). Donovan, Laudan and Laudan (1988), a collection of empirical studies of scientific change, also includes five case studies from chemistry. In Latour's seminal *Science in Action* (1987) there are more references to Crick, Mendeleev and Pasteur than to Einstein, Newton and Copernicus (and more references to engineers such as Diesel and Reynolds than to physicists). At least one case study appeared (Klüver and Müller 1972), taking up Habermas' (1968) *Frankfurter Schule* critique of 'positivist' approaches to science such as those of Popper and Kuhn, showing how the acceptance of the Kekulé structure of benzene was connected to dye manufacturing and the expansion of the German chemical industry.

Second an interest developed in the experimental side of science. This was a reaction against the bias within philosophy of science towards (formal) theory and has also favoured the appearance of more chemical case studies. Although initially this research concentrated on things like air pumps, electron microscopes, and atomic parity violation experiments, several detailed chemical case studies have been published. Hacking's *Representing and Intervening* (1983) discusses microscopes, 'making' quarks, etcetera. But there are *also* discussions of Boyle, Dalton, Davy, Lavoisier, as well as references to Paracelsus, Berzelius, Brønsted, Kekulé, Lewis, Pasteur, Prout, and von Liebig.[104]

Third on a smaller scale, the old interest in the logical reconstruction of scientific theories took on new life in what is called structuralism in the philosophy of science. Here the amount of interest in chemical theories is noteworthy.[105] Finally, and most recently, interest has developed in cognitive science of science, which focuses on the context of discovery rather than of justification. Here too, case studies from the history of chemistry play a role in the discussion; in particular, the chemical revolution of Lavoisier.[106]

These changes in the philosophy of science started to have some impact in restoring the balance between chemistry and physics in philosophical discourse. But although a range of detailed chemical case studies have been published since 1960 (Block 1-11), the more influential contributions to the debate have remained focussed on physical ones. Most of the chemical case studies, moreover, are not concerned with what might be specific for chemistry. Usually they 'test' positions in 'general' philosophy of science. For example, Ramsey (1990) uses a case study in chemical kinetics to discuss issues of realism and anti-realism and to argue against Latour's idea of blackboxing instruments; in his (1992) Ramsey argues that the 'contingency' does not disappear if the instruments are available 'off the shelf'; they merely become epistemologically 'translucent' rather than 'black'. Publications such as these assume the unity of science: scientists use instruments – question: are they 'translucent' or 'black'? If one looks for conclusions that may tell us something specifically about chemistry, they strike one as superficial. Few practising chemists will be surprised to hear that (Ramsey 1992):

> The epistemological instability of chemical kinetic data was due to a number of factors including the resolving power of the instruments, the recognized lack of clear theories about chemical processes, and the recognized lack of experimental data on many chemical reactions.

In judging to what extent chemistry was neglected in the philosophy of science, it should be noted that if in 'general' philosophy of science an example from chemistry is chosen, this may be of interest in its own right. An example is the exchange between Howson and Franklin (1991) and Maher (1993), in which Mendeleev's work figures as the major example of a Bayesian prediction model, drawing attention to the peculiarities of the periodic table as a 'law' or 'theory' and the relevance of the prediction of new elements for the acceptance of the table.

---

[104] See also [3.5] on Hacking's entity realism.
[105] See Hettema and Kuipers (1988), Kamlah (1984), Balzer, Moulines and Sneed (1987), and Lauth (1989, 1993). See Scerri (1997b) for a critique of the claim of Hettema and Kuipers (1988) that they have successfully axiomatised the periodic table.
[106] See Thagard (1989, 1990); various contributions in Giere (1992).

Block 1-11 Selected examples of case studies involving chemistry, published between 1975 and 1994.

Akeroyd (1986-1993)	using examples from the history of chemistry to 'test' various models of the growth of science (Lakatos, Laudan)
Akeroyd (1990a)	uses a chemical example to propose, by analogy, an oscillatory model of scientific development.
Balzer et al (1987)	stoichiometry
Bantz (1980)	theories of chemical bonding as a case study for the structure of discovery
Bechtel (1988)	fermentation theory
Carrier (1993)	criticises experimental realism, drawing on Stahl, Priestley, Gay-Lussac, Dalton
Diamond (1988)	polywater episode (testing the thesis that younger scientists are the first to shift; result: negative)
Frické (1976)	Dalton versus Avogadro
Gay (1976, 1978)	theory of chemical radicals and the asymmetric carbon atom to assess models of scientific method
Hettema and Kuipers (1988)	periodic table
Hoffmann (1990)	molecular beauty
Lauth (1989)	stoichiometry
Lauth (1993)	Avogadro's number
Mauskopf (1988)	19th century French crystallography
McEvoy (1989)	on the chemical revolution in the tradition of Foucault
Rheinberger (1992a, 1992b)	protein and RNA synthesis in the laboratory (experimental reasoning)
Rocke (1988)	molecular structure of benzene (Kekulé's theory)
Rothbart (1993)	intertheoretic analogies during the 19th century unification of organic and inorganic chemistries
van Brakel (1993a)	polywater
Zandvoort (1985)	on nuclear magnetic resonance
Zandvoort (1988)	polymer chemistry as a testing ground for Lakatos' notion of research programmes

and on the chemical revolution - oxygen replacing phlogiston (Lavoisier):
Toulmin (1957), Putnam (1975), Kitcher (1978) Laudan (1984:68-102), Nola (1980), Kamlah (1984), Shapere (1984:325-333), Donovan et al (1988), Perrin (1988), Zucker (1988), Thagard (1989,1990), Sankey (1991), McAllister (1993)

Scerri (1997a) has argued that Mendeleev's 'dramatic' predictions of new elements may have contributed little to the acceptance of his periodic system. For example, the citation which accompanies his award of the Davy Medal by the Royal Society of London makes no mention of these predictions. On the other hand, in his Faraday lecture in 1889, Mendeleev said

> When, in 1871, I described to the Russian Chemical Society the properties, clearly defined by the periodic law, which such elements ought to

possess, I never hoped that I should live to mention their discovery to the Chemical Society of Great Britain as a confirmation of the exactitude and the generality of the periodic law.

In a note he adds: 'I foresee some more new elements, but not with the same certitude as before.' Hence this is an example both in general philosophy of science (on 'novel predictions') and in the philosophy of chemistry (on the idiosyncracies of the periodic system of Mendeleev).

Chemical examples may also figure in other parts of philosophy. In the philosophy of mind Searle (1983) has suggested that mental states are realized in and caused by the brain in much the same was as the liquid character of water is realized in and caused by the movement of individual ('non-wet') molecules. Johnston (1997) uses the example of diamond and soot to argue that what some call the hard part of the mind-body problem is perfectly general and has nothing to do with special features of mind and body. Elsewhere, modern definitions of the chemical notion of valence figure, as analogies, in esoteric ongoing discussions about the possibility of reducing truth to non-semantic notions.[107] There are also formal analogies in which chemical examples play a role. For example there is a mathematical similarity in the description of neural nets (or connectionism) on the one hand, and immunology and autocatalytic networks on the other (Farmer 1990). And autocatalytic reactions have drawn the attention of process philosophers.[108] Further, enantiomorphism (the 'chemistry of space') has been discussed in connection with Kant's antimony that one hand is a reflection of the other whereas no rigid motion in infinite Euclidean space can map the one embedding into the other.[109] Finally, philosophers may 'borrow' some idea from chemistry, which may, in retrospect, bear on the philosophy of chemistry. An example is Frege's famous *Begriffsschrift* (a formal language for 'pure thought'), published in 1879, which is based partly on his knowledge of the language of chemistry. He 'borrows' the chemical concepts '(un)saturated' and 'disintegration' (Majer 1996). Frege's formal approach could then be applied to the language of chemistry.

The chemical example that figures most prominently in hundreds of philosophical publications of the past few decades is 'water is H_2O'. This will be further discussed in chapter 4.

1.8 birth of philosophy of chemistry proper

By 1990 mathematics, biology, and cognitive science occupy respectable areas within the philosophy of science and there is interest too in the philosophy of economics and astronomy. The philosophy of medicine is on the rise, though hardly interacts with mainstream philosophy of science. A vast literature on the philosophy of the social sciences and the humanities (philosophy of social science, of literature, of art, of history) has appeared. However most of it is not part of philosophy of science, as normally understood, partly due to the divide between

---

[107] Field (1972), Putnam (1978), Weatherall (1993).
[108] See Earley (1981, 1992, 1998)
[109] See Le Poidevin (1994) and Nerlich (1995). Cf. Van Cleve and Fredrick (1991).

Block 1-12 Emergence of philosophy of chemistry at international conferences

4th ICLMPS	1971	Bellu
6th ICLMPS	1979	van Brakel and Vermeeren
9th ICLMPS	1991	Ruthenberg, Ramsey
10th ICLMPS	1995	Eisvogel, Hendry, Lei, Ruthenberg, Schummer, Vihalemm, Zielonacka-Lis
PSA	1978	Bogaard
PSA	1990	Allchin, Hofmann, Ramsey
PSA	1992	Allchin, Hofmann and Hofmann
PSA (colloquium)	1994	Ramsey, Rothbart, Scerri
PSA (symposium)	1998	Earley, Rothbart, Woody, Scerri
WCP	1998	Schummer
APA (symposium)	2000	Ramsey, Scerri, Woody

ICLMPS : International Congress of Logic, Methodology and Philosophy of Science
PSA: biannual meetings of the Philosophy of Science Association
WCP: World Congress of Philosophy; APA: American Philosophical Association, Central Division

analytic and continental philosophy. The philosophy of law and the philosophy of religion belong to separate parts of philosophy. There is considerable interest in the philosophy of linguistics closely tied to the philosophy of language (or mind). Finally there is some interest in the philosophy of technology (including engineering and agricultural sciences) and considerable interest in environmental philosophy, a subject usually considered part of applied ethics.

Then finally, in 1994, philosophy of chemistry was born. That is not to say *nothing* happened before 1994 (see previous sections). But in March 1994 the *First International Conference on Philosophy of Chemistry* took place in London, followed a month later in Karlsruhe by the *Tagung Philosophie der Chemie: Bestandsaufnahme und Ausblick*.[110] In October of that year the American *Philosophy of Science Association*, organised for the first time in its history a colloquium on the philosophy of chemistry during its biannual congress. In November the *Second Erlenmeyer Colloquium* on the philosophy of chemistry took place at Marburg.[111] In December 1994 the meeting *Riflessioni Epistemologiche e Metodologiche sulla Chimica* was held in Rome.[112]

Shimmerings of what was to come can be seen in the years leading up to 1994 (see also Block 1-12). The 1994 Karlsruhe meeting was organised by the *Arbeitskreis Philosophie und Chemie*, which was founded at a meeting in Coburg (Germany) in June 1993. The first *Erlenmeyer-Kolloquium der Philosophie der Chemie* took place in November 1993.[113] Also, in response to the negative image of the chemical industry, a programme 'Chemistry and the Humanities' had been running for some time in Germany, leading to Mittelstraß and Stock (1992). In 1993 there were meetings on the history and philosophy of chemistry too in London, Bradford and Perugia. The Proceedings of the latter were published as *Atti del*

---

[110] The contributions were published in Psarros, Ruthenberg and Schummer (1996).
[111] The proceedings have been published in Janich and Psarros (1996).
[112] Papers were published in Mosini (1996).
[113] Papers were published in Janich (1994c).

Block 1-13   Post-1994 areas of interest in the philosophy of chemistry.

journals	*Hyle* (1995-), *Foundation of Chemistry* (1999-)
conference proceedings	Janich (1994c), Janich and Psarros (1996, 1998), Mosini (1996), Psarros, Ruthenberg and Schummer (1996), Psarros and Gavroglu (1999)
books, philosophy of chemistry	Schummer (1996a)
books, history of chemistry	Hoffmann (1995), Duncan (1996), Bensaude-Vincent (1998), Bensaude-Vincent and Abbri (1995)
chemical substance	see chapter 3
chemical thermodynamics	see section [5.2]
quantum chemistry	see section [5.3]
chemical laws	see section [5.1]
chemical models	see Block 6-1 and note 382.
language of chemistry	Bensaude and Abbri (1995), Schummer (1996c, 1998b), contributions in Janich and Psarros (1996), Crosland (1996), Weininger (1998), Laszlo (1998), Sanchez Perez and Sanchez Marin (1997)
chemical space, synthesis	Schummer (1995, 1997c), Cornforth (1993), Tontini (1999)
instruments, analysis	Cerruti (1998), Laszlo (1998), Rothbart and Scherer (1997), Rothbart (1999), Baird (2000)
some recent case studies	Akeroyd (1996, 1997), Allchin (1997), Brush (1999), Glas (1999), McEvoy (2000), Psarros (1997), Shelton (1995), Zollinger (1997)

V Convegno Nazionale di Storia e Fondamenti della Chimica (Marino 1994). As the title indicates, this was the fifth meeting of its sort (the first one taking place in 1985). Probably the first journal in which papers on the philosophy of chemistry were published with some regularity was the Italian journal *Epistemologia*.[114] In Italy the initial interest came from Chairs in theoretical chemistry, not philosophy.[115] In the Netherlands interest began in 1981, stimulated by A. Rip, H. Vermeeren and P. van der Vet.[116] There was however little follow up. Early interest in the philosophy of chemistry in various countries was often tied to meetings with a predominantly historical orientation.

By 1995 there were seven contributions to the philosophy of chemistry at the 10th International Congress of Logic, Methodology and Philosophy of Science held at Florence (Block 1-12).[117] The journal *Hyle: An international Journal for*

---
[114] Lévy (1979), Del Re (1987), Liegener and Del Re (1987a), Villani (1993), Mosini (1994), Psarros (1995b).
[115] For example: Del Re (1974), Barbier and Vincent (1981), Paolini (1981).
[116] See *Chemisch Magazine* (The Hague), October 1981, pp. 591-8. Cf. Vermeeren (1986), van der Vet (1979, 1987, 1989), Zandvoort (1985, 1988).
[117] For titles of presentations see *Hyle - Bibliographie (post 1990 literature)* at the *Hyle* website (http://www.uni-karlsruhe.de/~philosophie/hyle.html).

the Philosophy of Chemistry appeared in electronic form in 1995 and in 1997, its third, now full-fledged volume appeared in printed form. The *International Society for the Philosophy of Chemistry* was formally established at an international symposium on the philosophy of chemistry and biochemistry held at Ilkley (U.K.) in July 1997. Again in 1997, a special issue of *Synthese* on the philosophy of chemistry came out, bringing it to the attention of a broader philosophical audience.[118] In 1999 the first issue of *Foundations of Chemistry* appeared.

Since then work on the philosophy of chemistry has rapidly proliferated – so much so that there is already too much to review here (Block 1-13). Both traditional themes like 'reduction' or 'substance' received considerable attention from various directions and 'new' issues such as chemical synthesis were addressed. Subjects like the 'language of chemistry', which before 1994 had received a little attention gained momentum. Block 1-13 presents a crude division of current interests in the philosophy of chemistry. No doubt this will have to be adjusted soon.

---

[118] In 1986 a Symposium on the Philosophy of Chemistry appeared in *Synthese* (vol. 69, No. 3) with contributions of Hall (1986), Vermeeren (1986), and van Brakel (1986).

2   philosophical preliminaries

2.1   manifest and scientific image[119]

The terminology of manifest versus scientific imagery stems from Sellars. The former is the daily practice or common-sense-human-life-form; it refers to things like water, milk-lapping-cats and injustice-angry people. It also applies to sophisticated interpretations of 'people-in-the-world'.[120] The scientific image is concerned with things like neurons, DNA, quarks, and the Schrödinger equation, again including sophisticated reflection, and a promise of more to come. The distinction is often made by philosophers, even if the terminology of Sellars is not used. For example, Nagel (1970:117f) distinguished between

> the characteristic commonly attributed to things on the basis of everyday encounters with them, and the accounts of those things given by scientific theories that formulate some ostensibly pervasive executive order of nature. The more familiar and manifest traits of things are those which scientific theory attributes to them.

It is a distinction that in the modern world is all pervasive and has encouraged the dominance of naturalism and physicalism in analytic philosophy and beyond: 'the physicist's language gives us a *truer* picture of the world than does the language of common sense' (Smart 1964:47), a view of a world free of norms, a world such as scientists describe. Sellars' scientific image corresponds to Eddington's viewpoint of no one in particular, to Nagel's view from nowhere, and Williams' absolute conception of the universe, a conception of the world as it is, independently of our inquiries or experiences.[121] The same notion of objectivity occurs in 'non-scientific' contexts as well, for example in the style of the omniscient narrator or in Rawls' notion of judgements that occur behind a veil of ignorance.

In the first chapter of *The Many Faces of Realism*, Putnam (1987) castigates Sellars for eliminating the manifest image from his scientific realist world picture. According to Putnam, Sellars eliminates the world of middle-sized objects and their use by persons, replacing it by a superposition of particles, fields, and a space-time manifold (or however scientists would describe the world at the end of their inquiry). The following quotation from Putnam brings out the tension between the scientific image and the manifest image (1987:4):

> In the melodramas of the 1890s the Seducer always promised various things to the Innocent Maiden which he failed to deliver when the time

---
[119] This section is based in part on van Brakel (1996b).
[120] Sellars uses the expressions 'man-in-the-world', 'man-in-the-universe', 'discourse-about-man-in-all-discourse'.
[121] Eddington (1921), Williams (1978:236-49, 1985:136-9), Rawls (1971).

came. In this case the Realist (the evil Seducer) promises common sense (the Innocent Maiden) that he will rescue her from her enemies (Idealists, Kantians and Neo-Kantians, Pragmatists, and the fearsome self-described 'Irrealist', Nelson Goodman) who (the Realist says) want to deprive her of her good old ice cubes and chairs. Faced with this dreadful prospect, the fair Maiden naturally opts for the company of the commonsensical Realist. But when they have travelled together for a little while the 'Scientific Realist' breaks the news that what the Maiden is going to get isn't her ice cubes and tables and chairs. In fact, all there really is - the Scientific Realist tells her over breakfast - is what 'finished science' will say there is - whatever that may be. She is left with a promissory note for She Knows Not What, and the assurance there are some Dinge an sich that her 'manifest image' (or her 'folk physics', as some Scientific Realists put it) 'picture'. Some will say that the lady has been had.

This seems to dovetail with Sellars' (1967:143) comment that:

> I agreed with Kant that the world of common sense is a 'phenomenal' world, but suggested that it is 'scientific objects' rather than metaphysical unknowables, which are the true things-in-themselves.

At the outset I should stress that Sellars' and my use of the word 'image' is a literary conceit. It does not presuppose a mental representation or mirror of the world, nor does an 'image' name a concrete type. It is merely a 'useful fiction'. Sellars' terminology simply is a reminder of the fact that when confronted with a clash between science and what we already believe, there may be an 'automatic' inclination or demand in the modern world to favour the scientific image way of putting things. But I will use 'manifest image' in a different sense than Sellars, avoiding his associations of the manifest image with sense data or phenomenal (phenomenological) data. My use should rather be thought of as akin to the concept 'form(s) of life'. That is to say 'manifest image' is short for 'manifest form(s) of life, understood interculturally', and that is how the term will be used in the sequel (Block 2-1). In particular, the manifest image in my sense should not be associated with the view that there are basic-level categories ('natural kinds') like water and red, which are 'manifest' natural categories because of the way the world and our neurophysiology have been fine tuned by evolutionary processes.[122]

Although 'manifest' can often be identified with 'macroscopic' and 'scientific' with microscopic (and this tends to be Sellars' view), these terms should not be taken as synonyms. Assume a scientific theory that takes supra-macroscopic objects or 'middle-sized' unobservables as its basic objects. Such a theory would be a straightforward part of the scientific image. On the other hand, unobservables are postulated in the manifest image. Folk psychology particularly draws on an abundance of unobservables. Not only gods and angels, but also beliefs, desires, hopes, and meanings are unobservables. As to the physical world, very young children al-

---

[122] For a defence of this view see Lakoff (1987) and for a critique van Brakel (1991a).

Block 2-1   Manifest form(s) of life and congeners.[124]

manifest image	Sellars
form (s) of life	Wittgenstein
Lebenswelt, natural attitude	Husserl
Lebenswelt	Habermas
Dasein	Heidegger
folk psychology	Churchland
common sense	Austin
background	Searle
world of practical realities	James
natural attitude of everyday life	Garfinkel
everyday life	de Certeau
habitus	Bourdieu
praxis	Bakhtin
conceptus cosmicus	Kant
phronesis	Aristotle

ready invoke 'tiny invisible particles' in explanatory strategies, for example in understanding why water tastes like sugar when sugar has 'disappeared' in the water (Au, Sidle and Rollins 1993)[123] - the unobservable is scarcely the prerogative of the scientific image.

According to Sellars the task of the philosopher is to fuse the manifest and scientific perspective into one 'stereoscopic' or 'synoptic' view (Block 2-2).[125] He also says that the manifest image should not be 'overwhelmed in the synthesis' he aims for, and he stresses the folly of attempting to replace the manifest image *'piecemeal* by fragments of the scientific image'.[126] Nevertheless, the scientific image claims (or is defined as having as its goal) to give 'the *whole truth* about that which belongs to the image' and 'science is the measure of all things, of what is that it is, and of what is not that it is not'. Sellars also says that the critique he is engaged in is 'one which compares this [manifest] image unfavourably with a *more* intelligible account of what there is'.

One reason Sellars claimed primacy for the scientific image is because of the incompleteness of the manifest image. Sellars also found the two images incompatible.[127] The question that immediately arises is: From which perspective does Sellars claim such things? Although Sellars' writing sometimes suggests a third stance, the problem raised is more serious: it is a conflict between world pictures, each of which claims to be complete in principle.

---

[123] Cf. Quine (1960:223) about the solubility of sugar: 'men talked easily of solubility before those explanations (i.e. the explanations of chemical theory) were at hand'.
[124] For an elaboration of using 'form of life(s)' *at the same time* in the singular and the plural see van Brakel (1998a, 1999c).
[125] Compare Garfield's (1988) 'binocular vision' in which he tries to combine the two images as 'coequal and mutually complementary'.
[126] This acknowledges the pervasive holism of the manifest image and foreshadows the frame problem in artificial intelligence (van Brakel 1992b).
[127] It does not rest, *pace* van Fraassen (1975), on the incoherence of the manifest image itself (Thornton 1981; Sellars 1963:29).

Block 2-2   Sellars on the tension between the manifest and scientific image.[128]

> The aim of philosophy is to understand how things in the broadest possible sense of the term hang together in the broadest possible sense of the term.
>
> The philosopher is confronted by *two* pictures of essentially the same order of complexity, each of which purports to be a complete picture of man-in-the-world, and which, after separate scrutiny, he must fuse into one vision: the manifest and the scientific image.
>
> The manifest image limits itself to correlational techniques that can tell us about perceptible and introspectible events; the scientific image postulates imperceptible objects and events for the purpose of explaining correlations among perceptibles.
>
> Although the framework of perceptible objects, the manifest framework of everyday life, is adequate for the everyday purposes of life, it is ultimately inadequate and should not be accepted as an account of what there is *all things considered*.
>
> A world picture which included *both* the physical objects of the manifest image *and* complex patterns of physical particles would contain a redundancy.
>
> The fact that each theoretical image is a construction on a foundation provided by the manifest image, and *in this methodological sense* pre-supposes the manifest image, makes it tempting to suppose that the manifest image is prior in a *substantive* sense; that the categories of a theoretical science are logically dependent on categories pertaining to its methodological foundation in the manifest world of sophisticated common sense in such a way that there would be an absurdity in the notion of a world which illustrated its theoretical principles *without also illustrating the categories and principles of the manifest world*. Yet, when we turn our attention to 'the' scientific image which emerges from the several images proper to the several sciences, we note that although the image is *methodologically* dependent on the world of sophisticated common sense, and in this sense does not stand on its own feet, yet it purports to be a *complete* image, i.e. to define a framework which could be the *whole truth* about that which belongs to the image. Thus although methodologically a development *within* the manifest image, the scientific image presents itself as a *rival* image.
>
> There is an important sense in which the primary objects of the manifest image are *persons*. Thus the conceptual framework of persons is not something that needs to be *reconciled with* the scientific image, but rather something to be *joined* to it.
>
> To all of which, of course, the manifest image replies that the scientific image cannot replace the manifest without rejecting its own foundation.

Both the manifest and the scientific image are idealisations, in the sense that each 'is a conception of an integration of a manifold of images, each of which is the application to man of a framework of concepts which have a certain autonomy'.

The point is not, as Bernstein (1966:305) argued, that Sellars failed to clarify the ground rules for the metaframework in which to judge whether the redefinition of the manifest in scientific terms is *acceptable*. The point is that both pictures claim to provide the resources for whatever metaframework we might need. There-

---

[128] All unidentified quotations in this section are from Sellars (1963), in particular from pp. 4-9, 15, 18-20, 27-31, 38, 40, 302-3.

fore it is wrong to refer to Sellars' two images as embodying (or employing) 'distinct conceptual schemes', 'conceptual frameworks', or 'different types of theory construction'.[129] Terms like 'commitment' or 'stance' - although admittedly vague - give a better indication of the sort of problem Sellars has raised. Sellars suggests that there are no independent grounds to adjudicate between the two rivals: any arguments given are always arguments from within one image or the other. This, however, misses the point. The issue of rivalry is an external question (in the sense of Carnap),[130] which, to the extent it is meaningful, is a question within the manifest image.

The lacunae or incompleteness Sellars sees in the manifest image are only there from the perspective of the scientific image. Similarly, the manifest image is 'a sort of naive-scientific image' only when seen from the scientific image. Sellars is equally mistaken in thinking that within the manifest image the mental cannot live happily with the physical. Combining the mental with the physical is only a problem for the scientific image. As to the alleged incompleteness or vagueness of the manifest image, van Fraassen (1998) has argued that the scientific image 'is as replete with uncashed and ultimately uncashable promissory notes as the manifest image'. In support he gives a detailed analysis of the vagueness of the *scientific* concept of 'shape' [5.5]. Of course these 'uncashable promissory notes' in science should not be thought of negatively. It is because of these that there can be any change. This is implicit both in Kuhn's suggestion that rationality *requires* incommensurability, as well as in the open-ended concept of rationality that we find in the writings of James and Peirce. But if both the manifest and the scientific image draw on vague concepts, there is no clash of the sort that 'proves' that both cannot be right. As an example, consider Sellars' worry, that *as described* a pink ice cube cannot be identical with any object in the world described by science. As concepts and corresponding worlds are equally vague in the scientific and manifest image, there is no real clash. Hence, there is no incompatibility. The scientific image can describe humans who talk about pink ice cubes,[131] whereas the manifest image easily accepts explanations and stories about 'scientific' ice cubes as useful in certain circumstances.

Sellars' pink ice cube seems to present us with an unavoidable contradiction between the manifest and the scientific image. It *seems* something has to go. But, as van Fraassen (1986) argues 'nothing has to go'. This doesn't completely resolve the tension because the conclusion 'nothing has to go' is reached by argument. The scientific image would still have a *commitment* to the claim that the only true knowledge is scientific knowledge - no matter how the philosophical arguments run. Probably, Sellars would have agreed that *at the moment*, 'nothing has to go'.[132] But this doesn't change his commitment to the ideal of a Peircean end of inquiry, when the scientific image *will* account for everything. Looked at this

---

[129] As Aune (1990), Seibt (1990:234-7), and Sicha (1988) do.
[130] See Carnap (1956, 1963:929) and DeVidi and Solomon (1994).
[131] For an example of how the homogeneity of manifest colours can be explained using current scientific particulars see Clark (1989).
[132] Cf. Seibt (1990:231-270).

way, there are important consequences. Justification of such a commitment will be of a sort that is both pragmatic and grounded in the manifest image. Such a pragmatic justification is not in terms of scientific theories and arguments. Rather there is a certain asymmetry between the manifest and the scientific image that guarantees that if we really push towards the presuppositions and commitments of certain lines of argument, we will always end up in the manifest image. That is to say, the problem of the 'third stance' referred to above has a simple solution. The 'third stance' is *always* that of the manifest image.

When Sellars claims that the manifest image is the ordinary world of objects, events, and *persons*, one might suggest this is 'the real world if there [were] such things, and nothing otherwise' (van Fraassen 1998). But in the manifest image the latter reflection does not arise in this detached way. This is because the world being 'such things' simply is the world. Within the particular manifest image at hand, one cannot refer to the world *as described* by this or that rudimentary ontology. The thought that leads to this suggestion already presupposes 'a world of objects, events, and persons' *there*.[133] This manifest world is presupposed as given in any scientific practice. There is *no* 'sense in which the scientific picture of the world *supersedes* the descriptive ontology of everyday life' (as Sellars suggests).

Like Bernstein (1966), van Fraassen (1998) argues that it's not clear from which perspective Sellars is telling his story and concludes that he is located nowhere. But it is obvious which stance Sellars is adopting: He's telling the story from the scientific bench. Sellars addresses a real problem, despite the contradictions which arise when we try to formulate the issue 'clearly', that is, when assuming 'literal' ontologies in global domains where there are none - answering external questions as if they were internal questions. The images or stances Sellars calls up are like language, provided we have no mistaken ideas about language.[134] The choice of a stance is like the choice of a language in Carnap's (1956) sense. If such a choice or stance is analysed from the perspective of one allegedly universal discourse, paradoxes automatically arise. This point applies not only to Sellars' conception of the manifest and the scientific image. Other examples include Kuhnian incommensurability, Quine's web of belief, Goodman's worlds, and even the concept of truth (as Tarski showed).[135]

One might claim that there are numerous incompatible world pictures, but for present purposes all 'ordinary' world pictures are lumped together in the manifest life forms, because each clashes with the scientific image for similar reasons, in particular because of the problem of how, as Sellars puts it, 'categories pertaining to man as a *person* who finds himself confronted by standards (ethical, logical, etc.) which often conflict with his desires and impulses, and to which he may or may not conform, can be reconciled with the idea that man is what science says he is'.

Sellars' arguments for the primacy of the scientific image don't touch the manifest image because he offers them from the scientific bench. Even by scien-

---

[133] Doubts can of course be raised about the existence of *particular* (kinds of) 'such things'.
[134] Cf. note 183.
[135] Kuhn (1991), Quine and Ullian (1970), Goodman (1978), Tarski (1956).

Block 2-3 Primacy of the manifest image or form(s) of life.[137]

- MI is, in principle, homogeneous and complete; SI is fragmented and incomplete. Attempts to provide a unified science or one scientific method have failed.
- SI separates fact/value, subject/object, cognition/affect; MI is homogeneous in relation to rationality and morality.
- MI acknowledges differences of opinion (at least usually); SI not really.
- SI is dependent on folk intuitions and colloquial language: the notion of cause is grounded in MI; the idea of chance is grounded in MI; epistemic virtues are grounded in MI ; even logic and quantum mechanics cannot do without MI.
- This doesn't imply that MI is static or that MI wouldn't be influenced by SI or that MI-views are by definition better than SI-views.

MI = the manifest image; SI = the scientific image.

tific standards these arguments are not valid unless philosophical assumptions are used that lie outside the scientific image. Such arguments are in fact borrowed, in Sellars' terminology, from the 'perennial tradition', which includes for him the history of Western philosophy as part of the manifest image.[136] Moreover, the scientific image is dependent, at crucial points, on the manifest image, not merely historically, but at every important juncture of epistemic and ontological ratification (Block 2-3).

2.2 surfaces

In the previous section I introduced the distinction between the manifest and the scientific image. If we paint in broad strokes, 'macroscopic' corresponds to 'manifest', and '(sub)microscopic' with 'scientific', but in what follows finer distinctions are needed:

- The manifest/scientific distinction refers to the distinction between what is 'normally' called science and what not. This is a sociological, not essentialistic definition. The difference between physics and chemistry is again defined sociologically (chemistry being what one finds in a department of chemistry). On this criterion, physical chemistry is part of chemistry [5.2].
- The micro/macro distinction applies to levels of description in both physics and chemistry. Often 'macro' and 'micro' correspond to the distinction between phenomenological and fundamental laws, but not always. The prototypical example of a macroscopic theory is thermodynamics (both in physics and in chemistry). Microscopic theories introduce talk of molecules. A prototypical example is statistical mechanics. Quantum chemistry and quantum mechanics refer to an even 'deeper' level of description and I will refer to it as 'submicroscopic'.

This terminology will not cover all difficult border cases, but it suffices to make some crucial distinctions. I will now elaborate these distinctions using 'surfaces'

---

[136] Cf. van Fraassen (1995): 'whether or not the world exists is not settled by the success or acceptance of physical cosmology, *except* relative to certain philosophical points of view.'
[137] For more details see [8.2] and van Brakel (1996a).

Block 2-4   A few examples of surfaces or interfaces.

- ordinary solid objects in contact with a surrounding liquid and or vapour (gaseous) phase
- a solid deposit on another solid or solidified glue between two solid surfaces
- a solid surface in contact with a liquid or vapour phase in which a chemical reaction is taking place for which the solid is a catalyst
- a layer of oil on sea water, on sand particles, or on bird feathers
- a rain drop in air or on top of an 'oily' leaf, an oil drop in water, a soap bubble (internal and external surface)
- mercury in a barometer in contact with glass and 'vacuum'
- contact between various phases in heterogeneous systems, for example crude oil and water or steam in secondary oil recovery

or 'interfaces' as an example. For example: how should we envisage the surface of a sample of liquid water or the interface between liquid water and water vapour above it? Such an inter*face* can also be called an inter*phase*, because it is the interface between two phases or states of aggregation such as solid, liquid, and vapour. The vapour phase includes reference to a gas phase and vacuum (the limit of a vapour phase without any matter left). Some examples of surfaces or interfaces are given in Block 2-4.

Stroll (1988:53-65) has suggested that there are two different notions of surface, the common sense 'manifest' notion, which he calls *Leonardo surfaces*, according to which the interface between air and sea is neither air nor water; and the scientific notion of surface, which he calls *Somorjai surfaces*, according to which surfaces are physical objects. Stroll stresses that according to the manifest view it is by no means obvious that surfaces are material objects. Adams (1988) dismisses this view because Stroll 'regards philosophy as a self-contained "discipline"' and neglects the scientific side of the subject. Adams suggests that familiarising oneself with the mathematical discipline of topology will resolve all definitional questions about surfaces.[138] But the mathematical discipline of topology Adams invokes deals in abstract objects and abstract surfaces; is has *nothing* to say about things like an interface between liquid water and water vapour.

Stroll says that scientists identify surfaces with the topmost layer of atoms. But I don't think it's correct to say (ibid:54):

> There is virtual unanimity within [the scientific] community that any acceptable scientific account of surfaces must be about submacroscopic structures.

The study of surfaces belongs to interface chemistry, a part of physical chemistry and surfaces, like most other subjects covered in physical chemistry, can be looked at on a number of different levels. An engineer might be concerned to make a surface look especially smooth; any theory she uses is no more than an instrument relative to the practical goal. Here the technologist's (idealisation of a)

---

[138] It gives as the definition of the surface of an object that it consists of the points of the object that are closest to its outside, i.e. arbitrarily close to points not belonging to the object.

macrosurface merges with the manifest notion of a smooth surface and the theorist's non-existing 'frictionless surface'. At the other extreme, a quantum chemist may try to find the right sort of approximate mathematics to calculate the electron density at the surface of a particular oxide; so she's looking at something 'thinner' than a 'last layer of atoms'. Some scientists of course *do* look at the atomic level of the surface. Again others at the hydrodynamics and thermodynamics of macroscopic surfaces. There is no argument internal to science that makes one of these 'levels' more real than another. Moreover, scientists have great difficulty connecting the different levels of description in a satisfactory way.

If the question of priority *is* raised - which we don't have to raise, I suggest it is the macroscopic, thermodynamic surfaces - idealisations of manifest surfaces, that are much more real than surfaces at the submacroscopic level. If we don't look at the molecular/atomic level, we can ascribe an energy to a surface and have forces work along it, which can explain all kind of things. For example, why drops from a tap have a certain size; why if you take a liquid in a container out of the field of gravity it won't have a flat surface; why it is possible to blow soap bubbles; why a drop of oil creeps through every corner of your watch; why most liquids rise in a glass capillary (but mercury doesn't); the phenomena of surface tension waves, and so on. Though all these phenomena involve manifestly observable surfaces, it is not true, as Stroll suggests, that 'macroscopic-surface talk is, as it were, part of folk physics'.

On the other hand, at the molecular level it is not at all clear what the surface is. Consider the interface between liquid water and water vapour. There is a constant movement of molecules to and from the surface. The kinetics of this coming and going is described in statistical terms. A model might be: think of an ant hill which is overcrowded, its 'surface' covered with crawling ants. Moreover they are constantly sucked away from the 'surface' by an ant eater. A researcher is constantly dropping ants on top of the others as well. If we look from a sufficiently large distance no doubt there is a black surface there, but where is the surface exactly if we come nearby? The situation is similar for Stroll's Somorjai surfaces.

Also consider the experiment which was (allegedly) carried out by Franklin to estimate the size of molecules. On an absolute wind-free day he went to a small lake and started to put drops of pure oil on the water surface. He added oil until the whole lake had that typical rainbow-coloured-oily appearance. Assuming there was now a monomolecular layer of oil on the water surface, he could estimate the size of the oil molecules (as to the order of magnitude, the estimate happened to be correct). In this situation the 'head' of the oil molecule is 'in' the water and its 'tail' is in the air (because the atoms of its head have an affinity to water, and the atoms of its tail abhor water). Now consider the question: how many surfaces do we have here and where are they? Thermodynamically the answer is rather straightforward. There are three phases: a 'bulk' liquid phase, a 'bulk' gas/vapour phase, and an 'interface' or 'surface' phase. On the molecular level the only sensible answer would seem to be that there are two surfaces (one at both sides of the soap molecules), but this is inconsistent with the macroscopic view (both that of thermodynamics and that of folk physics). Quantummechanically speaking there is no surface or interface [5.5]. More important than the question how many surfaces or interfaces there are, is that the simple experiment ascribed to Franklin illustrates how the macroscopic and microscopic, as well as the manifest and scientific, interact to provide understanding.

2.3   reduction[139]

One way of reformulating Sellars' contention that the scientific image should have priority over the manifest image is to say that whatever there is in the latter has to be reduced to the former (or eliminated altogether). Before addressing the specific question whether chemistry can be reduced to physics in chapter 5, I'll review the concepts of reduction, supervenience, and emergence. Each of these aims to give a different interpretation of what I'll call 'interdiscourse relations'. For a start there's nothing wrong in understanding the latter phrase as 'interlevel relations', 'linking propositions', 'theoretical identities', or 'bridge laws'. Predicates labelling different discourses (or levels or domains or disciplines or theories) include: physical, chemical, behavioural, brain, genetic, cultural, moral, and so on.

The expression in scientific English that most often exemplifies reference to interdiscourse relations is 'to underlie' and its cognates. Its use can be illustrated with the following examples: A particular molecular structure underlies the property of being brittle. A conjunction of states representing the kinetic translational energy of the constituent molecules underlies the temperature of an ideal gas. Physical kinds underlie chemical kinds. In general, 'B' (the 'base') underlies 'S' (the reducible, supervening, or emerging facts, properties, kinds, laws or whatever). It is usually unclear what saying that B underlies S means, as there are many different ways to paraphrase this statement. In unreflective scientific language-use, talk of B underlying S seems to make no distinction between things being identical; being somehow related; one thing constituting, causing, controlling, or determining another; one thing making some sort of contribution to another; one thing being ceteris paribus statistically relevant for another; and so on. Again words in the last sentence suffer from the same sort of ambiguities. Still, what seems common to all occurrences is that 'to underlie' implies a form of hierarchy or asymmetry between the different discourses that are related. If B underlies S, apparently it is *not* the case that S underlies B. Theories of reduction, supervenience, and emergence are philosophical theories that aim to give metaphysical and/or epistemological content to the relation between B and S.

'Reduction' is related to a variety of other issues in the philosophy of science, for example the (dis)unity of science, models for the development of science and for the change of theories, the analysis of theoretical terms, the purpose of science, and types of scientific realism. Moreover, in one guise or another, reduction is pervasive throughout the history of philosophy (Block 2-5). Not only has it played a central role in the history of analytic philosophy and philosophy of science, but also the philosophies of Hegel, Marx, Nietzsche, Foucault, or Derrida, to name but a few, can be read as reductive programs.

---

[139] Earlier versions of sections [2.3] and [2.4] have also been published in van Brakel (2000a).

Block 2-5   Examples of proposed reductions.

S	is reduced to	B
platonism	'everything'	abstract concepts
materialism	'everything'	material entities
idealism	'everything'	ideas, spirits, monads
pantheism	'everything'	God
neutral monism	'everything'	sense data
anomalous monism	'everything'	events
naturalism	'everything'	natural science
naturalism	'everything'	physics
naturalism	'everything'	form of life
naturalism	values	facts
naturalism	reasons	causes
naturalism	philosophy	science
logicism	mathematics	logic
positivism	theoretical entities	observables
objectivism	colour	reflection of light
subjectivism	colour	activated neurons
eliminative materialism	folk psychology	neuroscience
behaviourism	psychological entities	observable behaviour
behaviourism	meaning	observable behaviour
intentionalism	meaning	intentions
regularity view	causes	regularities
Humean supervenience	causes	events
individualism	social processes	actions of individuals
atomism	substances ('stuff')	atoms
atomism	society	individuals
atomism	meaning	meaning of words
cultural relativism	their concepts	our concepts

The classical notion of reduction in the philosophy of science, which originated with the Unified Science programme of the logical positivists, had the following characteristics (Lévy 1979a, 1979b):
- It is an asymmetrical, non-reflexive, and transitive relation.
- The relation is linear in that it holds between not more than two theories.
- Its first aim is an explanatory one, usually that of connecting successor theories in the history of science – such 'synchronic' or 'intralevel' reduction will not however concern us further.[140]
- Its' second aim is the unification of scientific theories, either from the same or different disciplines - this will be our primary concern.

The classical view, exemplified most clearly by Nagel (1961), proposed two criteria for reduction: theoretical terms of S should be definable in terms of the lexicon of B; laws of S should be deducible from the statements of B. In a famous paper Oppenheim and Putnam (1958) proposed the following 'package deal':
- micro-reduction of the sort described by Kemeny and Oppenheim (1956);

---
[140] For the difference see for example Nickles (1973).

- levels of reality exist, with lower levels being more fundamental and composed of simpler elements;
- unity of science;
- cosmic evolution.

Early criticisms of this package deal tend to parallel criticisms of the Vienna Circle's unity of science programme: too much rational reconstruction towards the ideal of a universal language and ignoring scientific practice.[141] In reply to such criticisms various modifications were proposed. Notwithstanding their greater sophistication and rigour, they were scarcely different from earlier proposals.[142]

Without dwelling on details, there are two aspects of reduction that *must* be made more concrete. First the question must be posed what sort of thing it is that is being reduced. At least three types of reduction can then be distinguished. Constitutional reduction concerns the question of whether the domains B and S are ontologically identical, i.e. whether the S-entities are constituents of the same elementary substrates with the same elementary interactions as B-entities. Epistemological reduction concerns whether the concepts (properties, natural kinds) necessary for the description of S, can be redefined in an extensionally equivalent way by the concepts of B, and whether the laws governing S can be derived from those of B. Explanatory reduction concerns the question of whether for every event or process in S there is some mechanism belonging to B which (causally) explains the event or process. Note that this is a much stronger requirement than the observation that there are many phenomenological facts that can be given a plausible explanation in terms of a microtheory. I'll concentrate on epistemological reduction, because the other two make no sense without some sort of cognitive connection between the S- and B-domain.[143]

The second, more crucial issue is hidden in the phrase 'supplemented by suitable supplementary assumptions', which I omitted from the definition of epistemological reduction. Without this addition the reduction won't work, and undermines the idea that what is at issue is *reduction*, as acknowledged by Nagel (1961, 1974). Primarily, there are two kinds of 'supplementary assumptions': bridge laws and initial and boundary conditions. Bridge laws are usually taken to be nomological biconditionals which connect one term of the reducing theory with a corresponding term of the reduced theory or of equations which do the same job for a whole family of terms; either they translate kind-predicates in one science into those of a more basic one, or specify a metaphysical relation, like *being identical to* or *being a necessary and sufficient condition for*, between the kinds of one science and those of the reducing science.[144]

The obvious question to ask is: Aren't the bridge laws themselves in need of some explanation in terms of the reducing theory *alone*? The prototypical example

---

[141] For an early critique of Oppenheim and Putnam's 'sweeping' statements see Schlesinger (1961).
[142] See Lévy (1979), Sarkar (1992), Endicott (1998), Hooker (1981).
[143] For a detailed example focussing on the issue of explanatory reduction in the context of the philosophy of chemistry see Vemulapalli and Byerly (1999).
[144] Nagel did not require his bridge laws be biconditionals in form. Their purpose was solely to allow the derivability of the laws of the reduced theory from those of the base theory.

of a bridge law since Nagel (1961) is: 'temperature is mean kinetic molecular energy'. Another notorious example is 'water is H_2O'. The defender of reductionism will say that temperature is *nothing but* mean kinetic energy of molecules. However, putting aside the assumptions underlying the notion of averaging,[145] this bridge law doesn't apply to all (or even any) occurrences of temperature. Temperature is, in general, *not* the same as average mean molecular kinetic energy [5.2]; similar problems arise in the case of 'water is H_2O' [4.1].

Nagel (1961:372) was well aware that reduction was not possible 'unless a postulate is added relating the term 'temperature' to the expression 'mean kinetic energy of molecules'.[146] Later writers underplayed this problematic aspect of bridge laws. However, even if bridge laws could be understood as identities that justify elimination or replacement of the 'reduced theory', there would remain the rarely addressed question of initial and boundary conditions. One thing is clear: they are no part of the reducing theory.

2.4 supervenience and emergence

Due to the failure of reductionistic programs in the philosophy of mind and elsewhere, talk of reduction has been replaced by talk of supervenience. The most common claim is that a discourse S supervenes on a discourse B, *if*, in some sense of *necessary*

> it is necessary that if an S-truth changes, some B-truth changes;

and/or:

> necessarily, if two situations (objects, states, events) are identical in their B-properties they are identical in their S-properties.

Though it is notoriously difficult to give a satisfying definition of supervenience (Block 2-6), the connection with older ideas about reduction will be obvious: the supervenience relation is the generalised successor of the 'bridge laws' or 'linking propositions', limiting the relation to a token-token, instead of a type-type relationship. The hope was that supervenience relations would have all the advantages of reduction with none of its disadvantages.

Supervenience can be seen as a sophisticated version of the mirror of nature paradigm. The picture is sophisticated because more than one mirror is allowed in the world. Each mirror gives a different 'autonomous' picture of (part of) the world, but one mirror - the 'ideal' physical one - mirrors reality as it *is* (ontologically speaking). All other mirrors supervene on (part of) the ideal mirror. One could perhaps say that, on the one hand, the supervening mirrors have somehow emerged, but, on the other hand, they picture mere appearances, without cosmic significance. In all this, it remains an open question how far supervenience ex-

---

[145] And a host of other problems connected to the suggestion that thermodynamics can be reduced to statistical mechanics [5.2].

[146] Nagel (1974): 'the theory T is not derivable from (and hence not reducible to) the theory T', although T may be derivable from T' when the latter is conjoined with an appropriate set of bridge laws'.

Block 2-6 Variety of supervenience definitions.[147]

> KIM
> S *weakly supervenes* on B if and only if necessarily any two things that have the same properties in domain B have the same properties in domain S (that is, B-indiscernibility entails S-indiscernibility).
> S *strongly supervenes* on B if and only if the weak supervenience relation (defined above) applies across possible worlds.
> S *globally supervenes* on B if and only if any two worlds that are B-indiscernible are also S-indiscernible.
>
> ARMSTRONG
> If there exist possible worlds which contain an entity or entities B, and if in each such world there exists an entity or entities S, then and only then S supervenes on B.
>
> HARE
> Supervenience brings with it the claim that there is some 'law' which binds what supervenes to what it supervenes upon. That is, necessarily, if Sa then there is a valid inference of the form:
>   for all x, if $[(Bx => Sx) \& Ba]$ then Sa
> the two premises of which hold.
>
> DAVIDSON
> A predicate p is supervenient on a set of predicates S if and only if p does not distinguish any entities that cannot be distinguished by S.

tends; that is, what is to be included in the ideal mirror, and what supervenes on it? Presumably, in a thoroughgoing physicalist materialism, chemistry would supervene on the ideal mirror of physics and not be part of the *ideal* mirror (the subject of Kant's *eigentliche Wissenschaft*).

Much of the literature on supervenience is on such an abstract level that one wonders whether anything at all is being said (van Brakel 1996a). For example, most definitions of supervenience relations refer to objects or worlds or other things being 'indiscernible' or 'identical'. But what does it mean to say that two objects are 'exactly alike'? The problem passed over here is that in the *actual* world there are no two objects that are exactly alike - at the very least no manifest objects: 'identical' objects occupy different places, so their contexts (i.e. their boundary conditions) are different, if only slightly so. Lewis (1983:355) gives as an example of exact copies the output of a 'perfect' photocopying machine:

> Copy and original would be alike in size and shape and chemical composition of the ink marks and the paper, alike in temperature and magnetic alignment and electrostatic charge, alike even in the exact arrangement of their electrons and quarks.

But it is unclear what this means, because the arrangement of electrons and quarks is in constant flux and not well-determined when not being measured.

The most noteworthy characteristic of most supervenience definitions is that they do *not* state an asymmetrical relation, contrary to microreductionistic expec-

---

[147] Kim (1984, 1993), Armstrong (1989:103), Hare (1984), Davidson (1993). For discussion see van Brakel (1996a, 1999b).

tations. As Kim (1990:14), the most influential writer on the subject of supervenience, says:[148]

> Our mereological intuition, that macrophysical properties are asymmetrically dependent on microphysical structures, seems to be the major influence on our thinking.

That is to say, mainstream supervenience discussions, although claiming to be in the business of developing a form of non-reductive materialism, are wedded nevertheless to the intuition of a microreductionistic image of the world: the asymmetry is taken for granted, appealing to a long-standing atomistic intuition as part of the scientific image (Block 1-2). Armstrong, who makes symmetry an explicit part of the definition of supervenience, is a minority view. The dominant view is that the underlying discourse should be somehow prior to the overlaying discourse. But, though most writers on supervenience *suggest* there is a dependence relation, it has never been clearly specified - the implied asymmetry is missing from formal definitions (Block 2-6). If one likes the word 'supervenience' and one believes that water is H_2O, one can say that being water supervenes on being H_2O, but it is then equally true that being H_2O supervenes on being water.

Dieks and de Regt (1998) have pointed out how problematic reduction is *within* the physics of elementary particles. A text book will tell you that a neutron is composed of three quarks. But, if one digs a bit deeper, 'it proves to be impossible even to associate the description in terms of neutrons with any fixed number of quarks' (ibid:46). Therefore, they suggest that 'supervenience adequately captures the intuition behind reductionism', because sometimes specific links between levels of reality can be found and this is enough to provide empirical support for the metaphysical supervenience thesis. There is, however, a complication with the notion of supervenience that does not arise with most notions of reduction. If the supervenience thesis were true, for example for the relation between 'the physical' and 'the chemical', the following intuitively compelling argument applies:

- Chemical property X (say, 'electronegativity') causes another chemical property Y (say, a particular chemical reaction).
- X supervenes on physical story P.
- Y supervenes on physical story Q.
- That X causes Y supervenes on P causing Q.
- If P is a sufficient cause for Q, there cannot be another cause (principle of overdetermination).
- X does not cause Y.

As is not uncommon when one tries to set up a formal argument, it raises more questions than it answers - for example, why should one accept the principle of overdetermination and why does the conclusion follow [7.1]? However, the drift of the 'messy' argument will be clear.[149] Supervenience threatens to lead to a form of

---
[148] Part of the citation of Kim on p. 75.
[149] The type of argument stems from the philosophy of mind, where it has already led to scholastic disagreements about what the alleged argument *is*. See, for example, the exchange between Noord-

eliminativism. Although it protects the conceptual autonomy of the supervening discourse, ontologically (both in terms of entities and in terms of causes), this autonomy is bought at the cost of becoming, strictly speaking, epiphenomenal: a discourse that may have more practical relevance than the base discourse, but a discourse that is, strictly speaking, not about anything at all. Reduction scores better on this point, because *if* reduction would work then all chemical concepts, laws, causes could be tied via definitional correlations with physical concepts, laws, and causes. That is to say, on the reductionistic view, there is an 'exact translation' of X into P and of Y into Q. So there's only one cause X/P causing Y/Q and the problem of overdetermination doesn't arise.

After the initial hype about supervenience being *the* answer to a form of non-reductive physicalism, it was realised that few if any goods were delivered and in the past few years interest has shifted to emergence. This notion goes back to the 1920s, being the predecessor of the notions of reduction and supervenience discussed above.[150] At the time the terms 'supervene' and 'emerge' were used interchangeably. For example, Pepper (1926) said: 'an emergence [is] a change in which certain characteristics supervene upon other characteristics'. Roughly, emergent properties can be defined as properties of a 'whole' which are not possessed by its component parts. The following set of definitions, with P a property of a system x, is representative of contemporary discussions (Spencer-Smith 1995):

- P is emergent in x: P is novel in x, and no physical theory of the components of x can explain or predict P (radical emergence).
- P is emergent in x, relative to T: T is a theory of the components of x and P is novel in x, but T can neither predict nor explain x (epistemic emergence).
- P is emergent in x: P is novel in x, and P is explained by interactions between the components of x (interactional emergence).

Most scientists in 'postmodern' disciplines such as nonlinear dynamics, connectionist modelling, chaos theory and artificial life have, like Nagel (1961:375), no problem combining reductionism and a weak form of interactional emergence. As Wimsatt (1996) puts it: 'A reductive explanation of a behaviour or a property of a system is one showing it to be mechanically explicable in terms of the properties of and interactions among the parts of the system.'

Today's radical emergentism tends to agree with virtually all forms of reductionism and supervenience on the following points: the ultimate base is physical, systems have *systemic* properties, and a form of mereological supervenience applies. Talk of emergence brings in the question whether the systemic properties are *resultant* (i.e. reducible to the properties and relations of the parts of the system) or *emergent* (not reducible, not even to *relational* properties of the parts). The difference between 'resultant' and 'emergent' can be fought out over Broad's

---

hof (1999) and Kim (1999).
[150] Nagel (1961:366-380) gives an excellent survey of the early literature; for the recent re-emergence see Beckermann, Flohr and Kim (1992). For the history of 'emergence and reductionism' see also Stöckler (1991). For chemical examples in a recent discussion of emergence see Schröder (1998).

(1925:59) 'traditional' example of sodium chloride, or more complicated cases like fermentation, where an unexpected mode of organisation at the chemical level can explain that fermentation is more than an *ordinary* chemical process (Bechtel and Richardson 1992). Even Broad seemed to assume that emergent properties strongly supervene on microstructural properties (Beckermann 1992). That is to say: the emergent properties cannot be *defined* in terms of (deduced from) microstructural properties. But given both kinds of properties, they are connected by necessary laws.

If modes of interaction count as properties, emergentism becomes a truism because then all properties are emergent properties. This is particularly relevant to chemistry. A molecule (or atom) like sodium has a certain physical structure (nuclei and electron shells). The description of this system, for example the wave function of electrons, describes nothing but the modes of interaction of the molecule or atom, which can be realised in certain experiments (i.e. interactions with certain media). So, a phenomenon which in all possible circumstances behaves - interacts with any kind of environment - like a sodium atom is a sodium atom. It makes no sense to claim that a sodium atom has certain properties by itself, regardless of its modes of interaction. As there is no end to possible interactions, all properties of the system considered emerge from interactions with its environment (which also raises the question of the borderline between the system and the environment). I'll return to this issue in connection with the holism of quantum mechanics [5.5]. Of course, the strongly relational character of chemistry has often been noted before.[151]

The discussion about emergence is traditionally linked to that of downward causation. This notion will change in meaning with the specification of the kind of emergence. In the minimal sense of interactional emergence, 'downward causation' is the influence the relatednes of the parts of a system has on the behaviour of the parts (not the influence of a macro-property itself). If the notion of emergence is to make sense when considering interdiscourse relations between talk of chemical substances, chemical molecules, and quantum mechanical calculations, perhaps it is more a case of 'backward' emergentism. That is to say, it is the details of molecular structure or the introduction of quantum mechanics that is novel, not the properties of chemical substances. On the other hand, if one takes 'novel' in such a way that the most recent theory is the place from which everything is looked at, 'backward emergence' turns into 'radical emergence'. Then 'macrochemistry is not reducible to microchemistry' and 'microchemistry is not reducible to quantum mechanics' can be restated as: 'the property of being pure water is an emergent property relative to molecular chemistry' and 'being H_2O is an emergent property relative to quantum mechanics'.

---

[151] See for example Cassirer (1923:203-20), [1.3] on chemism, and [3.1] on chemical space.

2.5 natural kinds[152]

In two publications that reverberated through the philosophy of science, mind and language, Kripke (1980) and Putnam (1975) made gold, water and tiger famous as natural kinds and set the scene for numerous philosophical publications in which chemical substances, in particular water, figure as the prototypes of natural kinds. However, opinions among philosophers differ widely on typical examples of natural kinds (Block 2-7). Less common proposals for natural kinds include: dreams, good stories, well-formed arguments and illocutionary forces.[153] Discussion rages as to whether 'object', 'person', 'belief' or 'disease', or 'natural kind' is or is not a natural kind.[154] Disagreements about more 'obvious' examples abound as well. Kripke and Boyd mention yellow as a possible natural kind. But according to Hacking the colours are *not* natural kinds. In an influential article of Quine (1969), the primary example of a natural kind is red, though he eventually dismisses it because of the relative unimportance of colour for inductive success. Hacking thinks that social kinds (people, their behaviour and creations) should be kept apart from natural kinds, but Boyd (1991) argues that most social kinds are natural kinds too: any kind that functions in induction or explanation has a claim to be a natural kind as well.[155] On another track, Wilkerson (1988) disputes giving 'vegetable' or 'stone' (which are on Hacking's list of examples), the status of natural kind; such entities afford superficial knowledge, not detailed scientific analysis. He argues that biological kinds are prototypical natural kinds. In contrast, Dupré (1989) pointed out that Wilkerson's examples, do *not* name *scientific* kinds, i.e. biological species or higher taxa. Neither 'wolves' nor 'oaks', Wilkerson's prototypical examples, correspond to a biological species or higher or lower taxa. Moreover, many philosophers of biology argue that species are not natural *kinds*; they are rather to be seen as historical *individuals*.[156] As a consequence, in a later publication Wilkerson changed his view:[157]

> Biological natural kinds are determined by genetic real essences which are causally responsible for the behaviour of individual members of the kind. But, since there is considerable interspecific genetic similarity and intraspecific genetic variation, there are far more biological natural kinds than species.

Finally, concerning the received view that water, gold, and tiger are prototypical natural kinds, Churchland (1989:295) argues that these are merely *practical* kinds - *genuine* kinds are (fundamental) *physical* kinds:

> The familiar multitude of putatively natural kinds embraced by common sense, and by the many derivative sciences [which would include chem-

---

[152] This section is partly based on van Brakel (1992a, 1998c).
[153] Flanagan (1995), Bruner (1986:11), Searle and Vanderveken (1985:179).
[154] See on belief: Needham (1972), Ramsey, Stich and Rumelhart (1991:207,281), Chomsky (1995); on person: Clark (1991), Lowe (1991), Rovane (1993); on disease: D'Amico (1995), Margalit (1979); on object: Daly (1998); on the natural kind 'natural kind' Witmer and Sarnecki (1998), Shain (1993:290), Browning (1978).
[155] See also on social kinds Currie (1988), Wilkerson (1995:ch. 3).
[156] See Ereshefsky (1991), Ghiselin (1987), Kluge (1990), Splitter (1988), Williams (1985).
[157] Wilkerson (1995:133); cf. Dupré (1993).

Block 2-7   Some examples of natural kinds.[158]

Peirce	chemical elements, chemical compounds, all animals and plants, the sciences (ordered hierarchically), stoves, lamps (classified according to purpose), artists, business men, scientists, and classifications of works of art according to nature of composition
Russell	dogs, wood, atoms, molecules, electrons, positrons, neutrons, discrete series of energy levels
Putnam	water, gold, tiger, horses, electricity, tuberculosis
Churchland	mass, length, duration, charge, colour (of quarks), energy, momentum
Hacking	all sorts of animals, vegetables, minerals, insects, fish, stone, stomach
Bigelow	copper, gold, protons, electromagnetic fields
Buchwald	electric conductors, viscosity, metals, electrolytes (and everything else that can be measured)
Savellos	earthquakes, heart-attacks, high tides, photosynthesis, but not shootings, butterings of toast, and philosophical conventions
Kuhn	force, mass, element, ray (in wave theory, in emission theory)
Khalidi	quadruped, parasite, vitamins, hormones; diamagnetic, paramagnetic, and ferromagnetic materials
Boyd	salt of sodium, flame, yellow

istry], are at best merely practical kinds. Genuine natural kinds form a very small, aristocratic elite among kinds in general, being found only in the most basic laws of an all-embracing physics. And if there are no such laws, or if the human cognitive medium should turn out to be a representational cripple, then perhaps there are no natural kinds.

The lack of consensus about good examples of natural kinds is tied to the lack of consensus about what might be the criterion of being a natural kind (Block 2-8). For a start consider these four criteria: resemblance or similarity; figuring as projectible predicate or in a law of nature; origin; essence. That each of these four criteria makes some preliminary sense can be illustrated as follows, focussing on biological species. First, we may draw on morphological or perhaps functional similarity. This is nearest to the approach of grouping those things together that resemble one another in manifest characteristics. The second possible criterion is that of interbreeding, which might be taken as a typical example of a projectible predicate involving laws or regularities of nature, without much theoretical embedment.[159] Third, evolutionary theory in its neo-Darwinian form, has led to the descent criterion of species. Here, what matters is the origin of the entity. The fourth criterion, that of species as clusters of individuals bearing strong genetic similarities, is the archetype of a natural kind with an essence. Here what counts

---

[158] For Peirce on natural classes see van Brakel (1998c) and Hulswit (1997). Further Russell (1948), Putnam (1975), Churchland (1989), Hacking (1991), Bigelow, Ellis and Lierse (1992), Buchwald (1992), Savellos (1992), Kuhn (1993), Khalidi (1993), Boyd (2000).
[159] Note that coming of a common stock is not plausibly essential to a species, which could easily have been cross-bred from independent mutations.

is underlying deep structure, not manifest similarities, macroscopic scientific projectibles, or common origin.

This four-fold partition is for illustrative purposes only and is not representative of the variety of biological species concepts currently in vogue.[160] The example also shows that if we can't fit all four into one unified whole, the alternatives undermine the natural uniqueness of each. Ruse (1987) has argued for a 'consilience' of criteria, but this suggestion has gained little support. Major problems for finding *one* definition of biological kinds are: species evolve, are time and earth-bound; there is large intraspecific genetic variation; hybridisation and asexual organisms are difficult to fit in.

With some ingenuity the same four-fold division can be applied to other types of natural kinds, to obvious ones like gold and water, as well as to less plausible ones like schizophrenia.[161] Let's now look at some of the proposed criteria of characterising natural kinds in some more detail, starting with similarity.

As Quine (1969:116) says: 'there is nothing more basic to thought and language than our sense of similarity; our sorting of things into kinds.' I take it that nobody denies this. The problem is, it doesn't take us very far. Though we may have an innate flair for categorisation, it will apply equally to 'natural' and 'non-natural' kinds.[162] Quine also stresses that similarity has to be taken as a comparative notion; otherwise natural kinds could not be contained in other natural kinds. Moreover, as many have stressed, including Quine himself, it is not similarity, but similarity *in the same respect* that is required.

The notion of similarity can be considered at the manifest level ('what it looks like'), at the scientific or naturalistic level ('underlying similarities'), and at the metaphysical level ('resemblance of universals'). An example of the latter is Armstrong.[163] According to him:

> Resemblance can only be analysed in terms of respects of resemblance.
> Respects, in their turn, require to be explained by means of universals.

He would like to explain why our ordinary predicates which seem natural, like 'length' and 'red', *are* natural. However, he says that it is

> undeniable that different shades of red are different properties. It follows that *redness* is not a property common to all red things. To assert that a particular is red is to assert that the particular has some property, a property which is a member of a certain class of properties: the class of all the absolutely determinate shades of red.

So Armstrong's metaphysical approach removes *red* (and *colour* too) as real properties or natural kinds. 'Red' does not mark a natural similarity, though each

---

[160] Including: interbreeding, phylogenetic, ecological, structural similarity (genetic, chromosomal, developmental), cladistic, phenetic - some overlapping. For recent discussions on biological kinds see contributions in Ereshefsky (1991) and Wilson (1999).
[161] Witmer and Sarnecki (1998) list 13 different criteria that have been proposed to decide what is what is not a natural kind term.
[162] For a discussion and critique of Quine's notion of innate similarity spaces see van Brakel (2000b).
[163] Quotations in this paragraph are from Armstrong (1978a:57, 1978b:117,119,126). *Cf.* Armstrong (1983) for a somewhat different viewi

determinate shade of red does. These determinate colour shades are, contrary to introspection, complex properties: 'perceived qualities and relations are as much epistemological icebergs as any other aspect of reality.' This has the consequence that 'we will seldom, perhaps never, know what actual property the thing has, that is, what determinate shade of red it has.' Similarly, 'it may turn out, there is no such thing as length, but simply appearances of particulars having lengths, appearances founded upon other, genuine, properties.'

From a more strictly naturalistic perspective (like Quine's), Armstrong's approach is a lost cause from the start. It's clear that 'resemblance' should not be taken in a psychological sense (1978b:95): 'resemblance is always *identity of nature*' (also: 'is based upon nature'). At best perceptual similarities ('epistemological icebergs') will lead us to metaphysical resemblances, which, in the end, should 'reduce' to naturalistic universals. On a naturalistic view there cannot be a metaphysical world apart from the natural world. So it is science which should in the end encompass Armstrong's metaphysical universals (e.g. the absolutely determinate shades of red). But if the naturalistic ladder from psychology to physics is climbed we meet Quine (1969:138) again, who argues that advanced science transcends all superficial similarities without substituting deeper 'respects of resemblance':

> In this career of the similarity notion, starting in its innate phase, developing over the years in the light of accumulated experience, passing then from the intuitive phase into theoretical similarity, and finally disappearing altogether, we have a paradigm of the evolution of unreason into science.

An example of the scientific process of a step on the road of 'similarity disappearing' could be the change of chemical type of Dumas (who demanded similar fundamental properties of the members of the type) to the molecular type of Regnault where materials with very different properties are grouped using as criterion the substitution of functional groups.

We might be taken in by the argument to eliminate colour, because of the trouble of incorporating secondary qualities in physics – a major concern of Sellars for example. But if colour goes (because of what Quine calls its cosmic irrelevance), so too does water and ice cubes - pink, or otherwise.[164] Psychologically 'seeing similarities' works on anything. Quine's naturalistic approach in contrast shows that advanced science can do without natural kinds. Both his 'cosmic argument' and Armstrong's universals fail to support a substantial distinction between natural kinds and other kinds, at least for any kind of kind known by ordinary people.

Let's now consider the criterion of natural kinds having projectible properties or of figuring in natural laws. A modest formulation of this characteristic is (Dupré 1986): 'There is no harm in calling a set of objects that are found to have a substantial number of shared properties a natural kind.' But as Hacking (1991)

---

[164] The pink ice cubes are the ones Sellars' scientific Image would eliminate from the manifest image; cf. citation from Putnam on p. 42.

Block 2-8   A few proposed definitions of natural kinds.[166]

- a class of objects all of which possess a number of properties that are not known to be logically interconnected;
- each thing of that kind is of just that kind and not another kind by virtue of what that thing is on its own and apart from whatever decisions may be made by a sorter;
- a class of things that it seems in some sense natural to bring together under a general concept; a class of (all and only) things sharing some property or, perhaps better, some 'genuine' property;
- a type of property, process, state, event, or object studied by science, mentioned in scientific laws, and assumed to be a causal feature of the world;
- subjects of natural law;
- a spatiotemporally unrestricted or repeatable category ineliminatively presupposed by at least one true and explanatory law of nature;
- the sets that one picks out in giving explanations;
- characterized by *clusters* of properties which play an especially important *explanatory role* with relation to other properties and relations;
- not some defining essence or the fact that each participates in natural laws but the fact that each is always the same as itself relative to certain property ranges in accordance with natural necessity;
- distinguished from merely nominal kinds by virtue of the sort of generalisations we can warranty make about their members;
- natural kind predicates are inductively projectible, other predicates are not;
- to say that a particular predicate picks out a natural kind is to claim that making out the extension of that predicate would figure in the ultimate (ideal) practice;
- Natural kind terms are those which the omniscient spectator gives as names to the kinds which are most important in the cosmos' own quality space.

notes, there's a tension between requiring that the instantiations of a natural kind should have many properties in common (already advocated by Mill and Russell) and the aim of science to relate all these properties to one base property (as pointed out by Peirce). This tension might be resolved by defining a natural kind as a 'knot of similarities connected in a lawful way'.[165] The shared properties can then be given metaphysical work to do by playing a part in prediction, explanation, physical laws and causality: 'occupying the same causal role in nature' or 'having the same place in an ultimate scientific account of the world' (Levinson 1991)

If there are natural kinds then their properties, or at least some of their properties, are projectible. But having projectible properties doesn't make something into a natural kind, except when understood as 'a locally entrenched projectible

---

[165] Haack (1992), presenting this as Peirce's view.
[166] Russell (1948), Browning (1978), Hirsch (1982), Boyd (1991), Lowe (1991), Johnson (1990), Kitcher (1984a), Bigelow, Ellis and Lierse (1992), Kitcher (1993), Millikan (1984), Elder (1995), Wilkerson (1998), Dupré (1986), Shain (1993).

kind'.[167] I can predict/project how, in given circumstances, the following entities will behave: water, cyclopropenylindene,[168] oak, tiger, policeman, car, red paint, red-after-image, angry person, schizophrenic person, army, and so on. Of course some predictions may seem more reliable than others, but none are either absolutely guaranteed or completely arbitrary. In all cases there are ceteris paribus considerations and other uncertainties [6.2]. Similarly, a kind like *vovetas*, a category of the Native American Tsistsistas [Cheyenne] which includes most vultures, some hawks, two types of insects, as well as tornado's, has many projectible predicates.[169] The same applies to refrigerators or the Kwakw'ala word *lhenxa* which, amongst other things, refers to the kind of 'colour' a yellow banana and a green apple share.[170] We can think up the weirdest kinds which all have projectible properties (Hirsch 1993). That some kinds just *seem* more natural is merely because of deeper local entrenchment.[171]

So, if this is to lead anywhere for the natural kind theorist, we need more constraints. This is where the requirement comes in that natural kind properties aren't merely projectible, but function in 'proper' scientific laws. The point presumably is not that natural kinds or their properties are governed by scientific laws; after all, on the naturalistic view *everything*, however artificial or peculiar, is governed by scientific laws. The point must be that natural kinds figure prominently in the fundamental laws of science. But stressing the prominence of that role leads down to the slippery slope of Quinean ontological reductionism, along which natural kinds are eliminated by science. At best we end up with Churchland's view that the *only* natural kinds are 'mass, length, duration, charge, colour [of quarks], energy, momentum'.

To use an old Popperian terminology: on a thorough-going naturalistic view *all* natural kinds stem from a dogmatic attitude. A critical attitude however removes them: first the metaphysical, moral, affective, and secondary kinds go; then the biological kinds - think of them merely as historical individuals, haphazard results of natural selection. Chemical substances and elements are soon to follow, being framed on unwarranted anthropomorphic ideas about atoms and individuality ('molecules aren't *things*, they don't have shape' [5.5]). Finally we're left with a world of quarks (Quine 1992):

> My tentative ontology continues to consist of quarks and their compounds, also classes of such things, classes of such classes, and so on, pending evidence to the contrary.

---

[167] The terminology of 'entrenched' and 'projectible' stems from Goodman (1983); for discussion of projectable hypotheses and predicates see Stalker (1994).
[168] On this elusive C_3H_2 isomer see for example Schaefer (1986).
[169] This is a Tsistsistas [Cheyenne] word, the reference of which includes most vultures (*Cathartidae*), the common nighthawk (*Chordeiles minor*), swarms of green darners (*Anax junius*, a dragonfly), swarms of red skimmers (*Libellula saturata*), and tornado's (meteorological events), which, amongst other things, are perceptually similar in displaying the same kind of whirling movements in the air.
[170] For a discussion of *vovetas* see van Brakel (1991a) and for *lhenxa* Saunders and van Brakel (1997b).
[171] For arguments supporting the indeterminacy of kind terms see Wilson (1982), Li (1993), Dupré (1993).

And even they may have to go at the final pythagorisation of science. As Dirac (1939) suggested:

> If we express the present epoch, 2×10^9 years, in terms of a unit of time defined by the atomic constants, we get a number of the order 10^{39}, which characterises the present in an absolute sense. Might it not be that present events correspond to properties of this large number, and, more generally, that the whole history of the universe corresponds to properties of the whole sequence of natural numbers?

Notwithstanding the pull of eliminativism, there does seem to be some sort of consensus that natural kind, law, cause, induction, explanation, disposition, and a few others are syncategorematic terms, and part of these connections are often included in definitions of natural kinds (Block 2-8). For example, in the glossary of an anthology of classical articles in the philosophy of science, natural kind is defined as (Boyd, Gasper and Trout 1991:778f):

> A type of property, process, state, event, or object studied by science, mentioned in scientific laws, and assumed to be a causal feature of the world. The primary instances of natural kinds are objects of scientific taxonomy, such as electrons in physics, zinc in chemistry, and species in biology. Natural kinds are contrasted with phenomena that are assigned no such systematic, organising role, such as an event's occurring after I drop this pen, or an object's being located 34 miles west of the Liberty Bell.

But, as one might expect, no consensus on the meaning of terms like 'law' or 'explanation' exists; so it brings us no nearer to *the* definition of natural kind. It's surprisingly common to interdefine natural kinds and natural laws.[172] Moreover, there is also an inclination to avoid difficult questions (Block 2-9). In fact, natural kinds are often introduced by a philosopher of science at a crucial point in order to justify a not further argued preference or admiration, to ground induction in objective similarities for example, to explain why nature is lawful, or to ground the reference of scientific terms as in the causal theory of reference [2.6]. Hence Shain (1993:291) has concluded that:

> The concept of natural kind is brought in as a deus ex machina to save a metaphysical view of science which received emotional grounding, but not analytic support, from the instrumental success of science.

The threat of natural kinds as a grand illusion can only be countered by falling back on a form of metaphysical realism: natural kinds have essential or hidden properties, the properties that attach to an object in any possible world, or throughout any possible change in which the object endures. Consider Armstrong:[173]

---

[172] For example Carrier (1993:393), Churchland (1989:288), Johnson (1990:63), Bigelow, Ellis and Lierse (1992), Haack (1992:24f), Putnam (1990a:71). For discussion see Douven and van Brakel (1998).
[173] Armstrong (1978:66) and Armstrong in Bogdan (1984:261). Compare also the quotation of Kim on microreductionistic intuition on p. 75.

Block 2-9 Avoiding the issue of natural kinds.[174]

Currie	Of course I'm in no good position to say exactly what it takes for a property to be natural, though I assume that by and large we can agree on particular cases.
Hirsch	The notion of a natural kind is surely problematical. I want, however, to take this notion pretty much for granted in the present discussion.
Horgan	Providing an adequate philosophical account of natural kinds is no trivial task, of course, but I shall leave that task to one side here.
Miller	Scientists are our authorities for determining what is and what is not a natural kind. Science investigates underlying natures and the result of these investigations clarifies for us the limits of the reference of our familiar terms.
Pargetter	We have a fairly clear, but hard to explicate idea on what is to count as a natural property.

Behind the set of 'surface' properties there is a set of more deeply hidden properties: the *real essence* of the natural kind. The laws which a thing obeys are in a sense part of its essence.

Similarly, Wilkerson (1988) says that natural kinds

are characterised by real essences, intrinsic properties that make the individuals the kind of things they are, and which lend themselves to detailed scientific investigation The real essence of a thing not only determines its proper *de re* classification, but also directly determines many of its properties, irrespective of any system of classification we may find it convenient to adopt. Nature, as it were, provides the system of classification, whether or not we have the wit to use it.

An essentialist view leads us back through the whole discussion. Just as there are lots of truly alternative predicates that are projectible and display lawlike behaviour, so we may also contemplate truly alternative natural kinds, each with its own essence. Also note that carving nature 'at the joints' doesn't necessarily provide explanatorily significant categories. For example the cross-species distinction between male and female can be defined rather well in terms of the 'carving' between small and large gametes. But the gamete size by itself has little explanatory power (if any).

The idea that an essence is an underlying trait on which the more manifest properties supervene can be used in any scheme. The essence of 'tiger' could be the dynamic tiger gestalt, its functional organisation, its genetic structure, its ecological niche, all of this, or none. The essence of schizophrenia could be a genetic abnormality or a type of double-bind; or it could be the wrong dopamine level, a general synaptic loosening, a broken ego, or something else again. The essence of *vovetas* might be the form of their typical whirling movements. The essence of white is that it will always reflect most of the incident light or that there is no transparent white or that the reflectance of white objects (as contrasted with the

---

[174] Quoted from Currie (1990), Hirsch (1982:252), Horgan (1982), Miller (1992:434), Pargetter (1988).

Block 2-10 Stereotype and essence of green and *lhenxa*.[175]

> GREEN
> *defining characteristic*: the class 𝕲 which is called 'green' by most speakers of English
> *ultimate characteristic*: each element of 𝕲 causes in 'normal' human beings a negative signal in the LM-channel and no signal in the LM/S-channel the absolute value of which is larger than the signal in the LM-channel
>
> LHENXA
> *defining characteristic*: the class 𝕷 which is called *lhenxa* by most speakers of Kwakw'ala
> *ultimate characteristic*: each element of 𝕷 causes in 'normal' human beings a negative signal in the LM-channel and a positive signal in the LM/S-channel.

luminance) is the same throughout changes in illumination (Westphal 1987:12-39). Consider the specification of the stereotype and ultimate ('essentialist') character of green and *lhenxa* in Block 2-10. Both can claim to have a perfectly acceptable scientific essence, cutting nature at the joints. One response might be to say that this example strongly suggests the nature of the ultimate conception: in the ultimate conception there is no room for *either* green *or lhenxa*. What belongs in the ultimate conception is the LM- and LM/S-channel (or whatever is substituted in successor science). This leads us back on the eliminativist's path.

We are confronted with the following dilemma: Either we allow kinds like *lhenxa* or *vovetas* to be natural kinds, having projectible properties, essences, or whatever else goes with being a natural kind. Or, if we want to deny *lhenxa* and *vovetas* this status, we're pushed all the way down the primrose path of eliminativism. I see no way of stopping half-way or spreading into a multiplicity of *autonomous* schemes or domains which nonetheless are *dependent* on one physicalistic base (which is what supervenience aims to achieve). If one is drawn to microreductive essentialism, one ends up eliminating all substances, being left at best with 'quarks', 'superstrings', or whatever 'final physics' settles upon and something nobody, not even a few hundred theoretical physicists, can really grasp.

2.6 causal theory of reference

One reason for the increased interest in natural kinds in the philosophy of science is the concern about incommensurability. Incommensurability might be seen as a consequence of a Fregean theory of reference (according to which sense determines reference), combined with the Quinean idea of meaning holism. An antidote to the virus of incommensurability or relativism might be CTR - the Causal Theory of Reference, which allows sense to go all over the place, while stipulating reference to be stable. The suggestion is that though scientists may be wrong in the meaning (sense) they ascribe to their theoretical terms, in using these terms, they are *already* referring to *natural* kinds, existing independent of any human endeavours

---

[175] The description of the ultimate characteristics follows the dominant theory of colour perception. For details on, and criticisms of this theory see Saunders and van Brakel (1997a).

Block 2-11 Characterisation of causal theory of reference (CTR) cum essentialistic realism.[179]

> CTR operates under the assumption that things sufficiently distinguishable and re-identifiable to bear names have 'nature', that there is in principle a distinction between their true and apparent natures, that there is such a thing as what they *really* are which science attempts to learn and about which we may be ignorant or mistaken while yet succeeding in referring to them.
> The requirement CTR imposes on science is to discover what the objects into which we linguistically divide the world really are.
> We do not stabilise a term's reference by associating certain properties with continued use of the term. It is rather the world, independently of our linguistic predilections, that dictates the conditions for retention of reference. The essence on which reference depends is, then, metaphysical in that it need not be an experientially accessible feature of the referent.
> Every specification CTR makes of what it takes to be the referent of a natural kind term brackets the proviso that science is right. In effect, CTR appeals to 'final science' for determinations of essential properties.

like science. The meaning (reference) is thus stable across 'incommensurable' theories.

Roughly one might say: If we drop Fregean eternal universals (natural kinds, cognitive reference points or Platonic universals that all humans equally try to grasp), we end up with incommensurable Kuhnian worlds. All classifications and categories become language and theory dependent. CTR saves the one world or reality (and ideal thought as well) by replacing the human independent Third Realm of Frege with a human independent world of natural kinds and their essences. Another way of putting it succinctly is in terms of two options:

- Content determines object (the traditional Fregean view). Elucidation: The speaker's 'internal Intentional content' is sufficient 'to determine what he is referring to, either in his thoughts or in his utterances'.[176]
- Object determines content (CTR). Corollary: 'Cut the pie any way you like, "meanings" just ain't in the *head*!'[177]

Perhaps the staunchest contemporary defender in the philosophy of science of CTR is Leplin (1988).[178] The citations from his writings in Block 2-11 summarise the aims and promises of CTR, and should be compared with the criteria for natural kinds (discussed in the previous section) and with the work of Kripke and Putnam (to be discussed in [4.1]).

Arabatzis (1998:154) is a recent example of the use of CTR in the philosophy of chemistry, also drawing on what was later called entity realism [3.5]. He considers the chemists' and physicist' reference to electrons. Around 1890 chemists

---

[176] Searle (1983:199), who was one of the founders of the modern description or 'cluster' theory of meaning, which CTR aims to overthrow. I borrow the formulation of the two options from Devitt (1990).
[177] Putnam (1975:227); see further [4.1].
[178] The philosopher of science who wrote most extensively about the causal theory of reference is Boyd (1989), but he later modified his view substantially (Boyd 2000), as did Putnam.
[179] Direct quotations from Leplin (1988).

conceived the electron as a classical, static particle endowed with magnetic properties and not always subject to Coulomb forces. Physicists, on the other hand, regarded the electron as a dynamic non-magnetic particle, endowed with quantum properties, and being constantly in very high-speed elliptic motion around a positively charged nucleus under the influence of Coulomb-like attractive forces.

But says Arabatzis, from 1896 on there is 'referential continuity of the term "electron"'. There was an increasing variety of experimental situations 'involving' electrons. But because 'chemists and physicists shared the view that the electron manifested itself in the same experimental situations' and because 'they employed the same taxonomic criteria' they 'were talking about the same thing' (Arabatzis 1998:158).

An account like that of Arabatzis (and numerous similar references) sounds reassuring. But it doesn't always work. The electron case only works because we consider a brief time period and come at the example with the conviction that the existence of electrons cannot be disputed. Phlogiston is a simple and notorious example of a case where CTR does not work easily. If phlogiston doesn't exist, how is it possible that scientists believing in it were making 'good progress'? The only way out for CTR would seem to be to claim that the phlogistonists had been referring to oxygen all along (Carrier 1993). But that seems to place the claim to realism of CTR exclusively in the eye of the beholder. Alternatively, if one would say we need another theory to account for the success of non-referring terms, there is no reason to suppose that this 'extra' theory of reference wouldn't work for the referring terms as well (Douven and van Brakel 1998). The problems CTR is confronted with can be summarised as follows:[180]

- CTR assumes that the reference of natural kind terms is fixed by a baptismal event, when the term is first introduced, though the object to which I point can always be variously interpreted (as Wittgenstein famously remarked). In the CTR-literature the issue is referred to as the qua problem.[181] Concrete CTR-definitions of natural kinds always contain essential references to local circumstances and points of view. In the case of perceptual contact a certain degree of conceptualisation of the causal environment is necessary to fix the reference; perceptual contact underdetermines reference.
- CTR has difficulty explaining cases where science completely abandons the entity to which a term refers (as in the case of phlogiston or caloric particles).
- When scientific terms have their reference fixed theoretically (e.g. 'positron' or 'neutrino'), for example by mathematical desiderata, the explanatory theory overdetermines reference (and it may later turn out that nothing fits the specified causal role).

---

[180] See Devitt and Sterelny (1987), Devitt (1990), Hacking (1983:75-91), Johnson (1990), Kroon (1985), van den Brink and van Brakel (1988), Douven and van Brakel (1995), Stanford and Kitcher (2000), Weinert (1991).

[181] 'There must be something about the grounding situation that makes it the case that it *is* a grounding of a natural kind term, and not talk about, say, an artefact; something must pick the sample out *qua* member of a natural kind' (Devitt and Sterelny 1987:73). For a nuanced combination of CTR and descriptive elements see Stanford and Kitcher (2000).

- In interpreting the history of a term's usage CTR asks us to be charitable in assuming that the intention of all involved was to point to the same natural kind all along. But why is continuity more deserving of our charity than discontinuity? The defender of CTR doesn't only need a solution to semantic incommensurability (fixing the reference of theoretical terms), but also for the possibility of methodological or axiological incommensurability: we can only gain more knowledge about the same thing if this thing stays of interest to science.

A more general question is whether CTR or another theory of meaning can enlighten the nature of 'meaning' and 'meaning change' in science. This assumption has been attacked from various quarters. As Elgin (1983:41) points out:[182] 'Any body of discourse [for example scientific discourse] can be systematised in a variety of ways, and for different purposes different systems may be appropriate.' For the physicist (or chemist), to say that Rutherford's theory of the atom is wrong (or at least not sufficiently right), it is irrelevant whether the history of the use of the word 'electron' is described in Fregean or CTR terms. But then neither is there a fact of the matter as to *the* reason why scientists favour one theory over another (or of *the* rational reason for a theory change as reconstructed from the luxury of hindsight).

This leads to an even more general point - the reification of meaning. Although a view shared by many philosophers, scientists, and others, meanings are *not* things. They are not in a Fregean 'platonic' Third Realm. Worse, at least for CTR, they are not in the world either, other than as the subject of *passing* token-interpretations of token-utterances or token-inscriptions in a social setting. Meanings are assigned to bits of idiolect, not by appealing to a theory of meaning or interpretation in which the meanings of a language have been codified, but by a 'theory-in-flux' for interpreting idiolects.[183] The entrenchment of (scientific) natural kinds and other theoretical entities in science has to be seen in this light: as products of communicative interaction, continuously reconstructed.

I will now turn to subjects specific to the philosophy of chemistry. The causal theory of reference will figure in the background in chapter 4, issues of reduction of chemistry to physics will be addressed in chapter 5. The next chapter on chemical substances can be read as a case study on natural kinds. Throughout there will be references to the relation between micro and macro and the tension between the scientific image and manifest image, to which I will return explicitly in the last chapter.

---

[182] See, from a more naturalistic perspective, Nersessian (1991), Shapere (1991).
[183] See on passing theories of interpretation Davidson (1986), Callaway and van Brakel (1996). For 'identity' of meaning being merely pragmatic see Quine (1990) and van Brakel and Geurts (1988).

3 chemical substances

3.1 the science of stuffs

One area of research in the philosophy of chemistry is concerned with the nature of chemical substances (such as gold and vitamin C) and investigates the status and relationship of such concepts as pure substance (also called chemical compound, chemical species, chemical kind), molecule (often considered the microscopic essence of a pure substance), atom (building block of a molecule or chemical element), and of associated concepts such as valence (a measure of affinity between constituents of a substance), structure (in the sense of spatial distribution between microconstituents and their affinities) and phase (the state of aggregation of a substance such as solid, liquid, vapour). This notion of chemical substance is different from the metaphysical concept of substance that has figured in the philosophical tradition since Aristotle.[184] Schummer has argued that since Thales' suggestion that 'all things are made of water',[185] there has been a subsequent *Entstofflichung* ('de-stuffing') of philosophy, giving utter priority to form over substance or 'stuff': *Entstofflichung* of science (mechanical world picture) and of language (mass terms reduced to form terms - as in Quine's reduction of objects to quadruples of numbers); knowledge of substances reduced to that of secondary properties or to Kant's *Ding-an-sich*.[186]

Schummer suggests that chemistry is governed by an action-related conception of knowledge as distinct from the emphasis on formalisation and mathematisation of physics. This view doesn't merely emphasise the interactive aspects of the experimental side of science. The chemical praxis of making new things (new 'stuffs') is different from that of making careful measurements or carrying out 'crucial' experiments. There is a greater affinity of chemistry to technology or art than to physics. The fact that chemistry is constantly enlarging the world it studies, by making new stuffs and the differences between the chemical space of stuffs and the time-space of physics, make the interaction of the cognitive and material praxis of chemistry very different from that of physics (and biology as well). When compared with physics, most striking is perhaps the relative abstraction from primary qualities of (macroscopic) objects such as form and size. As long as an object can be placed in an experimental context, chemists do not care about spatial co-ordinates or the number of physical parts. In physics two objects are identical, if and only if they have the same space-time co-ordinates; in contrast two objects are chemically identical, if and only if they are found at the same place in chemical space (which means they enter the same chemical reactions).

---

[184] Witt (1989), Hoffman and Rosenkrantz (1994).
[185] Cf. Aristotle, *Metaphysica*, 983b20.
[186] See for this and the next paragraph Schummer (1995, 1996a:229-296, 1997a, 1997d).

Chemical space contains all possible substances.[187] Seen as a network, chemical space consists of the pure substances at the nodes; the relationships between the nodes are chemical reactions correlated to experimental practice. The dispositional properties of a substance include the interactions via all known and unknown chemical reactions (including reactions with as yet non-existing substances).

The stereotypical meaning of substance is: 'material from which something is made and to which it owes many of its characteristic qualities'. But substances can also be transformed into other substances. Such transformation or synthesis occurs naturally or intentionally. Chemical synthesis moves into synthesis or design of new material artefacts if it aims to control the properties of chemical substances that provide the artefact with certain required properties. For example, the molecular composition and structure of artificial leather doesn't need to be identical to that of natural leather, provided it has the same macroscopic properties. There has been an almost stable exponential growth of the number of chemical substances over the past 200 years; 1820: 10^3, 1860: 10^4, 1900: 10^5, 1960: 10^6, 1985: 10^7 (Schummer 1997b). Not only are more and more substances added, as it were, by more of the same, but substances are added with completely novel properties. For example 'starburst' dendrimers[188] are heterocyclic molecules that consist of a central polyfunctional core to which successive branched layers, called generations, are added. Each subsequent generation brings about a doubling of the end groups and a change of conformation until the dendrimer adopts a spherical shape. The volume inside is shielded from relatively large molecules while still remaining accessible to small ones.[189] Numerous other new types of substances have been synthesised,[190] creating new parts of chemistry such as organometallochemistry and new types of materials such as ceramic composites.[191]

Knowledge about material properties cannot be completed, because there's no end to making new stuffs. It makes no sense therefore to refer to 'intrinsic' properties of a substance 'an sich', apart from real interactions. In making new substances, unpredictable relations may occur, sometimes leading to chance findings - an 'impurity' that turns out to be something wildly new (e.g. Kronenether which led to a Noble prize).[192] The chemical elements span a space of possibilities beyond imagination. It has been said that reality is chemically unfathomable, an inexhaustible novelty of matter lying ahead (Tontini 1999), with the number of actualised properties vastly less than their possible number.

---

[187] Its relational structure can be described in terms of the operational definitions of 'element', the notion of chemical mass equivalent, and chemical reactivity (Schummer 1996a:182-223).
[188] They are also called cascade molecules, arborols, or micellanes.
[189] Potential applications of dendrimers include: synthetic models for enzymes and globular proteins, catalysts and template reagents, biosensors, drug carriers and transporters, unimolecular micelles and reverse micelles, synthetic membranes, molecular electronic devices, and photographic imagery.
[190] For examples of new 'weird' and 'beautiful' molecules see Hoffmann (1990).
[191] Modern composites are true 'mixts' (mixtures) in the Aristotelian sense, displaying properties that are more than the mere addition of the properties of their components (Bensaude-Vincent 1998).
[192] The issue of impurities introduces the question of (the distinction between) repeatability and reproducibility (Plesch 1999).

Hence to study the world from the stuff perspective is vastly different from the mechanistic study of primary qualities. Relative to the stuff perspective, talk of atoms and molecules is subsidiary (cf. protochemistry [6.3]). Any transformation of 'stuff' is first and foremost a qualitative change. No underlying quantitative description can fully grasp the 'emergent' property. Chemical systems are complex systems which cannot be reduced to their 'elements'.[193] A simple example is the difference between a liquid and a vapour. It is not difficult to blur this distinction under special experimental conditions (of pressure and temperature), but that does not undermine the qualitative difference between a liquid and a vapour or, speaking more scientifically, the sense of ascribing thermodynamic properties to the interface between a liquid and a vapour [2.2].

In this chapter I'll argue that the chemical notion of substance is wholly defined in terms of laboratory procedures and other experimental practices, and can be given no essentialist definition. Any identification of a particular substance may change under the influence of new observations, including observations using spectroscopic techniques and similar (sub)microscopic methods. But the final arbiter will, in the end, be observations at the macroscopic level. This is not merely because reading the instruments provides macroscopic data.[194] Poisonous water for example, would be impure water (an impure substance), independent of whether a (microscopic) cause can be found.

To set the scene I'll start with a discussion of manifest and scientific water [3.2], followed by sections on molecular structure and microreductionistic essences [3.3], a macroscopic definition of pure substance [3.4], and polywater [3.5]. The latter illustrates how manifest, macroscopic, microscopic, and submicroscopic considerations - be they experimental or theoretical - intermingle, interact, and intercalate to arrive at consensus about the properties of different substances.

3.2 manifest and scientific water[195]

The tension between the manifest and the scientific, introduced in [2.1], is illustrated well by Hare's (1984) defence of ordinary, liquid water as not supervening on (scientific) H_2O. Hare says there's no such supervenience or dependence. 'Water' and 'H_2O' are words with different meanings (senses) and the notion of natural kind or substance underlying this sort of supervenience is a recent phenomenon, related to the history of western science. Hence (Hare 1984:13f):[196]

> In the primitive sense, when we say that something is water, we do not imply that there is any (let alone any particular) chemical or physical structure such that stuff which has it is always water and that stuff has it. That is what we should be implying if it were a classic case of supervenience.

---
[193] Müller and Hörz (1996), Müller (1998), Mainzer (1997).
[194] In spectroscopic measurements macroscopic magnitudes (e.g. distances between spectral lines) are interpreted, using a model, to determine microscopic magnitudes (e.g. distances between atoms).
[195] Parts of sections [3.2], [3.3] and [3.4] were included in van Brakel (2000a).
[196] Cf. Hare's definition of supervenience in Block 2-6.

And as to the two meanings of 'water' he says:

> What has happened is that, when it was discovered that water was composed of H_2O, a new use of the word came in alongside the old, as the *OED* [*Oxford English Dictionary*] correctly records. But the old has survived as an alternative.

Psychological research is on the side of Hare. One might think that people (at least 'properly educated' people) use the presence or absence of H_2O as the primary criterion to decide what is to be called 'water'. However experimental research shows there is no ground for such 'psychological essentialism' (Medin 1989). Essentialist beliefs are not enough to fully explain category membership and word use. There is no single factor people use to identify the category boundaries of water (Malt 1994).

Still Hare's view seems to go against the grain of all common sense knowledge about chemistry or 'stuff'. For example, the philosopher Forbes (1985:199) says:

> There is an intuitive notion of substance according to which, to be of the same substance is to behave the same way in the same circumstances. Furthermore, it is part of this intuitive idea that the phenomenon of same behaviour is *explicable*. We expect the superficial and easily detectable differences between pieces of stuff to reflect fundamental differences which explain the superficial ones, and it is the fundamental differences which have the final say in classification; so someone who refuses to classify samples in this way may fairly be said not to understand what a substance is.

But what sort of criteria are to be used to assess whether Forbes understands what a substance is? Note that the point of discussion is not whether or not successful correlations (interdiscourse relations) can be found between microscopic and macroscopic descriptions. The question is, what is the right way to explain what (pure) substances are? Or what should the 'first' meaning of 'pure substance' be? It might be suggested that Forbes' suggestion also characterises the intuitions of the working chemist. However in the practice of chemistry, the intuition that observable (manifest) properties 'emerge' from (are caused by) unobservables is on a par with the intuition that (macroscopic) knowledge about chemical reactivities permits certain conclusions to be drawn about microstructural models. Hence, practice as such doesn't support the asymmetry of the relation between the micro and the macro that is usually implied in discourses of philosophy and common sense. Though many philosophers of science and scientists side with Forbes, the symmetry of the relation between the macro and the micro is acknowledged occasionally, as when Hempel (1999:339) says:

> Avogadro's number is the link between the macrocosm and the microcosm. Given the values of the macroquantities, values of related microquantities can be computed, and vice versa.

Hare's concern with the ordinary language use of the word 'water' and its accompanying referential intentions, is also very different from the 'intuition' of someone like Kim (1990:14), an authority in the literature on interdiscourse relations, who says:

> Chemical kinds and their microphysical compositions (at least, at one level of description) seem to strongly covary with each other, and yet it is true, presumably, that natural kinds are asymmetrically dependent on microphysical structures. Here our mereological intuition, that macrophysical properties are asymmetrically dependent on microphysical structures, seems to be the major influence on our thinking, cancelling out the fact that the converse strong covariance may also be present.

Kim takes the view that manifest objects are appearances of a reality constituted by microphysical structures. Such a view takes for granted that the macroscopic, manifest world is dependent on the microstructure of the world in such a way that it is underlying things that are more real and determine appearances. In crude jargon: Science uncovers the *Dinge-an-sich* that explain the phenomena we see. For Hare, the natural image is the manifest image of ordinary language and common sense; for Kim the natural image is the scientific image and its microreductionistic motivations. Often, though not always, Kim's view will be supported by scientists. For example, in volume 1 of a comprehensive treatise on water Kern and Karplus (1972:21) write:

> Since quantum mechanics provides an accurate description of molecular phenomena, a detailed understanding of the water molecule is available from theory. This implies that it is possible, in principle, to predict the structure and properties of water.

Although I will be arguing *for* the priority of the manifest image over that of the scientific image if, *and only if*, the issue of priority is raised, I also believe Hare's views illustrate what is wrong with the oversimplifications of an ordinary language approach. Hare (1984:13) says, in support of his view that (liquid) water and H_2O (in liquid, vapour, or solid form) are two different things, that

> if [chemists and ordinary men] were parched with thirst and begged for water and you directed a jet of steam at them they would not thank you.

But presumably they would not thank you either if you directed a jet of ordinary liquid water at them with a temperature of 98 °C. Dictionary definitions give descriptions of the 'normal' case in 'normal' circumstances. It is obvious that someone parched with thirst has as little interest in a jet of steam, as in being thrown in 'the liquid of which the sea is composed' (another *OED* description of what 'water' means). Similarly, a person who uses 'butagas'[197] to cook, needs a cylinder with 'butagas-liquid' - the same cylinder containing only butagas (that is 'butagas-gas') would be useless.

Another example Hare uses in his argument against the view that being water is a supervenient property, is that Harald, when crossing the Channel between Britain and France in 1066, did not mean H_2O when using the term 'water'. However Harald might well have believed that water is one of the elements of which

---

[197] Butane used as fuel gas.

Block 3-1   What is water?

water is what sustains the universe	(3.1)
water is one of the four elements	(3.2)
water is one of many elements	(3.3)
water is a mixture of inflammable air and vital air	(3.4)
water is H_2O (including D_2O and other isotope variations)	(3.5)
water is H_2O (excluding D_2O and other isotope variations)	(3.6)
water is the quantum mechanical composition of H_2O, H_4O_2, H_6O_3, ..., D_2O, D_4O_2, ..., T_2O, ...	(3.7)
water is the quantum mechanical composition of $(H_2O)_n$, OH^-, H^+, H_3O^+, ..., *H, *OH, ...	(3.8)
water is something like H_2O	(3.9)
water is some quantum mechanical superposition of a number of oxygen nuclei, twice as many hydrogen nuclei, and as many electrons as happen to be hanging around	(3.10)
water is whatever is water according to quantum mechanics	(3.11)

all bodies are composed (Block 3-1), and this would be sufficient, on Hare's own terms, for being water to be a supervenient property [2.4]. Of course *we* assume that the molecular structure of the North Sea in 1066 was primarily something like H_2O. Of course Harald did not know that. But nothing particularly interesting follows from just that. Perhaps one needs *some* theoretical perspective to consider apparently different things as the same (Wiggins 1980), but it doesn't have to be an appeal to microproperties.

Moreover, notwithstanding the authority of the *OED*, Hare is wrong about the history of the meaning of 'water'. The common sense stereotype of water (in liquid, vapour, or solid form) and the notion of water as a more or less pure substance, have kept much the same sense in the western culture since Aristotle, who says, for example:[198]

> The finest and sweetest water is every day carried up and is dissolved into vapour and rises to the upper region, where it is condensed again by the cold and so returns to the earth. Of solid bodies those that have solidified by cold are of water, e.g. ice, snow, hail, hoar-frost. Water freezes in winter. Ice is made up of water.

Though for Aristotle water is primarily a 'theoretical construct' (being one of the four primary bodies or elements), he clearly assumes that water is not merely a transparent thirst-quenching liquid, but a substance which can appear in different phases. Although he is 'confused' about the nature of gases or vapours, he knows quite well the processes of condensation and distillation.[199]

By the 18th century ideas about water as a substance had hardly changed (Block 3-2). After it had been decided/discovered in 1727 that air was not an element, but a mixture and (in 1781) that water was not an element, but a compound consisting of 'inflammable air' and 'vital air', the description of the stereotype did not

---

[198] The four sentences in the citation occur separately in the *Metereologica*, 354b27-30, 388b14, 347b36, 385b5; cf. 340a25-b3, 341b7-11, 383a112.
[199] See in particular *Metereologica*, Book 2, part 3.

Block 3-2 Eighteenth century dictionary definitions of water.

Chamber's Dictionary (1728)
Water, Aqua in Physicks, a simple, fluid and liquid Body; reputed the third of the four vulgar Elements. Whether Water be originally Fluid? We sometimes find it appear in a fluid, and sometimes in a solid form - that Water is ice. Water, if it could be had alone, and pure, would have all the requisites of an element, and be as simple as Fire. Whether Water be convertible into Air? But such a Vapour-Air has not the Characters of true permanent Air, being easily reducible into Water again. The Water may be rarefied into Vapours, yet it is not really changed into Air, but only divided by Heat, and diffused into very minute Parts; which meeting together, presently return to such Water as they constituted before.

Croker's Dictionary (1766)
Water, in general implies a pellucid fluid, convertible into ice by cold; naturally pervading the strata of the earth, and flowing or stagnating on its surface. The difference of common water, arising from the circumstances of stagnation of motion, or of its containing more or less of those stony particles, which it always contains in some degree, are not so essential, as to prevent the whole from being considered as of only one kind. Simple, or pure water, in a just sense of the word, is not met with any where.

change significantly.[200] Irrespective of whether the theory of matter was atomistic or something else, from Aristotle to the present day there has been one sense of water, which includes water as a pure liquid, as a liquid containing impurities, and as a substance in solid, liquid, or vapour form. Although it is correct that not all uses of 'water' imply that it is H_2O, all uses of 'water' *do* imply that it is a natural kind of the pure substance type. Water in all its modifications (liquid, solid, vapour) is the same substance. This supports the point that knowledge about 'materials and their transformations' is more robust than the local microphysical picture of the moment. If Aristotle says that water that evaporates may condense again, there can only be pedantic or silly reasons to raise the issue of incommensurability. Of course, if Aristotle's ideas about 'substance' (or 'movement', etc.) are at issue, or if it is said that he was or was not an atomist, there is a sense in which the divergence between modern views and those of Aristotle might be called incommensurable.

So 'when we say something is water,' we *do* imply that there is a particular something 'such that stuff which has it is always water and that stuff has it'.[201] But this essence doesn't have to be microscopic. Hare's conclusion that water and H_2O are two different things is wrong. Of course there is a sense in which 'water' refers only to liquid water and a sense in which it refers to water which may be in the form of vapour, fog, rain, or ice; equally of course, these two senses are connected in colloquial language.

The confusion between H_2O and liquid water is widespread in the philosophical literature. For example, Unwin (1984) says, making the distinction between the

---

[200] See, for example, the 1797 edition of the *Encyclopaedia Britannica*.
[201] Quotations from Hare (1984).

Block 3-3   Preliminary characterisation of (chemical) substances.

> - Common substances in the environment can be distinguished, roughly, into solids, liquids, gases and mixtures of these.
> - Different substances can be distinguished by manifest (or stereotypical) properties like colour, density, and so on.
> - It is possible to classify and name different types of matter because they have properties (both 'appearance' and 'use' properties) that are relatively constant with respect to time and place (assuming a 'reasonable' stable environment - this applies in particular to the temperature).
> - The best (but not the only) examples of types of matter with constant properties are chemical kinds, i.e. pure substances.

two explicit: '*Thus water* (meaning H_2O) is a pure mass-term, but water (meaning liquid H_2O) is not'. Nevertheless, in many publications, 'water is H_2O' is considered to refer to a liquid only. For example, Putnam (1975:232) says that x bears the relation same$_L$ to y just in case x and y are both liquids and x and y agree in microstructural properties [4.1]. But if there is anything important about this sameness relation, it is that x and y are both substances, which can occur in more than one state of aggregation.

However in stressing the autonomy of the manifest image Hare *does* have a point. Ordinary (somewhat vague, but no less objective) concepts like 'water', and even 'pure substance' are much better entrenched than 'atoms' and 'molecules' (Block 3-3). If the question of priority must be raised - I don't say it should - it is manifest water that is prior to scientific H_2O and not the other way round. This view does not deny that knowledge about the structure of molecules and atoms can vastly increase the instrumental understanding of the properties and behaviour of water. The point is simply that such scientific explanations refer to *water*, where the latter term refers to manifest water (the same water Aristotle speculated about). The term does not refer to an entity defined in terms of the currently most popular (sub)microphysical theories. To claim otherwise is to appeal to a science that is in the business of discovering the essences of things. I'll come back to this last suggestion in the next chapter. For now it may be enough to stress that 'essence' is not a theoretical term in chemistry, physics, or mathematics.

3.3   molecule structure and microreductionistic essences

Although Kim and Forbes take for granted that a substance is defined in terms of its microreductionistic essence, there's no need to postulate microscopic essences to establish that what *seems* to be different, is in fact the same substance. For example, one *could* say that liquid water and water vapour are the same substances because both are H_2O. But one could also say that they are the same because if water is evaporated and then re-condenses, the 'same' water comes back. There are good macroscopic reasons (in the manifest as well as the scientific image) to consider water as a substance, *independent* of leanings towards microreduction. Similarly one doesn't need to believe in DNA or anything similar to conclude that certain butterflies and caterpillars are different forms of the same species.

The response of the microreductionist will be to say that the microscopic theory *explains* that liquid-water and vapour-water are the same substance by pointing out that what remains the same are the molecules and what differs is their kinetic energy. It might be suggested that these variations of macroscopic properties depend on (microreductive) essences, which *are* indiscernible for the same substance. The argument would presumably run as follows:

- On the phenomenal level there are only appearances which arise out of complex interactions of real essences of substances. For example, although water is usually considered transparent, a thick layer of water may look blue. The visual appearance not only depends on the properties of what it is an appearance of (in this case water), but also on the unique context in which the water, its observer, and the background lighting are situated. Still, it would be 'the same' water, because the underlying essence (being H_2O) would remain the same; only the appearance is different in different contexts.
- At the macroscopic physical level there are properties like solidity and boiling point. If they appear to vary with context this shows that they are merely contingent properties, nominal essences in Locke's sense (Block 3-4). Variations with context of macroscopic physical parameters are due to differences in the way real essences interact.
- What in the end there is, is to be found at the level of molecules (or whatever the 'real' constituents of substances might turn out to be). At this level there *are*, for one substance, only identical entities.

But if we look at some details, it is not clear what molecular essences being identical or invariant amounts to. Take the philosopher's favourite: 'being identical molecule for molecule'. It leaves unclear what exactly is assumed to be identical (and indiscernible). Does the identity include the velocities and relative positions of the molecules? If it does, then it undermines the idea of macroscopic objects being identical 'molecule for molecule,' because the velocities and relative positions of the molecules are constantly changing. If it does not, then, say, temperature would be excluded as a relevant macroscopic parameter for two objects being (in)discernible. At the microscopic level everything is statistical and changing - hence no two things are ever the same. These variations are averaged out, at least by approximation, at the macrolevel, but that is not the level the microreductionist wants as a base.

Armstrong (1999:80) notwithstanding his positive attitude towards what he calls the Moorean corpus of common sense,[202] says:

> The stuff water is made up of molecules, and these molecules are made up of just three atoms, two atoms of hydrogen and one of oxygen. I assert, against some philosophers and crazies - ratbags as we say in Australia - that this is *known*. Quibbles can be made, but no serious epistemic assault can be made on this piece of knowledge.

---

[202] Armstrong (1999:81): 'First there is our Moorean knowledge. It is the epistemic background of our lives, without which no further knowledge could be supported or even, I think, acquired.'

Block 3-4   Locke on the nominal and real essence of substances.[203]

> The essences of the sorts of things, and consequently the sorting of Things, is the Workmanship of the Understanding, since it is the Understanding that abstracts and makes those general *Ideas*.
>
> The *nominal Essence* of Gold, is that complex *Idea* the word *Gold* stands for, let it be, for instance, a body yellow, of a certain weight, malleable, fusible, and fixed. But the *real Essence* is the constitution of the insensible parts of that Body, on which those Qualities and all the other Properties of *Gold* depend.
>
> That Men (especially such as have been bred up in the Learning taught in this part of the World) do suppose certain specifick Essences of Substances, which each Individual in its several kind is made conformable to, and partakes of, is so far from needing proof, that it will be thought strange, if any one should do otherwise.
>
> And yet if you demand, what those real Essences are, 'tis plain Men are ignorant, and know them not. From whence it follows, that the *Idea* they have in their Minds, being referred to real Essences as to Archetypes which are unknown, must be so far from being *adequate*, that they cannot be supposed to be any representation of them at all.
>
> The changes that that one Body is apt to receive, and make in other Bodies, upon a due application, exceeding far, not only what we know, but what we are apt to imagine.
>
> We can never be sure that we know all the Powers, that are in any Body, till we have tried what Changes it is fitted to give to, or receive from other Substances, in their several ways of application: which being impossible to be tried upon any one Body, much less upon all, it is impossible we should have adequate *Ideas* of any Substance, made up of a Collection of all its Properties.
>
> If an *English-man*, bred in Jamaica, who, perhaps, had never seen nor heard of *Ice*, coming into *England* in the Winter, find, the Water put in his Bason at night, in a great part frozen in the morning; and not knowing any name it had, should call it harden'd Water; I ask, Whether this would be a new *Species* to him, different from Water? And, I think, it would be answered here, It would not to him be a new *Species*, no more than congealed Gelly, when it is cold, is a distinct *Species*, from the same Gelly fluid and warm.

However, what does it mean to say that water consists of molecules 'made up of just three atoms, two atoms of hydrogen and one of oxygen'. Underlying microscopic essences vary as much with context or circumstance as the nominal essences. There are H_3O^+ and OH^- ions in liquid water. There are H_4O_2-molecules, as well as other H_2O-polymers in water vapour. And how much ionisation or dimerisation and polymerisation there is, depends on the temperature and other contextual variables. The pragmatic answer that water is *predominantly* H_2O is not sufficient if we are looking for essences. A few dissolved Na^+ and OH^- ions (a drop of sodium hydroxide) don't belong to water. Conversely the OH^- ions (of the electrolyte water), that were already there, do belong to 'normal' liquid water. It is not that we are unsure which (distribution of types of) microstructure is the correct one. The point is that there is no *one* correct microstructure, because the

---

[203] Locke, *Essay concerning Human Understanding*, III.iii.12, III.vi.2, II.xxxi.5, II.xxxi.5, II.xxxi.10, II.xxxi.8, III.vi.13. Cf. Leibniz (1981: 267,294,312,324f, 338,400-2). For the complexities of interpreting Locke's views on essences and natural kinds: Stanford (1998), Stuart (1999), Shapiro (1999).

microstructure depends as much on the context and functions just as another nominal essence would.

One might perhaps suggest that what samples of water have in common, if pure, is that only they contain hydrogen and oxygen in a proportion two to one, and this is what H_2O means. However, as stated, this is not a microscopic essence (no assumptions being made about atoms or molecules), and taken literally, it doesn't distinguish between water and a mixture of hydrogen and oxygen in a proportion of two to one. One might wish therefore to add that the *atoms* of hydrogen and oxygen interact in some way and that it is this interaction that constitutes the essence of water. However, in order to know what sort of interaction there will be and how this sort of interaction depends on the circumstances (such as the temperature), we have to know more about what hydrogen and oxygen are. This leads us to the problem of isotopes. Heavy water may seem chemically the same as normal water (having only an extra neutron in the hydrogen nucleus), but it doesn't boil at 100 °C (Block 4-1) and organisms have different 'fitness' when forced to live on/in it. At an even lower level of description there is the difference between ortho- and para-hydrogen. They have the same chemical properties, but differ in the orientation of nuclear spin. Because of their difference in molecular energy levels, some physical properties, in particular the specific heat, are different.

There is also the question how far the essence has to be specified. There are always new phenomena at lower levels which ask for explanation. A single H_2O molecule has two equivalent OH-bonds on any ordinary account of molecular structure, yet they have different vibrational energies (which turn up in its infrared spectrum). The OH-bonding orbitals have different energies though they are equivalent. The reason is that the identical bonds are not independent. They have coupling in their potential force fields. Still in some sense they are equivalent. Is all of this part of the essence of being H_2O?

Finally the interaction between hydrogen and oxygen in H_2O (or between composing quarks or whatever) is probably to be specified by a kind of superposition of quantum mechanical wave equations. Passing over the fact that it is not altogether clear what a superposition of wave equations means for a macroscopic system [5.5], the result will be quite complicated. Why would we choose one set of equations as the essence of water and not another? Because they are the equations that depend on being *water*. We may as well contemplate the quantum mechanical equations for blood or for the Second World War. What criteria could we use to decide which subset of equations represent essences of substances and which are the equations for heterogeneous mixture? Precisely: the substances (like ordinary water) that have already been identified at the manifest and macroscopic level.

That the manifest image determines which micro-essences are to be selected is also presupposed in sophisticated philosophical discussions. For example, Blackburn (1985:62) says in his discussion of 'being water supervening on being H_2O':

> If we had an argument that it [i.e. H_2O] does not have to be water, then

we would just change the basis for the supervenience. We would [argue] for a releasing property, R, and the true basis upon which being water supervenes would be (being H_2O) and being R.

3.4 a macroscopic definition of pure substance

A pure substance is often considered a collection of molecules of the same type. For example (Bunge 1985:222):

> Substance x is chemically pure if and only if x is composed exclusively of either atoms or molecules of a single species or kind.

Here presumably, lest the definition becomes circular, 'molecules of a single species or kind', is to be understood as 'molecules having the same composition and structure'.

This definition however only applies in rare cases - pure water does not consist exclusively of one type of molecule all having the same 'structure'. The definition doesn't work for metals, salts, electrolytes, or dissociating liquids. It also breaks down for enzymes, antibodies, viruses, or more generally isochemical compounds and homeomers.[204] It sounds good to say something like 'a molecule is the smallest particle of a definite compound which still has the same properties'. But 'smallest particle' at best makes sense for ideal gases and (perhaps) a few liquids, not for water, carbon (diamond, soot, buckminsterfullerine[205]), salt crystals, proteins, or cellulose. And even in cases like alcohol or helium it is unclear what could be meant by saying that the properties of an assembly of molecules is the same as those of the compound. Alcohol (ethanol) is transparent and may contain dimers; helium (gas) has a particular pressure. A molecule doesn't 'contain dimers' nor is it 'having a particular pressure'.

Crystals present their own problems. Some synthesised crystals, such as $Ca_{0.75}Nb_3O_6$, display internal structures that boggle the imagination (Hoffmann 1990); it seems arbitrary to ask how many molecules there are in a particular crystal. Then there is the problem of definite and indefinite compounds - a problem already recognised by Mendeleev. The variation can be considerable, for example Na_xWO_3 (0.93>x>0.32) or Li_xWO_3 (0.57>x >0.31). Consider too ruby and tourmaline: strictly speaking they are solid solutions, but this seems to stretch the notion of pure right out of its common sense meaning. Then there are polymers. In practice, nylon is considered one compound, although strictly speaking every molecule with a different chain length is a different compound. Molecular interpretations also break down in heterogeneous catalysis,[206] as well as autocatalytic and cyclic reactions (Manzelli 1996). Even in the simplest cases, and from the mo-

---

[204] Isochemical substances have the same gross composition, but with different average size of aggregates. Homeomeres are substances with identical chemical activity.
[205] Fullerines, first made in 1985, are a class of carbon substances which are assumed to exist of clusters of carbon atoms that form a quasi-rigid structure which does not fit existing models for amorphous clusters, tiny crystals or liquid drops. Buckminstefullerine consists of 60 carbon atoms in the form of a kind of ball.
[206] If the surface of a heterogeneous catalyst (or any disordered solid) is described as a fractal, there is no molecular picture.

lecular point of view, the majority of pure materials are tautomers:[207] they do not consist of identical molecules, but of an intimate mixture of different species of molecules (metamers or polymers) in statistical equilibrium with one another and inseparable under ordinary experimental conditions.

Structure theory in the sense of van 't Hoff studied the geometry of the fixed positions of atoms in a molecule without considering the nature of the binding forces. But already in Kekulé's benzene molecule it is not possible to specify exactly where there are single and double bonds. At different stages the structure of the benzene molecule was represented as 'intermediate' between two, three, and five structures. There is no end to the extent of this kind of hybridity. In an ab initio valence bond calculation for benzene 175 resonance structures were used. To explain the chemical reactions of anthracene over four hundred different diagrams have been utilised. If there is any 'it', to which these four hundred stories apply, it would seem to be anthracene and not 'molecule which depending on circumstances can have any of 400+ structures'. Bullvalene, $C_{10}H_{10}$, around since 1963, is said to have 1,260,000 electronic tautomers each of which is separated from the other by an energy barrier of 12 kcal/mole. Then there are quasi-molecular species such as van der Waals complexes, which appear only in homogeneous mixtures, for example $ArCl_2$ with an estimated life time of 10^{-12} seconds.[208] Though apparently new chemical species, they cannot be purified and put into bottles. Their identity conditions too are evanescent: their 'molecular structure' is extremely dependent on change of temperature and phase transitions.

The concept of molecular structure seems to derive its meaning more from the way molecules are represented in models than from anything else. It is often thought that one can investigate individual atoms and molecules with spectroscopic techniques or that X-ray techniques, electron microscopes, and similar techniques give direct information about molecular structure. However in the construction of all these instruments and in the interpretation of the data they produce, theoretical constructs are used that already contain assumptions about molecular structures. In using X-ray diffraction techniques, the experimental technique *presupposes* a model of the molecule and abstracts from all other properties of the chemical substance except the 'dynamical properties associated with energy transformation and energy enhancement' of molecules of 'relatively inflexible geometrical configuration' (Rothbart 1999:261-8). The instrument is designed on the basis of a thought experiment in which the chemical substance is envisaged in terms of this model long before any measurement is carried out. Also note that different spectroscopic techniques use different models. Different techniques (interaction with electron beams, infrared spectroscopy, X-ray or neutron scattering, magnetic resonance data) give slightly different measures of molecular dimensions and atomic dimensions may be defined differently (such as ionic radius, covalent

---

[207] The term is used here in a wide sense. In a narrow sense it may refer merely to compounds of which the keto- and enol-form of an isomer are in equilibrium. See also Block 1-9.
[208] They are called van der Waals complexes because they are held together solely by the relatively weak van der Waals forces. See for discussion Early (1992), Schummer (1998b).

radius, van de Waals radius).[209] Electron diffraction data can be used to calculate the 'distance' between atoms defined as the mean value of the distances between centres of electronic clouds. Microwave spectra can also be used to calculate the 'distance' between atoms, but now defined as the distance between nuclei, as calculated from their rotational spectra. Hence, different experimental methods may lead to different conclusions about molecular structure. Though one speaks these days of 'pictures' of atoms and molecules (using scanning tunnel microscopy), we only *see* computer graphics.

'Molecule' is an indispensable, but thoroughly theoretical concept. It belongs to theories that are impressively empirically adequate, but without giving a clear idea of what entities are thought to exist. More importantly, to the extent it makes sense to say things like 'a pure substance consists of identical molecules', any empirical evidence will depend on a prior understanding of what a pure substance is. Are molecules the bricks of the world? 'Yes, if You need them to explain the phenomena You create in Your laboratory and No, if You don't' (Psarros 1998a:100).

Though most philosophers will automatically assume that a pure substance is to be defined in terms of molecules, there are exceptions. For example Lowe uses the macroscopic directly observable characteristics of a phase change to argue that there is no need to appeal to Locke's 'internal constitutions' (Block 3-4) to characterise substances.[210] Instead of a definition in terms of composition or molecular structure, a pure substance can be defined as a substance of which properties such as density do not change during a phase conversion (as in boiling a liquid or melting a solid). Such a definition is independent of one's beliefs in atoms, in an atomistic hidden variable interpretation of quantum mechanics, or any other microphysical story.[211] Such a pure substance persists as a phase of constant composition when the conditions of temperature, pressure, and composition of the other phases present undergo continuous alteration within certain limits (i.e. the limits of the existence of this pure substance). This can be observed when distilling sea water or when a bottle of Coca-Cola is put into a freezer.

If the properties of two co-existing phases remain invariant during a phase change, the system is called hylotropic. If it is hylotropic over a limited range of pressure and temperature, it is a pure chemical substance. If it is hylotropic over all pressures and temperatures except the most extreme ones, it is a chemical element. These definitions go back to Ostwald (1907:166,170):

> The mode of phase change in which the newly formed phases have at every moment the same properties and the same total composition as the original system is called a *hylotropic* transition. An element is a

---

[209] And these values are not 'atomic constants'. For example, the covalent radius of hydrogen is not the same in H_2O and CH_4. Note that atomic size (and electronegativity) is derived from molecular – not atomic – properties.
[210] Lowe (1989:186) gives the following definition: 'A change to an individual substance, S, of kind K, is a phase change for S just in case it is a change which things of kind K survive as a consequence of the natural laws of development for K.'
[211] Also the definition of a pure substance in terms of a combustion analysis ('elemental analysis', going back to von Liebig and Lavoisier) is a macroscopic definition, based on weighing macroscopic quantities of material.

substance which cannot be transformed into another non-hylotropic substance within the entire range of attainable energy influences.

In short: A pure substance is a body which forms hylotropic phases in a finite range of temperature and pressure. Then chemical elements are substances which *never* form other than hylotropic phases.

It shouldn't be underestimated how theoretical this 'empirical' notion of 'pure substance' is, though no word has been said about molecules. The 'theoreticity' is most apparent from the use of the term 'phase'. The phase rule and the theory of chemical thermodynamics provided Gibbs with the theoretical background for the concept of phase.[212] A phase is a macroscopic continuum which, when in a state of thermodynamic equilibrium has constant and uniform properties throughout, such as density, electric conductivity, magnetic susceptibility, and so on - the so-called 'physical constants' (Block 4-1). It is the notion of phase that makes it possible to give definitions of 'solution', 'compound', 'pure substance', 'element', independent of any atomic hypothesis (Timmermans 1963).

A bubble of air, a piece of sugar, a drop of salt water, a fragment of glass, is a phase. A tiger, milk, and most paints are polyphasic aggregates. Mechanical methods, using mechanical forces (filtration, centrifugation, grinding), can be used to separate heterogeneous and homogeneous materials.[213] Physical methods using thermal energy and hydrostatic pressure (distillation, crystallising, melting) can be used to divide mixed and pure materials. Mixed materials can be further divided into solutions (of the gas/gas, liquid/liquid, and solid/solid sort), addition compounds (such as hydrates) and aggregates (emulsions, conglomerates, colloids, smokes). Thermodynamic methods (using energy or pressure at higher levels) can be used to divide pure substances into compounds and elementary materials.[214]

A material is pure therefore if it is perfectly homogeneous after being subjected to successive modes of fractionating which are as different as possible and when attempts at further purification produce no further change in properties. Ideally separation techniques have to be applied an infinite number of times. Later refinements may show that what was once thought to be a pure substance is, after all, not pure. Moreover, different separation techniques (crystallisation, electrophoresis, and so on) set different standards of purity. The ideal pure substance would pass all types of ideal purification tests, i.e. tests with unlimited resolution. Note that depending on its history, the same pure compound may display different properties at the same temperature and pressure: the precipitate of mercury oxide is yellow; as a product of calcination, it is red (the colour difference being due to a difference in grain size).

Spectroscopic methods for purity tests always remain secondary to traditional purity tests in terms of the inseparability by any separation technique (Schummer 1997c):

---

[212] Gibbs' phase rule says that a system of c components in p phases has $c-p+2$ degrees of freedom (i.e. number of independent variables).
[213] A material is homogeneous if it cannot be separated into different materials by external or capillary forces.
[214] Cf. the classification of separation methods in chemical engineering (Block 5-2).

- There is no pure 'standard' unless it has been made by conventional laboratory procedures (or in rare cases when 'naturally' pure products are available, when purity has been checked by conventional means).
- No spectroscopic method can be used unless one already knows what the spectra of a particular pure substance look like.

Although it is possible to make predictions of the electromagnetic properties of substances, there is no theory that provides a general criterion to distinguish spectra of pure substances and mixtures on the basis of spectrographic data alone.

Categories that cause problems for the definition of pure substance presented above include (Timmermans 1963): enantiomers,[215] azeotropic mixtures, dissociative compounds in equilibrium, certain types of mixed crystals or other polymorphic compounds (e.g. *d*- and *l*-camphoroxime), synthetic polymers, many biochemical compounds, systems that are not in 'pure' thermodynamic equilibrium, isotopes.

Each of these categories would warrant detailed discussion. Here I have to limit myself to a few comments. In biochemistry the concept of purity, with its chemical connotations of small molecules, starts to break down. Enantiomers present a special case because normally, they cannot be separated by physical means; still they are usually considered to be a mixture of two chemical substances. However it is often possible to separate enantiomers by first carrying out a chemical reaction with another pure enantiomer. The resulting products can be separated by physical means and, by carrying out another chemical reaction, one can obtain the two enantiomers of the original mixture in pure form. Many enantiomers can be separated by crystallisation and subsequent *manual* separation of the crystals. So one might suggest that physical separation of enantiomers is possible in principle, provided we embed it in a wider practice and take 'bulk' macroscopic handling of substances as the final arbiter.

Even if we restrict ourselves to relatively 'simple' substances, the thermodynamic notion of pure substance is an idealisation. Water after all, is *not* a pure substance - that is so only under 'ordinary conditions'. At high temperatures it changes into a mixture: above 500 °C its vapour dissociates partially and its two gaseous constituents can be separated. Similarly, at low temperatures a mixture of acetylene, C_2H_2, and benzene, C_6H_6, is a solution, but at high temperatures it behaves as a single substance (in particular in the presence of porcelain).

The macroscopic definition of a pure chemical kind, avoiding all reference to molecules, goes back to Wald[216] and Ostwald.[217] Neither Timmermans (1963), nor Prélat (1947), nor the more philosophical approach of protochemistry [6.3], add much to the programme laid out by Ostwald. However an important, though easily overlooked observation of van der Waals (1927) should be added. He addresses

---

[215] Enantiomers are species containing equal amounts of two optical isomers, like *l*- and *d*-tartaric acid. Cf. Block 1-9.
[216] Wald (1897) credits Ostwald for motivating his research. Both Mach and Ostwald can be considered Wald's mentors. See Ruthenberg and Psarros (1994) for the 'positivist', 'operationalist', phenomenological chemistry of Wald and his correspondence with Mach and Ostwald.
[217] Ostwald (1902, 1904, 1907). Cf. discussion of his work in [1.1].

the issue only briefly, but with great rigour and notes that in order to define the notion of pure substance in macroscopic terms (as presented above), (Gibbsean) thermodynamics is drawn upon, which *itself* presupposes an *undefined* concept of pure substance.[218] Although van der Waals doesn't mention Ostwald, he gives credit to Wald for an earlier rigorous approach (van der Waals 1927: 229n2).

Historically, the concepts of 'substance' and 'element' were developed while 'interacting' with the notions of molecule and atom. This, however is a contingent fact. If everything we know about molecules, atoms, and the whole of corpuscular physics and quantum mechanics would turn out to be false, it would not change the observations that led to the operational definitions of chemical substance and chemical element. Any talk of atoms, molecules, and valences will be relative to this macroscopic scientific definition of pure substance. Notions like temperature, density, phase conversion, and so forth, are used to give a definition of pure substance *within* the scientific image, against the background of thermodynamics. Moreover, these scientific *macroscopic* definitions *presuppose* a notion of pure substance that can only be justified by appeal to its 'vague' meaning in the manifest image. When we are presented with 'difficult' cases pragmatic decisions have to be made as the notion of pure substance cannot be essentialised. There is not one essential definition of 'pure substance' that can avoid the need for 'inspired adhoccery' to deal with difficult cases.

3.5   polywater[219]

In previous sections I raised doubts about forms of reductionism and essentialism when applied to chemical substances. Here I want to illustrate how arguing against reductionism and essentialism, doesn't imply anti-realism. What I have argued *against* is that the molecules, atoms or quarks are *more* real than the pink ice cubes. Although my discussion of the polywater episode below is directly relevant to the issue of natural kinds and essences, my main concern is to show how decisions made by scientists as part of 'ordinary' scientific practice, are a 'co-production' both of macroscopic and manifest observables *and* theoretical interpretations, models and their intercalations.[220]

The case study I discuss bears on what is called entity realism in the philosophy of science. This species of realism is somewhat similar to the causal theory of reference (CTR [2.6]), in developing the baptising, reference fixing events of CTR into detailed accounts in terms of experimental scientific practice. The essentialism of CTR which says that the essence of a natural kind is tied to its phenomenal properties by physical laws, is replaced in entity realism by a form of causal realism. Entities have causal (dispositional) capacities, and it is these

---

[218] Van der Waals (1927:14,227): 'Wir setzen also bis auf weiteres den Begriff einer „chemisch reinen Substanz" als bekannt voraus' and 'die Betrachtungen stützen sich auf die Annahme, daß wir wissen, was wir unter diesen Worten [„chemische reinen Substanz" and „chemischen Individuums"] zu verstehen haben.'
[219] A shorter version of section [3.5] was published in van Brakel (1993a).
[220] For overviews and bibliographies of the polywater episode see Allen (1971), Everett, Haynes and McElroy (1971b), Franks (1981), Gingold (1973, 1974). For a review, placing 'polywater' in the more general context of water in capillaries and thin films, see Clifford (1975).

causal capacities that allow them to affect other things. The main proponents of entity realism, Cartwright and Hacking suggested that an entity is rightly thought to be real if we know how to manipulate it so as to create new phenomena.[221] Hacking's entity realism was developed in response to criticisms of the causal theory of reference, because 'the language game of naming hypothetical entities can occasionally work well even if no real thing is being named' (Hacking 1983:87). Hacking thinks this problematic aspect of CTR can be solved by changing the criterion of baptising an entity or natural kind (by pointing to its observable presence), to baptising an entity because its dispositional capacities can be manipulated to affect other entities. In his slogan, which might replace Putnam's 'meanings ain't in the head': 'if you can spray them then they are real'.[222]

However as Carrier (1993) has shown, phlogiston passes this test: 'Stahl employed phlogiston so as to intervene in and actively change other processes.' Nevertheless, it was later abandoned. Stahl's experiments worked because combustion (by oxygen) and calcination (by phlogiston) are alike, except that the two entities move in opposite direction. Stahl's account created a classification among the phenomena that is still taken to be correct. Therefore, Carrier concludes that where theories and entities may only fail the 'retention test', kind-structures pass the test: they are often retained after a major theoretical change like that of phlogiston to oxygen. This would favour a kind of structural realism: Stahl's broad taxonomy of classification is 'real', though his specific account in terms of phlogiston is not.

However I am not concerned primarily to argue here for or against whatever sense entity realism and Hacking's famous slogan might make as a theory of scientific practice – chemical synthesis would provide a better case study to analyse Hacking's distinction between experimenting *on* and experimenting *with* (Zeidler and Sobczynska 1995/96). I use the example of polywater to show that any microstory always comes *after* a problem has turned up in macroscopic terms, but, that being said, there is a constant interaction between descriptions at different levels (discourses) to arrive at a coherent and empirically adequate description of the manifest phenomena.

In asking in the philosophy of science whether certain entities exist or not, prototypical examples one thinks of are 'protons, photons, fields of force, and black holes' (Hacking 1983:21). Because we're all familiar with water, it may be exciting when we hear about the possible existence of *poly*water (even if the latter can only exist in microgram quantities). But what is so special about it? There is water and there is such a process as polymerisation (chaining of molecules), which are both well entrenched phenomena. Saying that polywater exists simply means that water has, in certain circumstances, the capacity to form polymers. So what 'polywater' means seems straightforward. Whether polywater *exists* is a different matter and is what the following case study is about.

---

[221] Cartwright (1983:87-99) and Hacking (1983:262-5).
[222] Hacking (1983:23). On Putnam's 'meanings ain't in the head' see [4.1]. On Cartwright's *'when you can spray them they are real'* see [6.2].

What was later to be called polywater or anomalous water was first reported by Fedyakin (1962) in Russian, in the journal *Kolloidnyi Zhurnal*:[223]

> It has been shown that water vapour close to the saturation point condenses on the walls of capillaries, forming an unstable film that passes over into continuous columns of liquid. The latter posses a structure differing from that of ordinary water, which explains the condensation of vapour at lower pressures than called for by Kelvin's equation.

Fedyakin's work became widely know in the western world in 1966 when the Russian physical chemist B.V. Derjaguin was invited to give a Faraday lecture in London in 1966. Until then there had only been publications on 'anomalous' water in Russian; after 1966 the number of scientific publications rose sharply. By 1970 the total number of publications had risen to more than 100 that year (of which only 10 appeared in Russian). By 1973 when the hype was over, about 500 scientific articles in total had appeared.

The term 'polywater' was first used by the American spectroscopist Lippincott, whose spectroscopic 'evidence' for polywater hit the headlines (Lippincott et al (1969).[224] Donahoe (1969) suggested in *Nature* that all water on earth might turn into a viscous solid.[225] The suggestion was immediately rejected by well-known English physical chemists, but that *Nature* published Donahoe's letter is indicative of the hype.

The first thing to note is that in the polywater case (as distinct from, say, quarks) there is no doubt that *something* exists (Block 3-5). Under certain conditions a condensate forms in capillaries, which *should* be water, but it has the wrong properties. On this much realism even an agnostic empiricist such as van Fraassen could agree: the condensate is an 'observable' - even to a non-scientist. There is no doubt that in a certain space-time region there is some material, most probably in liquid form. The question is: what is it *more in particular*?

Before considering this question it may be good to point out that interface and colloid chemistry - the domain to which the polywater issue belongs - is full of anomalies and each anomaly has generated numerous speculative interpretations [6.4]. Polywater is not essentially different from lesser known anomalies and speculations. The only difference is that it got the attention it did. Numerous articles on polywater appeared in general science magazines and the general press, including *Der Spiegel* and *The Huntsville Times* (Alabama).[226]

McKinney has suggested that[227]

---

[223] The Kelvin equation, mentioned at the end of the citation, says that a thread of liquid in a fine capillary whose walls are wetted by the liquid should have a lower vapour pressure. At the end of their detailed review of the polywater events, Everett, Haynes and McElroy (1971b:305) write: 'But the Kelvin equation still lacks experimental confirmation!'

[224] Other terms used for anomalous water include water II, specific water, orthowater, superwater, superdense water, (maximally, ultimately) modified water, cyclimetric water.

[225] Donahoe (1969), Bernal et al (1969), Everett, Haynes and McElroy (1969).

[226] Apart from extensive discussions in general science magazines, polywater was news in Frankfurter Allgemeine Zeitung, The Guardian, New York Times, Saturday Review, The Times, Wall Street Journal, Washington Post and also in The Plain Dealer (Cleveland OH), The Morning Herald (Sydney), and so on (Allen 1971, Franks 1981).

[227] All quotations of McKinney in this section are from his (1991).

> Scientists were justified in believing, for a short time, in the reality of polywater without ever using its causal properties to experiment on other more hypothetical parts of nature.

This is wrong. At no point were scientists *justified* in believing in the reality of polywater. I disagree even more strongly with McKinney's suggestion that many scientists 'believed in the reality of the polywater model,' because this 'is evident from the frequent use of the term "polywater" as standard nomenclature'. The staunchest defender of it, Derjaguin, used the term 'anomalous water' or 'water II', not 'polywater'. Of the papers presented at the First International Conference on Polywater (Bethlehem PA, June 1970), nine have 'anomalous water' in the title (of which one occurrence in quotation marks), three 'water II', one 'water polymers', and one 'polywater-water' in quotation marks. Though two contain the word 'polywater' in the title, in full these titles were: 'An Alternative Explanation for Polywater' and 'Polywater - A Search for Alternative Explanations.' I think the following quotation is representative for the vast majority of scientists who have published on the subject: 'this term ['polywater'] is used only for convenience and does not imply any assumptions on the part of the authors as to the nature or structure of this substance.' [228]

What physical scientists worried about in particular was how to obtain accurate, reliable, and reproducible measurements of very small samples obtained in circumstances where it was well known that inexperienced researchers *always* obtained 'anomalous' results, because of contaminants on the one hand, and the subtleties of the experimental techniques on the other. As to the sophisticated techniques used, the Raman technique used by Lippincott et al (1969), is notoriously difficult; their interpretations were criticised at once (see for example Everett et al 1970). At the above mentioned conference, Lippincott et al (1971:443) reported:

> Attempts to reproduce the Raman spectrum previously reported have been unsuccessful. It was concluded that there is reason to seriously question the concept that water exists in a stable polymeric condition and that contaminants, both inorganic and organic, may account for a number of the physical properties and other phenomena associated with 'polywater'.

As it was also acknowledged that the Lippincott et al (1969) work 'was taken in part from a [Ph.D.] Dissertation', one may speculate that that publication was perhaps prematurely submitted to *Science*.

Scientists often make bold conjectures about the existence of some entity or effect: if the conjecture turns out to be true, reputations are to be gained, and few 'peer group risks' are involved. In the case of polywater this effect was aggravated many times over, due to the hold the popular media got over it, triggering cold

---

[228] Petsko and Massey (1971:508). Also compare Page and Jakobsen (1971:427): 'The anomalous component itself is called "polywater", but this term is used only for convenience and does not imply any assumptions on the part of the authors as to the nature or structure of this substance'. Also (Brummer et al 1971): 'This description [viz. "polywater"] is controversial but, for simplicity, this terminology will be used here.

Block 3-5   The 'given' which had to be interpreted.[229]

> The crucial observation is that when freshly-drawn capillary tubes of diameter in the range 1-50 μm are placed in a closed space containing water vapour at 95-99 per cent relative humidity, a liquid condensed in some of the capillaries. The reduced humidity can be produced by allowing the vapour to equilibrate with a solution of sulphuric acid (the procedure used by Fedyakin), with saturated salt solutions, or with water held at a temperature a few tenths of a degree Celsius below that of the chamber containing the capillary. Fedyakin employed glass capillaries of unstated composition; subsequently borosilicate glass and glass capillaries have been widely used. The phenomenon is observed both when air is present and when air has been completely removed by careful vacuum techniques. The liquid obtained by these various procedures is usually called *anomalous water*.
>
> In their most recent experiments the Russian workers employ an apparatus constructed of pyrex and molybdenum, and draw the capillaries from high purity (99.995 per cent pure) quartz tubing. The main apparatus incorporates greaseless valves, and contamination from vacuum pump oil vapour is guarded against by efficient cold traps. Before each preparation the chamber containing the capillaries is baked out for 48 hours at 450°C high vacuum, and vapour from triply-distilled and degassed water is admitted through a break seal.
>
> It is clear that some of the recent work outside the USSR has been carried out without sufficient regard being paid to the supreme importance of scrupulous cleanliness at every stage in the preparation and handling of the material, and some workers have now apparently dropped work on the subject in despair.

war and other politics. Several scientists obtained money to investigate polywater, not because they or their grant givers believed in its existence, but because they couldn't risk missing the boat.

McKinney suggests that, in the polywater case, we can make a clear distinction between data and interpretation. He says: 'we can still obtain valid experimental results from any sample of anomalous water - just measure its density and viscosity.' But that's not true. Let's assume that it *is* a fact that there is a film of liquid above a meniscus of ordinary water in a capillary. To say that it boils at about 250 °C, has a much higher viscosity and density than ordinary water, and so on, is 'loaded' with interpretation. This is first because of the complexity of the experimental techniques, and second more fundamentally, because of the ceteris paribus character of the laws used to process the data [6.2]. McKinney quotes Derjaguin saying:[230]

> The velocity of a column's movement is inversely proportional to its length and viscosity. Therefore, one can readily calculate the viscosity.

Note that the viscosity is *calculated* from a law which contains, among other things, the ceteris paribus condition that the observed velocity is not influenced by contact angle hysteresis (whereas it is plausible that this condition was not met in Derjaguin's experiments).[231] That polywater would have a higher viscosity than water has been quoted in numerous (respectable) publications; and, in retro-

---

[229] Slightly paraphrased from Everett, Haynes and McElroy (1971:280ff,300).
[230] The name 'Derjaguin' is sometimes spelled 'Deryagin'; I use the first version throughout.
[231] For the notion of contact angle see [6.4].

spect, there *seems* little doubt that the unknown substance has a much higher viscosity than water. However, as far as I have been able to establish, careful measurements have only been reported by Derjaguin. Moreover:[232]

> Calculations show that under the probable circumstances of this [Derjaguin's] experiment, a difference of about 0.01 in [the cosine of the contact angle] between leading and trailing menisci is sufficient to account for the observed fifteen-fold increase in apparent viscosity.

Many otherwise competent scientists writing about polywater didn't know the first thing about contact angle hysteresis (another anomaly of sorts). It was on Derjaguin's authority alone that they accepted the high viscosity value. The problems about what *it* is start with mundane properties such as having a particular viscosity - a mundane, macroscopic almost manifest property, even before we start contemplating any molecular structure.

One more reason why research money was granted for 'polywater' was because of its potential practical significance. Some researchers simply got the brief to 'produce large quantities' of it (whatever 'it' was). Imagining this development to be successful, it is easy to see how polywater could have been used as a tool to do all kinds of things *with*, without necessarily knowing what it *was*.

So I don't think the distinction McKinney draws between the epistemology and ontology of experiment can be made. McKinney also seems to suggest that the anomaly was solved ('no new theory was required to explain the properties of the material') when it was established that the water was 'loaded with contaminants.' Still, on McKinney's rendering of the history of this episode, it remains an unexplained anomaly that if we start with triple distilled water and quartz capillaries,[233] we end up with a liquid film which contains less than 50% water. Not only does it contain '20-60% sodium by weight, 1% calcium, 3% potassium, 15% chlorine, and 15% sulphate', it also contains acetate and lactate. If contamination is the explanation, isn't it utterly implausible that the same contaminants were present in the capillaries in different laboratories which used a variety of techniques to make anomalous water? And even if 'anomalous water' were a liquid consisting primarily of water and '20-60% sodium [and] 15% chlorine ', it remains an anomaly why salts like NaCl (sodium chloride) didn't precipitate.

Actually McKinney's explanation in terms of contaminants does not represent what was going on at the time. First numerous other possible contaminants had been proposed, for example: borates, hydrosols, silica gels, nitrates, bicarbonates, carboxylic acid salts (Rousseau 1971:434). Second in the polywater rush not everyone used freshly drawn quartz capillaries:

> The latest Soviet work has been conducted with such care that it is difficult to believe that there can have been anything present in significant amounts other than water and silica. The same cannot be said for other workers who prepared their capillaries from tubing 'cleaned by conven-

---

[232] Haynes in JCIS (1971:564).
[233] McKinney, quoting Franks (1981), says about silica that the 'material is pure and does not contain residues of potentially water-soluble substances'.

tional methods' or 'boiled in *aqua regia* and thoroughly rinsed with distilled water'. The tenacity with which conventional acid cleaning reagents are adsorbed by glass and quartz surfaces is well known and stringent cleaning with steam or long soaking in distilled water is needed to remove the last traces of contamination. A significant omission from nearly all published work is any mention of the methods of handling the capillaries; some of the contaminants reported recently could have come form perspiration or from breath aerosol.

Third, in physical chemistry long experience is necessary before there is a chance of working in a reasonably contaminant-free way. Derjaguin never denied that *some* contaminants were always present in 'polywater.' The presence of '1% calcium, 3% potassium' and traces of sodium acetate and lactate is irrelevant for the assessment of the polywater hypothesis. Of course these data *were* relevant insofar as they offered an alternative explanation for say, particular spectral bands that had been ascribed to lengthening of O-H bonds (an alleged polywater property).

Probably the first good assessment of the 'real' anomalous water (excluding data that could be ascribed to straightforward contaminations) was given by Everett, Haynes, and McElroy (1971:1035), who concluded:

> As the only materials present in the capillary are silica and water, and as the properties outlined above appear to be consistent with those of a silicic acid sol, it is not necessary on the basis of the above evidence to seek an explanation in terms of 'polywater'.

This interpretation did not fit all reported data,[234] but a good reason to go with the interpretation of Everett, Haynes and McElroy, is that it was later endorsed by Derjaguin (Block 3-6). If it is accepted that anomalous water is a solution of an anomalous component in ordinary water forming a colloidal sol (on concentration taking on a crystalline form), then all the reported anomalous properties get straightforward explanations (Everett, Haynes, and McElroy 1970). For example, what first looked like a 30-500 times lowering of the vapour pressure as predicted by the Kelvin equation, is simply due to the presence of a relatively involatile anomalous components in ordinary water (in accordance with Raoult's law). The variability too in the reported properties of anomalous water (including the variable success rate in making it) can be explained by assuming a reaction between water and silica (ibid:304):

> The explanation of this variability is to be sought in the surface chemistry of silica, and in this particular problem in the influence of the precise conditions under which the quartz capillaries are drawn: factors such as the temperature profile along the tube, the ambient atmosphere, and the rate of cooling could all be important.

---

[234] Lippincott et al (1971:443): 'It was established both by infrared and microprobe analyses of "polywater", that neither silicon nor silicon compounds were present in significant quantities.'

Block 3-6   Final word of Derjaguin.[235]

> In all probability, the anomalous component is in general a mixture of silicic acid colloidal particles and molecular-dissolved compounds of Na, Si, O, C, K, Cl, and S. When produced in quartz capillaries under the cleanest conditions attainable, the samples chiefly contain O, Si, and Na. This [i.e. the occurrence of Na] may be explained by the enrichment of quartz surface layers with sodium impurities when drawing the capillaries. In the case of samples not obtained under strictly 'sterile' conditions, the presence of atoms of K, C, Cl, and S may rather be attributed to the surface diffusion of impurities from an ambient medium.
>
> The anomalous properties of condensates may be explained by the peculiar features of a reaction taking place between the vapour and solid surfaces in the process of condensation.
>
> The particles of a sol converting itself on evaporation in a gel impart increased viscosity, density, and refractive index to anomalous condensates.

The conclusion should be that there were no contaminants in the normal sense of the word. The product of an unexpected chemical reaction is not a 'contaminant' in the ordinary sense, though it does show the presence of a hitherto unknown ceteris paribus condition. The fact that for some reason the water and silica 'reacted' to form a silicic acid sol is still an anomaly and 'a newly created material' (McKinney's words for polywater). Why such a gel-like water-silica structure would form is not very clear, though less exciting than if 'it' had been one of the 30 proposed molecular structures of polywater.[236]

Moreover it is not the case that the disappearance of the concept of polywater has done anything to remove anomalous thin water films from the scientific scene. For example, twenty years later, we find Gee, Healy and White (1990:450) writing in the *Journal of Colloid and Interface Science* (the same journal in which the papers of the 1970 polywater conference had been published):

> Water films on fully hydroxylated quartz are much thicker than expected. The existence of different amounts of solute on various silica or quartz surfaces, or that certain silica samples dissolve to varying extents, appears to be the most obvious explanation The only certainty is that a great deal more experimental and theoretical work is necessary before the mystery of this and similar systems is finally unravelled.

Gee, Healy and White also point out that many of the earlier reports of such phenomena were not taken seriously by many scientists because

> the anomalous results obtained from these [earlier] investigations were often attributed to the presence of a gel-layer on the solid surface formed by dissolution of the silica, and also to the inherent rugosity of the surface. However, the experiments on mica [of which very many had been reported in the period 1978-1990] negate this notion since such a layer does not exist on the surface of mica in aqueous solution.

---

[235] Quotations from Derjaguin and Churayev (1973a, 1973b).
[236] Molecular structures proposed for anomalous water include tetrahedral $(H_2O)_4$ clusters, two-dimensional sheets in square array, $(H_2O)_{14}$ rhombic dodecahedra, $(H_2O)_4$ and $(H_2O)_6$ ring structures, linear chains linked by hydrogen bridges, various cyclic polymers.

There is no reference in this paper to any literature from the polywater episode (though there are six to publications by Derjaguin and co-workers covering the period 1935 to 1986). However, more recently one can find in the top journal *Langmuir* an article with the title 'Long range attraction in water vapour: Capillary forces relevant to "polywater"', in which the author writes (Yaminsky 1997):

> The old polywater experiments were sophisticated. The original experimental observation has never been dismissed. The great mistake of B.V. Derjaguin, interpreted properly, it seems, does have positive future development.

Newman (1989:142) observed: 'This whole research area [of surface chemistry] was considerably side-tracked during the 1970s by the "polywater" episode which is now happily laid to rest.' But research wasn't 'side-tracked' at all. It was just a bit of normal practice that drew the attention of outsiders. Of course, scientists were slightly embarrassed by the media coverage of an exciting hypothesis that turned out to be wrong, but this had little or no lasting effect on the investigations of the anomalies of interface and colloid chemistry.

Derjaguin (and collaborators) published hundreds of papers on surface phenomena from the 1930s to the 1980s. He was assisted by some of the best instrument makers in the world and was renowned for his ingenious experimental techniques.[237] One of the theories mentioned in every text book on colloid chemistry is the DLVO theory - the Derjaguin-Landau-Verwey-Overbeek theory. Derjaguin was justified in *speculating* about polywater, because the 'contamination' or 'surface dissolution' required to explain the observed physical characteristics was implausible. As Derjaguin and Churaev (1973b) retrospectively note:

> We earlier proceeded on the basis of the assumption that the dissolving ability of condensates cannot differ from that of the volume water because experiments showing that liquid water does not change its properties when introduced into the same capillaries in which the anomalous product was obtained were considered evidence of the absence of an influence of dissolution of the surface and its contamination.

Furthermore, the polywater hypothesis was not as revolutionary as might seem. Observations suggesting an unusual behaviour for water when contained in small capillaries had been reported for 50 years; anomalies involving capillary forces had been reported since 1870.[238] Since the 1920s there had been claims that liquids, in particular water, condensed in capillaries, exhibit vapour pressures substantially lower than those predicted by the Kelvin equation. Numerous other anomalies too had been reported of water in contact with materials like sand and coal and it had often been speculated that water near solid surfaces may be structurally different from bulk water (either liquid or ice). The only thing Derjaguin added was that the effect of the walls extended somewhat further and somehow stabilised. As it hap-

---

[237] 'No research group has tried to reproduce many of Derjaguin's careful experiments [on "polywater"]' (Allen 1971:557).
[238] For references see Allen (1971:556), Freundlich (1922:215-9), Haynes, and McElroy (1970), van Brakel and Heertjes (1977).

pened, anomalous water turned out to be, for the time being (Latour 1999), a 'polymer' of water and silica, not of water alone. Everett, Haynes and McElroy (1971) describe the silicic acid sol as a 'polydisperse polymer' (of silica and water), which may coagulate to a gel and 'age (possibly by further polymerisation or cross-linking).' Hence talk of water plus polymerisation being involved wasn't entirely wrong.

I conclude therefore, contrary to McKinney's suggestion in defence of Hacking's (1983) entity realism, that scientists were not led to believe in the existence of polywater because of their experimenting *on* polywater. There was, to be sure, much experimenting going on with water in capillaries. The causal effects of the anomalous condensates provided strong evidence that the effects weren't caused by ordinary water. It was right to conclude, provisionally at least, that the particular condensates were 'anomalous'. But the experimental part of the story stops at the anomaly. The data by themselves do very little to support one hypothesis or another. The condensates were anomalous because their manifest properties were not the properties of water, whereas that was really the only liquid that could turn up there. Any microstory whether provisionally accepted or not, will be about a substance that has already been identified relying on manifest and macroscopic observations.

Eventually, using information about *all* its (alleged, measured, possible) properties, complete consensus was reached that polywater, as envisaged, did not exist. Aspects considered included manifest properties of liquids such as viscosity and boiling point, as well as more 'hidden' or 'theoretical' macroscopic properties such as thermal expansion behaviour or entropy of evaporation; further microproperties, in particular concerning molecular structure, using X-ray diffraction and various spectroscopic techniques (infrared, Raman, proton magnetic resonance); and finally various quantum mechanical calculations methods, both ab initio molecular orbital models and semi-empirical methods, though at the time of the polywater episode, ab initio techniques to make quantum mechanical calculations for the stability of various types of O-H bonding were always inconclusive (Everett, Haynes and McElroy 1970):

> Whether polywater is eventually proved to exist or not there is no doubt that by an appropriate selection of models and parameters, a theoretical basis could be found for whichever turns out to be the experimentally established situation: theoretical calculations cannot have any reliable predictive value in the present case.

Accounts of what it was that caused the (apparent) anomalies were generated from and justified by the general theoretical background considered relevant for the phenomena at hand *and* for the evaluation of the experimental techniques used. It is because of what physical chemists know - theoretical knowledge about entities like quartz, water, interfacial tensions, disjoint pressures, sols, gels, and polymerisation, as well as practical knowledge like the mundane fact that air in laboratories is full of oily molecules - that some explanation of what is going on are considered more plausible than others.

Block 3-7   Quotations from papers presented at the First International Conference on Polywater (1970).

in support of the existence polywater
The molecular weight of water II was determined from the vapour pressure lowering of its solution in water I and found to be 180 ± 40. Corroborating results were obtained when the molecular weight was determined by the melting point and from the curve of phase separation at sub-zero temperatures. The only explanation that fits all our data is that water II consists of associated water I molecules.
A material forms in the capillaries during the anomalous preparation, whose formation cannot be explained on the basis of impurities. Thus no interpretation, other than a form of water (which is stable to 350-400°C), has yet been found to explain the infrared spectra of the products from 'cleaned' capillaries. All results presented have been reproduced.

dismissing polywater
Anomalous water is commonly supposed to be produced by a catalytic process, and the cessation of growth after some time is attributed to poisoning. An examination of the poisoning according to generally accepted ideas in catalysis indicates that the concentrations which would result would be far smaller than claimed. This discrepancy can easily be resolved by attributing the phenomena to condensation of water on soluble impurities.
A preparation of poly-H_2O results in material with the same infrared spectrum as poly-H_2O. The failure to observe the predicted isotope frequency shifts indicates that the reported infrared features do not originate from H_2O units. On the basis of these experiments it appears very unlikely that a polymer of water exists. We believe the polywater properties result from a complicated mixture of salts.
Three new results invalidate our original model. Because our original model did so well in consistently correlating such a wide variety of experimental data, our present disproof of this model leads us to conclude that polywater does *not* exist. Considering these results and other theoretical studies we conclude that there are no plausible structures for polymers of water more stable than the ordinary hydrogen-bonded species.
Our theoretical study reveals that the symmetric and asymmetric hexamers are not of major importance in the formation and stabilisation of water-II, if indeed water-II exist.

miscellaneous observations
'Polywater' is observed not to exchange rapidly with normal water, and demonstrates an Arrhenius line width temperature dependence in the range of 100 °C to minus 80°C.
That any substance, claimed to be polymeric in nature, should have so low an entropy of evaporation is, however, so startling that it is prudent first to question the validity of the interpretation of the experimental data.
Once again I should point out that extremely careful trapping must be carried out in order to avoid back-diffusion of trace organic material and the installation of a liquid nitrogen trap may be insufficient.
It is shown that the product [i.e. anomalous water] does not arise from impurities carried into the capillaries by creep of a liquid film from the water source over the surface of the cell. Rather, it results from a vapour-capillary reaction.
The mere presence of some salt is not sufficient to disprove the presence of Water II.

The form of life that science is, is such that experimental scientists are instrumental in making things and meanings by experimenting both on and with. But when consensus is reached about the existence of something, the whole bag and baggage of that form of life comes with it. It would be an artefact to consider some of the judgements expressed in Block 3-7 as *inherently* more compelling than others in reaching the verdict that no support for a water polymer exists in the circumstances under investigation.[239] Scientific practice is involved in a co-production of macroscopic and manifest observables *and* 'theoretical' interpretations, models, and their intercalations.

---

[239] Quotations in Block 3-7 are from, respectively Derjaguin and Churayev (1971:415), Page and Jakobson (1971:427), Taylor (1971:543), Rousseau (1971 and in JCSI 1971:563), Allen and Kollman (1971b:469), O'Konski and Levine (1971: 547), Aldrich et al (1971), Petsko (1971), Everett and McElroy (1971:529), Cadenhead in JCIS (1971:565), Brummer et al (1971:489), Fowkes, Lovejoy and Chow (1971:522).

4 essentialistic realism

4.1 Kripke and Putnam on water

Chemists may be surprised to hear how popular it is among philosophers to discuss the properties of water. To give the flavour of some of these discussions let me first give a somewhat hilarious example. In the prestigious journal *Mind* eight pages were reserved for a discussion of the relevance of Chomsky's (1995) observation that tea and Sprite are not called water although they contain roughly the same proportion of H_2O as tap water (Abbott 1997).[240] On a somewhat more sophisticated track, LaPorte (1996) when arguing that other writers greatly overestimate the role of microstructure in advocating theories of causal reference, refers to the difference between H_2O and D_2O, that between ruby, topaz, and corundum; jade and nephrite; diamond and charcoal; and similar examples. The chemical composition of coal and diamond is the same, but, presumably its 'microstructure' is different. Is it to count as one or two natural kinds? More generally, as Locke already realised (Block 3-4), any natural object will instantiate many different natural kinds. Such publications touch on the general issue of the tension between the manifest and scientific image and the 'dilemma' concerning whether physics or chemistry should be claimed the final arbiter of substances, even if the main focus of these publications is in the philosophy of language or mind.

It is common in these discussions to combine or even identify reductionism and essentialism: '(the concept, the laws governing) water can be reduced to (the concept, the laws governing) H_2O' becomes the same as 'the essence of (the natural kind) water is H_2O'. On such a microreductive picture, water really is H_2O. H_2O is the essence of the substance which, at the manifest level, is called water. The essence of being H_2O is invariant and determines (together with essences of other substances) the variation of manifest properties in different contexts. On this view the manifest properties are not epiphenomena; they are real, but supervene on the microproperties (or essences of natural kinds).[241]

It is this view that has been given a new twist in seminal publications of Kripke (1971, 1980) and Putnam (1975), which have had an impact on various domains of analytical philosophy. Most discussions have taken place in the philosophy of mind and language, but here I will only consider their suggestion that natural kinds have essences and focus almost exclusively on the example that is most prominent in their writings and the vast secondary literature:

---

[240] And the debate didn't stop there, moving on to 'living water' (LaPorte 1998); 'babies, chickens, and tomatoes, contain a higher proportion of water than Utah's Great Salt Lake' (Abbott 1999:145). There's also the question of tears and holy water.
[241] See quotation of Kim on p. 75.

water is H₂O (4.1)

Passing over their differences, in brief, Kripke's and Putnam's views concerning (4.1) were:[242]

- (4.1) is a theoretical identity - a kind of bridge law [2.3].[243]
- (4.1) states an essential property of water: science discovers essences [2.5].
- If (4.1) is true, then (4.1) is necessarily true.
- Both 'water' and 'H₂O' are rigid designators [244] - rigid designators are expressions that refer to the same entity in all possible worlds.
- The (Fregean) sense of 'water' does not determine its reference. A baptism event identifies the reference of water.

The last characteristic lies at the bottom of the causal theory of reference [2.6]. In what is called a dubbing ceremony, referent and word are tied to one another (Putnam 1975:225):

> The body of liquid I am pointing to bears a certain sameness relation (say, *x is the same liquid as y*, or *x is the same$_L$ as y*) to most of the stuff I and other speakers in my linguistic community have on other occasions called 'water'. The relation same$_L$ is a *theoretical* relation: whether something is or is not the same liquid as *this* may take an indeterminate amount of scientific investigation to determine.

Future generations remain causally connected to the dubbing event and hence causally connected to the referent of the word and specialists will eventually find out what 'essence' fixes the 'intended' sameness relation. The causal theory of reference blocks incommensurability of successive theories, because the term's referent supplies the stability needed for rational comparison of successive theories.

Talk of rigid designators is associated primarily with Kripke, but both Putnam and Kripke agreed that *if* (4.1) is true, it is true in all possible worlds. At the same time, (4.1) being true, is a posteriori (Putnam 1975:233):

> Once we have discovered that water (in the actual world) is H₂O, *nothing counts as a possible world in which water isn't H₂O*.

Contrary to most received views, (4.1) is a posteriori *and* necessary - a metaphysical, though not an epistemic necessity.[245] Dummett (1978:428) has suggested that Kripke's arguments that there are statements which are *both* necessary *and* a posteriori might be strengthened by considering that it is necessary that exemplars of a kind are of common origin: 'I suppose that anything produced in a laboratory would be only artificial silk, and not real silk, whatever its chemical structure.' However, being artificial or natural silk is a simple question when assessed in chemical terms: artificially produced substances are the same as those occurring in

---

[242] I write 'were', not 'are', because Putnam has changed his mind on almost everything listed here and Kripke turned his interest to other matters.
[243] Cf. 'It is going to be *necessary* that heat is the motion of molecules' (Kripke 1971:160).
[244] Also (Putnam 1975:231): 'when I give the ostensive definition "*this* (liquid) is water", the demonstrative 'this' is *rigid*.'
[245] See Kripke (1980:38,116-9,123-28,138); Putnam (1975:233).

nature, if they fulfil the operational criteria of being the same substance [3.4]. Of course it is possible to consider natural silk in the specific sense of being the product of a particular animal, or to consider silk in the specific sense of the material from which threads and then woven materials are made. But, as a substance, silk is silk, if it is the same chemical substance – whatever operational criterion for 'same chemical substance' is preferred. The 'common origin' considerations work well when applied to proper names as rigid designators, but not when applied to natural kinds (Forbes 1997). The extension of the notion of rigid designator to natural kinds has other consequences as well. Rigid designators are a kind of deus ex machina; they allow us to believe in essences without having to know what they are (de Sousa 1984). Moreover, even if we grant that water is necessarily composed of H_2O, we should not accept that the rigid designators 'H_2O' and 'water' refer to the same thing (Barnet 2000). Compare SiO_2, the molecular constituent of sand, quartz, and glass. If both quartz and SiO_2 are rigid designators, then they do *not* designate the same thing.

According to Kripke and Putnam, in discovering that (4.1) is the case, science has discovered the essence or nature of what it is to be water, by establishing the fundamental structure or the underlying trait of water: 'science attempts, by investigating basic structural traits, to find the nature, and thus the essence (in the philosophical sense), of the kind' (Kripke 1980:330).[246] Now consider:[247]

water has minimum density at 3.98 °C (4.2a)

water boils at 100 °C (4.2b)

water has a viscosity of 0.8904 centipoise (4.2c)

and similar characteristic properties of water (Block 4-1). Here manifest properties and macroscopic scientific properties are closely connected. It is part of the manifest image that oily liquids are more viscous than water and much hotter than water when they are boiled. The corresponding scientific properties make these manifest properties more precise by tying them to operationally well defined (and intercalated) measuring procedures. Even for those properties in Block 4-1 which have no obvious manifest counterpart, once they are introduced in terms of a scientific measurement operation, it is easy to point to their manifest effects. Surface tension is not a manifest property, though its effects are [2.2]. Of course this is a gliding scale; heat of formation perhaps only makes sense in a scientific context. In addition statements concerning the macroscopic but quite abstract thermodynamic properties of water like free enthalpy and entropy should be considered (Needham 2000). But here I restrict myself to the less abstract manifest properties of Block 4-1.

---

[246] Cf. Harré and Madden (1975:102): 'We follow the scientific tradition in identifying the real essence of a kind, material or individual, with its nature, which is progressively revealed a posteriori by empirical investigation.' Cf. discussion on essential properties of natural kinds in [2.5].
[247] Each of these statements is ceteris paribus [6.2]. For a start, for each a 'standard' pressure has to be specified and a temperature for (4.2c).

Block 4-1   Properties of water, heavy water (D$_2$O), and 'polywater' [3.5] at 25°C and atmospheric pressure.

	(H$_2$O)$_n$	(D$_2$O)$_n$	'polywater'	units
density	0.99707	1.10445	1.4	g/ml
dielectric constant	78.54	77.936	5	-
viscosity	0.8904	1.096	15	centipoise
boiling point	100	101.42	250	°C
freezing point	0		-40	°C
index of refraction	1.3325	1.3388	1.48	-
heat of formation	285.89	294.59	240	kJ/mole
specific electrical conductivity	<10^{-7}	<10^{-8}		ohm^{-1}cm^{-1}
magnetic susceptibility	negative		positive	
surface tension (at 20°C)	72.6		74.9	dynes/cm
molecular mass	18	20	180	dalton

Because (4.2) involves descriptions (instead of rigid designators) such as 'has a boiling point of 100°C', Kripke's arguments would not apply to (4.2). As I'll argue at length there's no ground for this. If there is sense in Kripke's concept of rigid designator then it depends on the context (in our world) in which an expression is used whether it is meant to be a rigid designator. Davies and Humberstone (1980) suggested reading (4.1) as an a posteriori true identity between two entities denoted by descriptions - removing the 'necessity' of the statement. Kripke gives few argument beyond stipulation why such a view would be wrong. He simply appeals to some vaguely entrenched Aristotelian intuition about essences and assigns the role of discovering these essences to modern science.

In explaining Kripke's position, Putnam (1981:47) writes

> It is 'metaphysically necessary' (true in all possible worlds) that water is H$_2$O but this 'metaphysical necessity' is explained by mundane chemistry and mundane facts about speakers' intentions to refer.

Mundane facts about water fall apart on the one hand, in the common sense knowledge about water - this is what is called the stereotype by Putnam (1975) - and on the other hand, the 'tests known by experts' – though neither Putnam nor Kripke give much of an explanation of what they mean by the latter. The stereotype is usually specified by saying that water is a colourless, transparent, tasteless, thirst-quenching liquid, and so on. However, what the extension is of such a stereotype seems very vague indeed. Apparently ice is not water, a thick layer of water is not water (because it is blue), whereas fluoridated drinking water is a stereotypical case of water. Putnam has pointed out that what he calls the stereotype is by no means supposed to include everything the speaker knows about a substance. If an average speaker were asked to produce a sample of something that is stereotypically water she would not exhibit a block of ice, although she may know perfectly well that ice is a form of water. How to delineate the stereotype is of some psychological interest, and of some interest to linguistics, but it has nothing to do with the questions about necessity or how reference is fixed.²⁴⁸ I'll follow Putnam in taking the stereotype to be a rather loose part of the meaning; in particular not everything that is contained in the stereotype needs to be true.²⁴⁹

I'll argue below however that it is of importance to relate the delineation of the stereotype to the tests known by experts. This point, viz. the interdependence of ordinary and scientific language, is strongly supported by Putnam in his (1990a), and has been stressed before particularly by Dummett (1978:427) in his discussion of Putnam's conception of the division of linguistic labour:²⁵⁰

> The meaning of the word 'gold', as a word of the English language, is fully conveyed neither by a description of the criteria employed by the experts nor by a description of those used by ordinary speakers; it involves both, and a grasp of the relationship between them.

If we were to agree with Putnam's law of the sociolinguistic division of labour,²⁵¹ we would have to give some account of how the stereotype (which governs the colloquial use of a term) is related to tests known by experts (which fix or define reference). In the present case it is particularly relevant what sort of markers are part of the stereotype. Surely, if it were a psycholinguistic habit to distinguish between rain, tap water, spring water, snow, and so on, without there being any inclination to call it all water, this would not provide a basis to relate the stereotype to tests known by experts who presuppose or confirm that these samples consist of the same liquid (take or leave a few impurities). On the other hand, if it is a marker of the use of the term 'water' that it refers to a liquid (with certain specific properties), then it is part of the stereotype of being a liquid that it can boil and solidify, that it is more or less pure, and that it has innumerable stable properties. Moreover, as Beattie (1993:323) has argued, if it is allowed that experts are wrong then there is a tension between the presupposed structural theory and Putnam's thesis of the linguistic division of labour:

> Even in our structurally attuned times, it's far from clear to me that the central meaning of our ordinary term 'water' (as opposed, perhaps, to the term 'H_2O') is anchored in structural rather than manifest properties.

As Ayer (1992:456) puts it in his reply to Putnam's (1990a):

> It is only because I have already identified my writing table, in this case ostensively, that I am able even to make any sense of talk about the atoms which composed it at the time of its origin.

Putnam (1975) seems to argue that extension is determined jointly by ostension and by scientific investigation²⁵² and he stresses that

---
[248] Putnam, personal communication, September 6, 1983.
[249] See on the relation of Putnam's stereotype to Frege's terminology Wiggins (1980:79) and Salmon (1982:12,153).
[250] The example of gold has a long history, starting with Locke and the detailed response of Leibniz (1981:304-28). The latter argued (ibid:338): 'As a rule, only the experts have sufficiently accurate ideas of a given material.'
[251] For arguments that Putnam's advocacy of the linguistic division of labour is inconsistent with the results of the Twin Earth thought experiment, see Beattie (1993).
[252] As when he says: 'corresponding to the stereotypes of the lay person is the technical sense of the expert' or 'intension and extension reciprocally determine each other and are mutually correlative

Traditional semantic theory leaves out only two contributions to the determination of extension - the contribution of society and the contribution of the real world!

But when the chips are down the microreductionistic essence does the real fixing of reference in Kripke's and Putnam's account.

Most philosophers will argue that the question whether (4.1) is true in the actual world, is completely separate from the question whether, if true, it is necessarily true. For example Farrell (1983) says: 'Let us suppose - it does not matter whether we are supposing truly or not - that water is composed of H_2O molecules'. But, in discussing these matters, it *is* relevant whether (4.1) is in fact true and what the meaning is of the terms occurring in (4.1). The relevant philosophical literature is full of phrases such as 'dubbing events', 'rule according to which the reference is fixed', 'tests known by experts', 'science has shown ', and so on. If the meaning of all these phrases is ambiguous - if it is left completely open what in fact the rules, tests, and things shown by scientists are, then it is unclear what the function of such phrases is in explaining or arguing for a particular philosophical opinion.

For example, it is important to know whether the extension of water includes ice, a layer of blue water, an isolated H_2O molecule, and so on; and whether 'H_2O' means 'producing two volumes of hydrogen and one volume of oxygen when subjected to electrolysis', or 'two atoms of hydrogen and one atom of oxygen in a particular spatial relation'; otherwise the discussion is not about substances, but is an exercise in a logical game based on extremely vague intuitions. It is not sufficient to say that (4.1) is a theoretical identity statement involving natural kinds, because nothing follows from this as to what sort of a posteriori statement (4.1) is. It is one thing to discuss the difference between 'a = a' and 'a = b'. It is a completely different question whether (4.1), 'Hesperus is Phosphorus', and 'the standard meter in Paris is one meter long'[253], are or are not, substitution examples of a = b in one sense or another. Moreover, it would be a mistake to think that in moving to a logically regimented language that would remove ambiguity about the meaning, scientific or otherwise, of 'water is H_2O' (Block 4-2).

The 'scientism' of Kripke and Putnam has of course been attacked by Wittgensteinians and other ordinary language philosophers. The Kripke/Putnam view leads to absurdities such as (Stroll 1991): Water = H_2O, Ice = H_2O; therefore Water = Ice.[254] Opponents like Stroll will also oppose any theory of direct or causal reference that presupposes a reference fixing ostensive baptism event, because what a word (in general) means is how it is used. What is of interest is the difference between the use-meaning of water and ice, not what might be hidden

---

terms' (1975:187-8). The next quote in the main text is from p. 245; cf. also p. 271: 'Ignoring the division of linguistic labour is ignoring the social dimension of cognition; ignoring what we have called indexicality of most words is ignoring the contribution of the environment.'

[253] According to Kripke the latter statement is both contingent and a priori. For discussion of this view see van Brakel (1990).

[254] The 'scientistic' reply would be that the Kripke/Putnam account has to be supplemented with a hierarchy of kinds (Elders 1994).

Block 4-2 Alternative logical 'translations' of water is H_2O.[256]

> $(\forall x)$ [water $(x) \to H_2O\ (x)$]
> $(\forall x)$ [water $(x) \leftrightarrow H_2O\ (x)$]
> $(\forall x)\ (\forall y)$ [{water $(x) \land H_2O\ (y)$} $\to x = y$]
> $(\forall x)\ (\forall y)$ [{water $(x) \land H_2O\ (y)$} $\to (P)\ \{P(x) \leftrightarrow P(y)\}$]
> $(\forall x)$ [water $(x) \to \Box\ \{(\exists x)$ water $(x) \to H_2O\ (x)\}$]
> $\Box\ (\forall x)$ [water $(x) \to H_2O\ (x)$]
> $\Box\ (\forall x)\ \Box$ [water $(x) \to \Box\ \{(\exists x)$ water $(x) \to H_2O\ (x)\}$]
> $\Box\ [(\exists x)\ \{$water $(x) \land H_2O\ (x)\} \to (\forall x)\ \{$water $(x) \to H_2O\ (x)\}$]

underneath.[255] Moreover, both Kripke and Putnam extend a theory about proper names to names of substances, introducing the indexical aspect of the baptising event, but such an extension is by no means obvious, if only because '"I described it as water" makes sense; whereas "I described him as Cicero" does not' (Hanfling 1984:199).

Other questions that automatically suggest themselves if one doesn't take the 'science is right' approach for granted include 'whether Putnam's account might not be guilty of a certain cultural and philosophical parochialism',[257] and why it is that 'the meaning of "water" is determined to no small part by the *impurities* of water' (Margalit 1979:35). Distilled water, which is quite pure water, is not 'typically' water (you may get it in special bottles at the drug store or the petrol station).[258] From a more scientific point of view, Mellor (1977:305-6) argues against Putnam's essentialist theory that in case of elements high in the periodic table (which do not occur on earth - only in laboratories), there is no 'ostensive reference, to just *this* archetype in *this* world'; and therefore Putnam's theory of causal reference fails. But that seems to miss the point. It is only in a scientific context that there is interest in these elements, and when one scientist has produced the element, even if its life time is very short, that would be sufficient to secure the ostensive reference to it.

Below I won't evaluate the account of Kripke and Putnam on 'general' philosophical grounds, but investigate whether their appeal to 'science is in the business of discovering essences' makes any *scientific* sense. Though the original publications are now a quarter of a century old, they are still referred to regularly and every year publications appear in which the Twin Earth thought experiment, to be discussed in [4.3], is rehearsed yet again.

More important than the specific discussions that still go on,[259] is the issue of the reference of natural kind terms grounded in the sameness and difference of hid-

---

[255] Wittgenstein (1953:§126): 'For what is hidden, for example, is of no interest to us.'
[256] For discussion see Ackerman (1983) and Forbes (1985:193). It is not implied that all formalisations are independent.
[257] Cassam (1986:97); *cf.* Hanfling (1984:203) and Beattie (1993).
[258] For a recent discussion on how to distinguish impure water from things that are not water, but contain more water than certain kinds of impure water see Abbott (1997, 1999), Brown (1998), LaPorte (1998), which show that the distinction can only be made on pragmatic grounds.
[259] Most of these discussions are taking place in the philosophy of mind and in analytic metaphysics, less so in the philosophy of science. A recent discussion of the issues of this chapter in the philosophy of science is Boyd (2000). For a recent discussion of the issue of rigid designators see LaPorte (2000).

den structure. This idea was addressed in a general sense in [2.5] and details undermining its obviousness in the case of water were given in [3.3]. In this chapter I consider Kripke's and Putnam's work as one particular influential historical case study that strengthened philosophers' intuitions about science's discovery of essences - particularly those philosophers who are concerned with issues otherwise remote from the philosophy of science.

4.2 molecules and atoms

Philosophers tend to take for granted what 'H_2O' means. However, it is not always clear whether the term 'H_2O' refers to one molecule,[260] to a set of similar molecules, to 'the' molecular structure, to the chemical composition, or to something else again. Also the often used term 'internal structure' is vague: does it refer to the internal spatial arrangement of atoms in molecules or to the internal arrangements of molecules within a substance (or both)?

Sometimes it is added explicitly that what is meant by (4.1) is that H_2O is referring to the molecular *structure* of water, i.e. it is shorthand for a *particular* topological arrangement of two hydrogen atoms and one oxygen atom.[261] On the face of it, one would think that 'chemical composition' and 'chemical structure' mean something different, but Salmon uses them interchangeable:[262]

> From these three ingredients - the ostensive definition of water, the fact that the paradigm has the chemical structure H_2O, and the fact that consubstantiality consists in having the same chemical structure - we easily generate the necessary a posteriori truth that water is H_2O, or more accurately that every sample of water has the chemical composition of two parts hydrogen and one part oxygen.

Whatever meaning for H_2O one chooses, as we have seen [3.3], statement (4.1) is *not* true, because of the problem of isotopes and the fact that water is not 100 per cent H_2O. It almost seems as if the meaning philosophers ascribe to 'water is H_2O' is precisely that which makes 'water is H_2O' if true, necessarily true, but 'water is H_2O', in fact, not true. This might further suggest that there are no truths in science; so scientists discover no fundamental structures of substances either. This cannot be the intention of the defenders of essentialistic realism, however.

Confronted with the falsity of (4.1), Putnam (1983:63) chose to replace (4.1) by:

> water is the quantum mechanical super-position of
> H_2O, H_4O_2, H_4O_6.... plus D_2O, D_4O_2, ... (4.3)

---

[260] Putnam (1975:239): 'one may refer to a single H_2O molecule as water'.
[261] For example: 'Molecules will be said to match if they contain atoms of the same elements in the same topological combinations' (Quine, 1977); or: 'Thus "molecule" of H_2O denotes a natural kind, membership in which depends on whether a given molecule is made up of certain kinds of atoms related in certain ways' (Hirsch, 1982:266).
[262] Salmon (1982:163; cf. 81,86,178).

because that is what scientists allegedly have discovered the chemical composition of water in the actual world to be (and not H_2O). But this still doesn't say anything about other isotopes or about the dissociation or dimerisation products of H_2O that can be found in water and the essentialism is watered down significantly: 'the "essence" that physics discovers is better thought of as a sort of *paradigm* that other applications of the concept must *resemble* than as a necessary and sufficient condition in all possible worlds' (ibid:64).

One may also wonder why Putnam takes for granted that being partly D_2O is part of the essence of water. Prima facie it would seem more reasonable on the view Putnam and Kripke are defending, to consider D_2O and H_2O different natural kinds.[263] The fact that water as it occurs on earth consists of a *particular* mixture of H_2O and D_2O and some other isotope combinations of hydrogen and oxygen,[264] is due to the history of the earth (which might well have been different). The substance with composition H_2O is quite different from the substance with composition D_2O (Block 4-1).

Moreover it seems too easy to take for granted that there can be radical changes in what scientists consider to be the essence of things and assume that scientists will make the essentialistic right choice when confronted with a dilemma, as they were when confronted with isotopes (Zemach 1976). Following Paneth, the IUPAC defined a chemical element in 1923 as a substance of which all atoms have the same nuclear charge.[265] Re-structuring chemistry on the basis of isotopy would have led to the demise of the periodic system. Here chemistry did not follow the path to reduction because of developments in physics.[266] That is to say the business of isotopes was relegated to physics. As de Sousa (1984:571) puts it in a critique of the Kripke/Putnam theory of natural kinds:

> H_2O is a *chemical* characterisation of water, not a *physical* one. Physically it turns out that water is a mixture of several sorts of molecules: ones containing Oxygen-16 and one containing the isotope Oxygen-18, as well as ones containing isotopes of hydrogen (deuterium or tritium).

But nowhere in the writings of Kripke and Putnam is it stated that H_2O is the *chemical* essence of water. They claim it is *the* essence. So anybody arguing that water has an essence has to address the isotope issue.

Such issues have been discussed at great length in the literature.[267] I think at least two problems are hidden in this discussion. First there is Locke's suggestion that our natural kind classifications are as interesting and informative about ourselves as about the natural orders of things (Block 3-4). If the essential properties

---

[263] As Zemach (1976:120-2) points out, Putnam will have to say that it turns out that after all water is not a natural kind, but a mixture of natural kinds. If the example of heavy water is too exotic, consider chlorine instead (Mellor 1977:303) in which there is not one isotope that predominates.
[264] There are eighteen isotopic variants of water, combining three hydrogen isotopes (H-1, H-2, H-3) and three oxygen isotopes (O-16, O-17, O-18). The issue of isotopes has also entered discussions about the causal theory of reference (Stanford and Kitcher 2000:104).
[265] IUOAC = International Union of Pure and Applied Chemistry.
[266] See quotations from MacKinnon, Hundt, Kragh, Heisenberg, and Pauli in Scerri (1993:50-1).
[267] See Zemach (1976), Mellor (1977), Platts (1983), Wiggins (1980:211-2), and more recently Boyd (2000:77).

have an explanatory function, wouldn't their relevance be determined by the interests of the investigators? Sometimes atomic number is the 'essential explanatory property'; at other times the isotope number is. Second, there is the question whether, if it is accepted that substances *do* have essential properties, how do we know that scientists have actually discovered any of these properties?

The point that views about the structure of matter might change completely has been levelled against the theories of Kripke and Putnam by several writers. For example, Averill (1982) has argued that nothing in Kripke's and Putnam's theory rules out the possibility that there are 'zits', very small particles lurking in and around the nucleus of the atom. The basic zit-structure determines all the properties of the atom, including the atomic number. Zit-structure is more basic than atomic structure: it allows variations on the same basic zit-structure such that gold can also have atomic number 283. Because of accidental properties of the Big Bang in this world, gold with atomic number 283 does not occur. Therefore Kripke's arguments only support the following: if a feature of a substance is basic according to a scientific theory that is both true and the best, most comprehensive theory possible of a substance, then this feature is an essential feature of that substance. However neither scientists nor we can ever find out whether the prevailing or any other scientific theory is a theory with such properties.[268] A suggestion like that of Averill may seem far-fetched, but it is no more far-fetched than the extremely influential effect of Putnam's invocation of Twin Earth [4.3].

If the philosophical concern is primarily to 'save' a statement that is both a posteriori and necessary, a solution might be to consider: 'Water contains hydrogen atoms' (Tye, 1983). But as there are many things that contain hydrogen atoms, this statement is extremely uninformative about what water is. So perhaps we should resort to something like (Block 3-1):[269]

the molecular structure of water is something like H_2O (4.4)

as a paraphrase of (4.3). However, as Putnam (1983:63f) acknowledges, then it becomes unclear what it means to say that two samples of liquid are both samples of the same liquid if they share 'relevantly very similar' molecular structures.[270] It would also seem that it's difficult to take (4.4), or (4.3) for that matter, as an identity statement in the sense of Kripke. In accepting something like (4.4) as the embodiment of the essence of water that scientists have discovered, those scientific statements which, if true, are necessarily true become both very abstract and very vague in their empirical or operational meaning. There seems to be little difference between (4.4) and saying:

water is whatever the molecular structure of water will turn out to be (4.5)

---

[268] Also consider the thought experiment of 'a universe where matter is continuous' (Forbes 1997:520).
[269] Putnam, personal communication (October 29, 1982).
[270] Putnam (1983:62): 'we do not call any other actual or hypothetical substance "water" unless it is *similar in composition* to this. But "similar in composition" is a somewhat vague notion'.

Even if we put aside extravagant suggestions like Averill's zits and the uncharitable readings of Stroll and other ordinary language philosophers, take science at face value, and accept Putnam's hand waving that water is 'something like H_2O', there remain problems for the essentialist programme. If the essence of molecules is to be defined in terms of constituting atoms and their interactions, clearly this will only work if atoms have/are essences. Let's look at this in some more detail. One problem of defining the essence of an atom has already been hinted at: Isn't it part of the essence of a particular element that it forms certain chemical bonds with other elements (atoms, ions, radicals) with certain properties? Even if one considers an element severed from the rest of the world (forgetting about relational properties), it's not clear what its essence could be. The problem encountered with finding the essence of substances in terms of molecular structure just repeats itself at a lower level.

Consider as an example Krypton - chosen because it figures in the one time definition of one meter and in discussions about Kripke's suggestion that there are not only statements which are a posteriori *and* necessary, but also statements which are contingent *and* a priori.[271] According to Kripke and Putnam the essence of an element is its atomic number. On an essentialist reading:

Krypton is the chemical element with atomic number 36 (4.6)

if true, is necessarily true, just like (4.1). Now let's look somewhat further into the meaning of the right-hand side of (4.6). The narrow meaning of

having atomic number 36 (4.7)

is that it refers to an atom with 36 protons in the nucleus and 36 electrons around it. The number of neutrons varies for different isotopes of the same chemical element. For example, the isotope Krypton-86 has 36 protons and 50 neutrons in its nucleus (as well as 36 electrons around it). Hence (4.7) is ambiguous because it rigidly designates a number of different isotopes. So let's focus on one isotope. One of the many properties of Krypton-86 is:

Krypton-86 may have an electron in the $2p^{10}$ or $5d^5$ energy levels (4.8)

However Krypton is not the only element for which (4.8) is true. But what does however seem unique for Krypton-86 is that it[272]

has the property that radiation of a *particular* wavelength is emitted when there is a transition of electrons between energy levels $2p^{10}$ and $5d^5$ (4.9)

which property is determined by the whole configuration of elementary particles which constitutes a Krypton-86 isotope. In Kripke's terminology: there are no possible worlds that contain Krypton-86 which do not have property (4.9). Hence

---

[271] The remaining part of this section draws substantially on van Brakel (1990).
[272] This particular example was chosen because of the definition of the meter as equal to 1,650,763.73 times the wavelength of the orange-red radiation emitted by Krypton-86 atoms kept in a vacuum and excited at the triple point of nitrogen (-210 °C), which radiation is emitted when Krypton-86 electrons excited to the $5d^5$ energy level fall back to the $2p^{10}$ level. In 1983 the meter was redefined in terms of time and the speed of light.

not only (4.7), but also (4.9), rigidly designates the (same) substance, viz. Krypton-86.

One might wish to argue that the only necessary and essential property of Krypton-86 is that it consists of 36 electrons, 36 protons, and 50 neutrons. But this will not do, because, say, one copper-65 isotope and seven hydrogen-3 isotopes together consist of the same numbers of elementary particles. What distinguishes Krypton from the weird mixture of copper plus hydrogen is that in Krypton the 36 electrons etc. are related to one another in a certain way. If (4.6) is necessary (if true), then most if not all properties of the Krypton electron configuration etc. must be necessary properties (if they are true properties), including (4.9). This simply follows by fleshing out Kripke's crude examples with some scientific detail. Similar questions can be raised if one asks for the essence of oxygen and hydrogen if the essence of H_2O is reduced to being a particular arrangement of these type of atoms.

If one enters the discourse of essences, one discovers not one essential microproperty, but a whole range of such 'essential' properties. Moreover, such essential microproperties are not independent of context - at best they are constant over a certain range of 'ordinary' contexts. We started out being interested in the 'essence' of water, but we seem to end up with the 'essence' of matter - which itself evaporates on further scientific scrutiny.[273] The whole notion of a *particular* and *pure* substance is eliminated in the process of reduction. It is taken for granted that 'chemical substance' *must* be a well defined category in terms of atomic primitives, and if not, it is simply eliminated during the reductive process. However, that the notion of '(chemical) pure substance' is methodologically prior to notions like atom or molecule seems indisputable [3.3]. To see the notion of pure substance as merely a ladder to reach the true reality of atoms and molecules might therefore be called the essentialistic fallacy – and similarly for the ladder from molecules to quantum mechanics [5.3].

4.3 possible worlds

According to Putnam, (4.1), or perhaps (4.3), serves to determine what is and is not to count as water in a way facts like (4.2) do not. They allow us to imagine a hypothetical planet called Twin Earth (TE), where we might find a liquid
- which fits the stereotype of water;
- which fits (4.2) - it boils at 100 °C;
- but which does not fit (4.1), or even (4.4) - it is not something like H_2O.

The Twin Earth story can be taken as referring to Twin Earth as somewhere remote from Earth, but in 'our' universe, or it could refer to another possible world altogether. The first version emphasises the importance of acquaintance in the context of referring terms. The second version emphasises that water is a rigid designator. In Putnam (1975) the distinction is not clearly drawn and as I will argue below that is how it should be.[274]

---

[273] Cf. citation of Quine on p. 63.
[274] Retrospectively Putnam (1990a:60) said: 'Far-away planets in the actual universe were playing

Twin Earth is '*exactly* like earth' except for the fact that (Putnam 1975:223)

> the liquid called 'water' is not H_2O but a different liquid whose chemical formula is very long and complicated. I shall abbreviate this chemical formula simply as XYZ. I shall suppose that XYZ is indistinguishable from water at normal temperatures and pressures. In particular it tastes like water and it quenches thirst like water. Also, I shall suppose that the oceans and lakes and seas of Twin Earth contain XYZ and not water, that it rains XYZ on Twin Earth and not water, etc.

That is to say, on Twin Earth:

water is XYZ (4.10)

And Putnam suggests that whereas on earth the word 'water' means H_2O, on Twin Earth it means XYZ. This is the causal theory of reference again: reference determines meaning. There is a demonstrative element in the fixing of the extension or reference of 'water'. For example, the reference of 'water', is fixed in such a way that $Oscar_E$ and $Oscar_{TE}$ may be in the same psychological state in 1750 with respect to the sense or stereotype of the term 'water', whereas $Oscar_E$ is living on Earth where (4.1) is the case and $Oscar_{TE}$ is living on Twin Earth where (4.10) is true, not (4.1).

A common sense approach to the suggestion that water on Twin Earth is not water is to ask: 'Wouldn't we be licensed to say that the extension of "water", is any liquid which is *either* H_2O *or* XYZ' (Margalit 1979:34). As Mellor (1977:303) puts it, there was water on both planets alike, and there still is. We simply discovered that not all water has the same microstructure. Why should it? Because its microstructure is an essential property of water? Well, that is the question. Consider too the following concern. Putnam appeals to the linguistic division of labour. It seems to follow that if speakers on Earth and Twin Earth both speak English, then 'most of the stuff I and other speakers in my linguistic community have, on other occasions, called "water"' (Putnam 1975:224) is the stuff 'either H_2O or XYZ' (Zemach 1976).

Such responses had little impact however and a whole industry developed discussing Twin Earth, taking the outline of the example for granted. For example, in a recent book, Jackson (1998:38) says Putnam

> invited us to agree with him that what counted as water on Twin Earth was not the stuff on Twin Earth with the famous superficial properties of water - being a potable liquid and all that; for short being water. We agreed with Putnam.

And he goes on to discuss the intricacies of the Twin Earth example for theories of reference throughout chapters 2 and 3 of his book. In a note, he acknowledges that 'in the mouths and from the pens of the folk it is indeterminate whether it is

---

the very same role in my own discussions that hypothetical situations ("possible worlds") were playing in Kripke's.' But it is not obvious that Putnam (1975) is unambiguous in this respect (if only because of the ambiguity of 'hypothetical situation').

H$_2$O or the watery stuff on Twin Earth that counts as water on Twin Earth', but (ibid:38n12):

> For simplicity I will suppress that complication in what follows and will suppose, with the majority, that the stories did not resolve an ambiguity but rather made conspicuous a hitherto unremarked feature of our use of the word 'water' and like natural-kind terms.

However, if we only consider chemically possible worlds, it is not possible to find a liquid with a different chemical composition (or different molecular structure) than water, which would be 'indistinguishable from water at normal temperatures and pressures'. How would the biochemistry of organisms work on Twin Earth if it is assumed that the contents of all cells remain the same except for H$_2$O being replaced by something which has a 'very long and complicated' formula. This is stipulating a world which is to any stretch of the imagination chemically, physically, and biologically impossible. Choosing water as the example, makes the suggestion of Twin Earth even more hilarious, but the same sort of counter argument would apply if a rare earth element had been chosen in the example. The only remotely possible way that Twin Earth could be in a physically possible universe, would be if we were to assume it to be located in a universe with a different history.[275] Putnam's hypothesised XYZ cannot be *water*. That suggestion would remove any plausibility of Twin Earth resembling earth in any way whatsoever. To suggest otherwise is simply nonsense. The more so if the suggestion would be that no matter how far statement (4.2) is extended with similar descriptions of water (Block 4-1), (4.2) applies to both water and water$_{TE}$, though water$_{TE}$ is not H$_2$O.

If we want to consider all logically possible worlds, we can stipulate worlds in which anything resembling 'our' physics and chemistry is different. But, even if these differences are assumed to be small (whatever that means in this context), it is difficult to imagine (in our world) what would or could be the case in the other world (apart from stipulating it in *our* language). Philosophers often dismiss such problems. But as thought experiments depend on intuitions, they better not be in conflict with self-evident intuitions that life would not be possible on Twin Earth. To the extent the intuitions underlying the Twin Earth example are not self-evident, it takes away the self-evidence of appealing to microstructure as the essence of water.

Let me go over this again. Consider:

$$\text{water is H}_2\text{O if and only if it has a boiling point of 100 °C} \qquad (4.11)$$

This is an equivalence, not an implication. Of course, in saying this we have to take (4.11) at face value; (4.11) is not a strict law in any sense because it applies 'approximately' under 'normal circumstances'. But these ceteris paribus conditions apply to both the left- and right-hand sides. One may stipulate other possible worlds, in which both (4.1) and (4.2), or one of these, are not true. In that sense

---

[275] Assuming each universe is the result of different symmetry breakings. Cf. quotation of Redhead on p. 201.

neither (4.1), nor (4.2) is necessary. If the statement is not true, of course it is not necessarily true either. There would not be any water in these worlds. In any possible world, either there is water, or there is no water. If there is water, it boils at 100 °C (ceteris paribus) and is ascribed (ceteris paribus) a molecular structure, which is crudely abbreviated as H_2O. If some of the properties of XYZ are different from those of H_2O, XYZ is not H_2O. Lots of liquids on first view may look like water, but turn out not to be. Perhaps some very clever imposters are possible.[276] But there is no philosophical problem involved in this case.

Again: assume XYZ and H_2O share many stereotypical properties, but when electrolysed XYZ fails to yield two volumes of hydrogen to one of oxygen. Then XYZ obviously is not water. Perhaps we have an interesting scientific problem here, but not a philosophical problem. On the other hand, if all properties we know of are the same, it is speculative, to put it mildly, to say that some water is XYZ instead of H_2O. We might as well say: on Twin Earth there is stuff that looks like water, but it does not have a molecular structure - which is a meaningless statement unless some sort of empirical support or other context is provided - for example the publications of Woolley discussed in [5.5], which might be used to set up an argument leading to the conclusion that water on Earth doesn't have a molecular structure either.

Since his (1975) Putnam has changed his position with respect to the 'metaphysical realism' of essentialistic realism. In a way he keeps essences, but 'pragmatises' them by tying them to 'our' intentions.[277] By 1983 Putnam does not support essentialistic realism anymore, but he still upholds the causal theory of reference. For example, instead of XYZ on Twin Earth, what is called water there is now 50% H_2O and 50% grook.[278] (This mixture, it is stipulated, will pass all the lay tests for water.) No reference is made to their being essential properties; there are not even references to microstructures (instead of 'molecular structure' only 'chemical formula' is used). But it is still the case that:

> The reference is partly fixed by the substance itself (through the use of examples). The word 'water' has a different extension on earth and on Twin Earth because the *stuff* is different, not because the brains or minds of Twin Earth speakers are in a different state than the brains or minds of earth English speakers in any psychologically significant respect.

Though it may seem that the 'true' identity conditions are not given anymore by what is H_2O, but by the 'tests known to experts' of the type listed in Block 4-2, the thought experiment is as nonsensical as the XYZ one, because it still assumes that if you replace 50% water in the universe by grook, every organism would just go on living as usual.

In Putnam (1988:30-3) it is still argued that 'reference is partly fixed by the environment itself' and the Twin Earth example is used. Here Putnam says that it

---

[276] Cf. Leibniz (1981:312): 'a way might be found of counterfeiting gold so that it would pass all the tests we have so far'.
[277] See Putnam (1978, 1981) and the Preface in Putnam (1983).
[278] Putnam (1990b:287), first published 1986, manuscript available 1983. He suggests that the chemical formula for grook is '$C_{22}H_{74}$...'. The next citation in the main text is from p. 288.

is a mistake to suppose that 'if XYZ plays the role of water on Twin Earth, then it must exhibit *exactly* the same behaviour as water on Earth, at least at the "observable" level.' And he goes on to argue that differences might well turn up between the two in chemical reactions 'average speakers' are never confronted with. However, he doesn't accept the suggestion that 'it is constancies in the "phenomenological" properties - e.g., the melting point and the freezing point - that fix the reference of substance terms' (ibid:129n13). But this still doesn't make the fairy tale about Twin Earth more convincing.

There is another problem with this possible world talk and the related notion of necessary properties and conditions. Putnam agrees that there is no such thing as a necessary condition for membership in the extension, because any 'discovered' necessary condition is a fallible result - hence it is always stated that (4.1) is necessary *provided* it is true. For example (Putnam says), we may always find out in the future that it is not true that fish breathe through gills, and if we do, breathing through gills is not a necessary condition for being a fish. But note that what we would consider a necessary condition depends on the prevailing background. Fallibly, a necessary condition, depends on what we know about the thing. Being a necessary condition only makes sense relative to a background. Consider:

There is no temperature below zero degrees on the Kelvin scale. (4.12)

A scientist in 1920 would say that if we try to imagine a possible world in which there are temperatures below absolute zero, this is very difficult because some of the most fundamental physical laws would have to be different. Such a world would not have been considered another physically possible world, but a possible world in which the term 'temperature' would have no reference. However, fifty years later, the concept of temperature has been enriched so far that it is no longer true that there are no temperatures below zero degrees Kelvin.[279]

This example illustrates that the distinction between what is a different possible world and what is a different context, depends on our knowledge and the contexts we have actually investigated. According to Putnam, water is what bears a similarity relation to local matter. However, what we have studied locally, may have been studied under boundary conditions that do not apply everywhere. 'Elsewhere' the boundary conditions may be different from 'here'. Most people can boil an egg in local water, but not all. This has (usually) something to do with the local boundary conditions, not with water being this or that. What is the case (here and in other possible worlds or contexts) not only depends on the theories we have, but also on the parts of the world we're considering and the prevailing boundary conditions.

The Twin Earth example has spawned a vast stream of publications of which, a quarter of a century after its introduction, the end is not in sight. Bhargava (1992) made the more 'plausible' suggestion that on Twin Earth water is X_yZ; but on the whole, philosophers have been absolutely rigid in sticking to the original example. Occasionally other possible worlds were introduced such as Dry Earth: 'in

---

[279] Cf. discussion on p. 125.

ESSENTIALISTIC REALISM

which there is no water, in which all the rivers and streams run dry, but everyone suffers an illusion making their experience similar to ours'.[280] And as one might expect discussions about Dry Earth take for granted that (Stoneham 1999:122)

> some concepts, such as *water* and *gold*, latch on to objectively real kinds or categories and consequently have their extension determined by those kinds rather than by features of the thinkers who use those concepts. The extension of *water* is determined to be that of a natural property, namely H_2O.

In 1991 Crane wrote: 'My excuse for adding another paper to the already vast literature is that if my arguments are right, they will help not just to solve, but to *dissolve* the Twin Earth problem.' But his arguments to dissolve the issue didn't work. Philosophers just continued writing about it.

Without exception, in the ongoing Twin Earth chronicles,[281] microreductionism is presupposed. For example, Jackson (1998:49) takes for granted that the extension of some words is 'settled by underlying nature' and writes in a revealing passage:

> Even beginning chemistry students, whose only way of picking out the acids is by the superficial properties of turning litmus paper red, may know that 'acid' applies to something by virtue of its having an underlying nature that plays a specified, significant role in chemical theory that they hope to learn about in future classes.

Perhaps the most disturbing fact about this literature is the insistence of sticking with an example that simply *makes no sense at all*. Occasionally this is recognised, as when Falvey and Owens (1994:109n) say:

> We follow tradition in ignoring here the obvious fact that the twins cannot be physically identical if there is no water on Twin-Earth. The example can be changed to accommodate this fact.

But why continue for years discussing the same example if it is incoherent and leave unclear how the accommodation of the example would work? Does it mean that if Falvey and Owens refer to 'molecular structure' the accommodated example won't contain any reference to molecular structure? They also write in the middle of a long argument that somebody

> may say 'water is H_2O'; but this determines water as the referent of his word 'water' only on the assumption that he uses the symbols 'H' and 'O' to denote hydrogen and oxygen, respectively.

One would like to know what would replace this sentence in the accommodated example.

It might be better if the philosophers so enchanted with Twin Earth would focus on some real-world examples. For example the Chinese counterpart of 'elephant' is 'xiang' (Li 1993). There is no doubt that 'xiang' originally referred

---

[280] Quoted from Stoneham (1999:120). Dry Earth was invented by Boghossian (1997:170).
[281] The title of a collection of papers on Twin Earth is called *The Twin Earth Chronicles* (Sharpe 1996).

to the Asian elephant. Today it refers to both the Asian and the African elephant. Was the essence of 'xiang' always already contained in the sameness relation that encompasses the Asian and the African elephant? Or, more specifically directed at the Twin Earth case: Just as sulphuric acid and hydrochloric acid bear a certain sameness relation to being acid, there is reason to say that XYZ and H_2O bear a certain sameness relation for being water. Though all philosophers writing about Twin Earth claim to be naturalists, it is a very strange naturalism indeed. Papineau (1993:19-20) in a book aimed at showing why a 'naturalist perspective is an inescapable consequence of certain physical truisms' writes:

> Consider Carl, who wants a glass of H_2O and Lrac, his physically identical Twin Earth counterpart, who wants a glass of XYZ [because] Carl is surrounded by H_2O, where Lrac is surrounded by XYZ [19-20]

How is one to engage in a discussion with a naturalist if the naturalist doesn't accept the inescapable consequence of the scientific truism that talking about XYX is nonsense, empty humbug, or worse?

4.4 proliferation of essences

Confronted with the fact that (4.1), 'water is H2O', is not true, we might drop (4.1) altogether, and continue the discussion about the reference of 'water' and whether there are necessary a posteriori statements, with reference to (4.2): water boils at 100 °C, and so on. Such statements are, beyond doubt, a posteriori: they are the result of empirical inquiry. They are used to fix the reference of substances and give operational meaning to relations such as 'being the same liquid as'. Furthermore, they exemplify the assumption that two samples of the same substance will behave in the same way in the same circumstances.

Instead of talking about science discovering essences, the Kripke/Putnam story, would then run as follows. Experts fix the reference of 'water' by using properties such as water boils at 100 °C. If such knowledge is true, then, on the Putnam/Kripke account, the experts fix the reference necessarily; (4.2), if true, is necessarily true. If somewhere else we would find XYZ, which superficially resembles water, then we don't follow the broadly Fregean conclusion that water refers to either H_2O or XYZ. XYZ is something different than H_2O and the experts Putnam's division of linguistic labour relies on, will tell us how to fix the reference of XYZ and what the difference is.

Putnam (1990:69) acknowledges that no other substance has the same boiling point at sea level or the same freezing point as H_2O. But he still assumes that some of the 'secondary' properties might be different: 'we might, surprisingly, find another liquid whose minimum density was 3.98C, but whose chemical composition was not H_2O'.[282] But this suggestion is ad hoc unless a suggestion is made how to envisage two chemical substances with very different molecular structure and *exactly* the same boiling and freezing point. Campbell has suggested that if (4.2) holds for water, it might still be the case that the underlying structure

---

[282] Putnam, personal communication, September 6, 1983.

flickered through being H_2O one minute, XYZ the next, something else after that, and on again unpredictably.[283] Hence, no matter how far the list of reference fixing properties (4.2) is extended, it wouldn't fix the underlying molecular structure. But this speculation can only be empirically meaningful in one of two ways. Either there is a scientific explanation such that the same natural kind, on a micro level, flickers through various molecular or other microstructures - cf. the earlier discussion about tautomers and resonance structures [1.6]. Or, H_2O and XYZ, or whatever else is flickering, refer to different things, in which case there will be macroscopic properties that change from moment to moment.

Let me put the tension between type (4.1) and type (4.2) statements more generally. In order to understand what natural kinds or chemical substances are, there is no need for a 'structural' theory of matter. The theory may change, but the boiling point of water will remain lower than the melting point of gold. Hence, *if* priority has to be given, it has to go to the manifest and macroscopic properties, not to the underlying micro- and submicro-properties. The molecular structure of H_2O is full of riddles, but no future developments will change the manifest difference between ice and liquid water or will change the (approximate) boiling point of pure water in 'normal' conditions. I must therefore conclude that the stability in the use of natural kind terms depends entirely upon manifest properties and macroscopic regularities such as the ones expressed in (4.2) for water. The primacy of manifest water over scientific H_2O is implicitly acknowledged in the most sophisticated Twin Earth debates as when Jackson (1998:52) says that 'water is the watery stuff of our acquaintance' is a priori, whereas 'water is H_2O' is a posteriori. The reference of water in all worlds is settled by what is watery and the subject of the relevant acquaintance in the actual world (ibid:39).

The reference of a pure substance can be fixed by means of an innumerable number of physical properties. The reference of terms like 'water' is provisionally determined on the basis of our acquaintance with samples. In the course of time, experience (including that of experts of various kinds) may yield more and more refined reference fixing observational characteristics. In fact, some of these reference fixing properties may link the stereotype and the tests known by experts directly. For example, property (4.2b), 'water boils at 100°C, is both a test used by experts to fix the reference of water, and, as Wittgenstein remarked, something everybody knows to be true for certain.[284] The fact that (4.2b) is not true on the top of the Mount Everest does not change the 'defining' certainty of it. Moreover, (4.2b) is also a partial definition of the centigrade scale, in which sense it might be considered as an a priori statement. These aspects are all intertwined. Depending on which aspect is foregrounded 'water boils at 100°C' can be considered a posteriori (being the result of measurement), as a priori (setting a standard for what water is); it can be considered necessary (in all possible world water boils at 100°C) or contingent (on the top of Mount Everest ...).

---

[283] John Campbell (Oxford), personal communication 1983.
[284] 'In a court of law the statement of a physicist that water boils at about 100 °C. would be accepted unconditionally as truth. If I mistrusted this statement what could I do to undermine it? Set up experiments myself? What would they prove?' (Wittgenstein 1969:§604).

So called essential properties are only so, relative to our knowledge or way of looking at the world. If the extension a term has is relative to its use within a given linguistic community, then in 1750 the extension of 'water', as used by Oscar$_E$, and Oscar$_{TE}$, is the same (sense determines reference), no matter whether we now like to say that in 1750 the molecular structure was different on earth and on Twin Earth and whether when confronted with this impossible situation we would decide to call both H_2O and XYZ water or decide that they are two types of water (analogous to what happened with isotopes).

If we had discovered the 'true' natural kinds (at least some of them), then they would exist 'metaphysically' on Kripke's account. If the scientific theory about such natural kinds were true, then both macro- and microproperties, and manifest properties for that matter, would be equally necessary - a point perhaps first stressed by Mellor (ibid:311). It may seem that 'being H_2O' is more essential than 'being a liquid having minimum density at 3.98 °C', because the first has more explanatory power and tells us about underlying structure. Nevertheless, although intuitively the explanatory power of (4.2) seems much less than that of (4.1), the arguments about necessity, rigid designators, and reference apply equally to (4.2).

This line of argument can be generalised further. As Witt (1989b) comments, comparing the 'new' essentialism with Aristotle's views:

> If one counts as essential the necessary properties of objects quite generally, however, there is no obvious justification for restricting the things which have essences. The choice of substance-like objects as their primary examples exploits our Aristotelian intuitions about substances and essences, but it neither justifies those intuitions nor does it justify the exclusion of less palatable examples.

The problem of being dependent on the (contingent) context arises for all natural kind properties, both microscopic and macroscopic. The marking of boundaries at the microscopic or 'real essence' level is just as natural or conventional as the marking of boundaries on the macroscopic or 'phenomenal' level. Perhaps Locke was wrong to say that we cannot know 'underlying' properties (Block 3-4), but it doesn't follow that the microproperties have a higher status metaphysically speaking, even if they have wildly heuristic value. To paraphrase Goodman: if there is an essence there are many, and if many none.[285]

---

[285] It is a paraphrase of 'if there is one world, there are many, and if many none'. Cf. Goodman and Elgin (1989:49-53).

5 the alleged reduction of chemistry

5.1 the idea that chemistry can be reduced to physics[286]

According to Primas (1991:163) 'the philosophical literature on reductionism is teeming with scientific nonsense'. In support he quotes the philosophers of science Kemeny and Oppenheim (1956), who said that 'a great part of classical chemistry has been reduced to atomic physics'. Perhaps it was not philosophers who invented the story about the reduction of chemistry to physics, because philosophers of science were no doubt familiar with the influential 1929 statement of Dirac on this issue (Block 5-1). The famous quote of Dirac is referred to in lectures for a general audience of scientists in a way that suggests everybody knows it. Usually the reference is uncritical, e.g. by Noble laureate Mulliken in *Physics Today* (1968); only rarely the reference to Dirac is more critical, as when Hoffmann (1998:4) says in a 'viewpoint paper' presented at a meeting honouring pioneers of computational quantum chemistry: 'only the wild dreams of theoreticians of the Dirac school make nature simple'.

As a specific example of the reduction of chemistry to physics, it is often suggested that the periodic table can be 'derived' from quantum mechanics – incorrectly, as Scerri has shown in a number of publications.[287] Such a reduction was already ascribed to Bohr, for example by Popper. But, contrary to his own claims (and those of Popper) 'Bohr populated the electron shells while trying to maintain agreement with the known experimental facts'. Later developments too in quantum mechanics cannot strictly predict where chemical properties recur in the periodic table. Pauli's explanation for the closing of electron shells does *not* explain why the periods end where they do: the closing of shells is not the same as the closing of periods in the table. Unknown electronic configurations of atoms are not derived from quantum mechanics, but obtained from spectral observations. Hundt's rule states an empirical finding and cannot be derived.[288] Moreover, the experimentally obtained filling rule has about 20 exceptions, for example:
- nickel, palladium, and platinum show a different outer-shell configuration, yet they are grouped together;
- helium, beryllium, and magnesium have two outer-shell electrons, but they do not fall into the same group;
- chromium doesn't follow the rule for the standing filling order;

---

[286] Parts of chapter 5 were published as van Brakel (1997).
[287] This paragraph is based on Scerri (1991b, 1993a, 1997a, 1998a, 1998b, 1999b); see also Löwdin (1969).
[288] Hundt's rule says that when multiple electrons occupy a set of orbitals having equal energies, they will occupy as many different orbitals as possible.

Block 5-1   The spread of Dirac's views.[289]

> DIRAC
> The underlying laws necessary for the mathematical theory of a large part of physics and the whole of chemistry are thus completely known, and the difficulty is only that exact applications of these laws lead to equations which are too complicated to be soluble.
>
> REICHENBACH
> The problem of physics and chemistry appears finally to have been resolved: today it is possible to say that chemistry is part of physics, just as much as thermodynamics or the theory of electricity.
>
> HEISENBERG
> Physics and chemistry have become fused in quantum chemistry.
>
> VAN VLECK AND SHERMAN 1935 – in the *Reviews of Modern Physics*
> To quote some now classic sentences of Dirac ...
>
> EYRING, WALTER AND KIMBALL 1944 – a well-known text book
> In so far quantum mechanics is correct, chemical questions are problems in applied mathematics.
>
> KEMENY AND OPPENHEIM 1956
> A great part of classical chemistry has been reduced to atomic physics.
>
> OPPENHEIM AND PUTNAM 1958
> The possibility that science may one day be reduced to microphysics (in the sense in which chemistry seems today to be reduced to it) ...
>
> NAGEL 1961
> Certain parts of nineteenth chemistry (and perhaps the whole of this science) is reducible to post-1925 physics.
>
> FEYNMAN 1965
> The Schrödinger equation has been one of the great triumphs of physics. By providing the key to the underlying machinery of atomic structure it has given an explanation for atomic spectra, for chemistry, and for the nature of matter.
>
> SCHWEBER 1997
> The laws of physics encompass in principle the phenomena and the laws of chemistry.

- in the case of scandium, quantum mechanics can predict which of three conceivable configurations has the minimum energy, but it cannot derive the correct configuration from first principles.

A more correct way of putting the relation of chemistry to quantum mechanics is that of the physicist Gell-Mann (1994:109ff, emphasis added):

> When Dirac remarked that his formula explained most of physics and the whole of chemistry of course he was exaggerating. In principle, a theoretical physicist using QED [quantum electrodynamics] can calculate the behaviour of any chemical system *in which the detailed internal*

---

[289] Dirac (1929), Reichenbach (1978:129), van Vleck and Sherman (1935:168), Heisenberg (1972:112), Eyring, Walter and Kimball (1944:iii), Kemeny and Oppenheim (1956:7), Oppenheim and Putnam (1958:35), Nagel (1961:362), Feynman (1965:19-18), Schweber (1997:177). Some more references to citations of Dirac's view: Jordan (1957:19), Bader, Nguyen-Dang and Tal (1981:897), Bogaard (1978:354n1), Scott (1992), Levine (1991:606), Gell-Mann (1994:109), Frenking (1998:106), Tomasi (1999:84f).

structure of atomic nuclei is not important. [But:] in order to derive chemical properties from fundamental physical theory, it is necessary, so to speak, to ask chemical questions of the theory.

But even Gell-Mann tends to suppress the 'chemical questions to be asked', or at least invites misquotation when he says: 'The laws of chemistry can in principle be derived from QED, provided the additional information describing suitable chemical conditions is fed into the equations'.

Though Heisenberg is often quoted as endorsing Dirac's view, he also said:[290]

> Although the chemical laws – or the laws of quantum theory that are placed above those of physics and chemistry – are timeless (because otherwise one couldn't speak of a 'law' at all), the special events, which we consider as chemical, only gradually emerged during the development of the universe.

which again is a more nuanced view. Nevertheless, the general consensus seems stuck with Dirac's view as propagated by Reichenbach, Nagel, Feynman, Schweber, and others.

The question of reduction is tied to that of autonomy. Janich (1998), in the introduction to the proceedings of a philosophy of chemistry conference concludes that there was a surprisingly broad consensus among the participants about a reduction-critical attitude.[291] He says: 'It is obvious that none of the contributions represents a form of reductionism, neither a crude, nor a subtle-careful one' (ibid:29). But such a conclusion by-passes the hard philosophical issues. Janich points out that one can distinguish between two foundational issues: that of methodological (historical, genealogical) autonomy, and that of ontological (metaphysical, closed system, natural laws) autonomy.[292] At a higher level of abstraction I might oppose this dichotomy, but given this way of speaking, it serves a useful purpose in the following way. Janich is surely right to say that chemistry is and will remain autonomous in its development (relative to other natural sciences). Nobody would expect it to be otherwise, given the long-standing and well-entrenched institutional separation of physics and chemistry (Hiebert 1996), fuelled by the 'imagined communities' of identity formation. And it is no doubt true that physical comprehension is *psychologically* different from chemical comprehension (Buck 1998). However such observations have little bearing on discussions about the *ontological* autonomy of chemistry. The latter question is a much more tricky one, only addressed, and then only indirectly, in a few contributions to this conference on the autonomy of chemistry.[293] Other contributions are either neutral or display a soft or crude reductionism (as this term is

---

[290] Heisenberg (1942:255): 'Während die chemische Gesetze - oder die der Physik und Chemie übergeordneten Gesetze der Quantentheorie - zeitlos "gelten" (denn sonst könnte gar nicht von einem "Gesetz" gesprochen werden), sind die speziellen Vorgänge, die wir als chemisch bezeichnen, erst allmählich in der Entwicklung des Weltalls aufgetaucht.'
[291] In his introduction to the *Proceedings of the 3rd Erlenmeyer-Colloquy for the Philosophy of Chemistry* (Janich and Psarros 1998) on the subject of the autonomy of chemistry.
[292] Corresponding to Janich's 'abschließende Form theoretischer Darstellung' and 'Abhängigkeiten im Zugang zu einem Wissen'.
[293] Psarros (1998a), Schummer (1998a) and Hanekamp (1998).

normally understood in the philosophy of science). For example, Frenking (1998:106-7), a theoretical chemist, says about the autonomy of chemistry that:

> Chemistry would become revolutionised in 1927, now that the very basis of all chemical phenomena, i.e. the chemical bonding, was understood for the first time. Chemistry as true science is still in a developing stage because quantum chemical research of the many chemical phenomena is still in an infant stage.

These statements, echoing Kant, are supported with the obligatory quote from Dirac. Similarly, Arabatzis (1998:155) suggests that the conflict between physicists and chemists over the electron was 'fully resolved' with 'the advent of the exclusion principle, spin, and eventually quantum mechanics'. The crudest form of reductionism can be found in the contribution of Mainzer (1998). He argues that chemistry has become a *science* in the sense of Kant, because it uses ever more mathematics. In saying this he takes for granted the non-autonomy of chemistry. For example he says (ibid:42-3):

> Molecular states are described by wave functions. The emergence of a chemical phenomenon is reduced to a physical symmetry breaking.

To paraphrase it in Kant's terms, what Mainzer is saying is that to the extent that chemistry can be reduced to quantum mechanics, it is a proper science.[294] Of course, chemistry as a practice is autonomous, but it is not an autonomous *science*. The physicist Dingle (1949a) put it more aggressively:

> Chemistry rightly figures prominently in the history of science; in the philosophy of science it should not figure at all.

The question 'can chemistry be reduced to physics?' is, in its generality, not meaningful. First, it is unclear how one would delineate and separate chemistry and physics - what for example, about chemical physics or mechanical and physical separation methods in chemistry and chemical engineering (Block 5-2)? Second, more than two discourses, not only 'chemical' and 'physical', need to be distinguished: there is manifest water, there are physical, physico-chemical, chemical, and biochemical *macro*properties of water, there is (chemical) molecular *micro*discourse and (physical) quantum mechanical discourse. And that's only the beginning. A whole variety of possible interdiscourse relations have to be addressed and the question of reduction has to be made much more concrete than talking about chemistry and physics. For example, by asking:
- Can chemical thermodynamics be reduced to statistical mechanics?
- Can the notion of chemical bond be reduced to quantum mechanics?
- Can 'being water' be reduced to 'being H_2O?
- Can heat and mass transfer in chemical engineering be reduced to 'fundamental' phenomenological equations such as Fourier's law for heat conduction and Fick's law for molecular diffusion?

---

[294] In his own words (Mainzer 1998:49): 'Chemistry is involved in a growing *network of mathematical methodologies and computer-assisted technologies* with increasing complexity. Thus, chemistry, is a *science* in the sense of Kant, but with changing frontiers.'

- Can the laws of Fourier and Fick be reduced to the equations of irreversible thermodynamics or to statistical mechanics?
- And so on.

Third, the notion of reduction has to be spelled out [2.2]: what is it that is being reduced (theories, concepts, properties, natural kinds, laws, explanations), how does it differ from related notions such as replacement, elimination, integration, supervenience, and emergence, and what *sort* of relation is it (dependence, identity, ...)?

To set the scene let me stress that Primas is absolutely right when he says that, if reduction is used in the sense that 'higher-level theory together with its interpretation can be deduced from the basic theory', then it is a *brute fact* that:[295]

- Chemistry has not been reduced to physics: 'how are the non-linear differential equations of chemical kinetics derived from linear quantum mechanics?'.
- Chemical purity is not a molecular concept: 'pure water does not consist of H_2O molecules'.
- The theory of heat has not been reduced to statistical mechanics: 'how can the zeroth law of thermodynamics be derived from statistical mechanics?'
- Temperature is not a molecular concept.
- Classical physics is not a limiting case of Einsteinian physics, nor a universal limiting case of quantum physics.

5.2 thermodynamics and statistical mechanics

For various reasons, physical chemistry (including quantum chemistry, chemical thermodynamics, interface chemistry, colloid chemistry, electrochemistry)[296] might well be considered a separate science, originating just before 1900.[297] Schummer (1998a) argues it fills the gap between physics and chemistry. I suggest that the best (pragmatic, not essentialistic) choice is to consider physical chemistry as part of chemistry. Physics only provides explanations for material behaviour in general. Physical chemists have to explain the specific behaviours of the 16 million plus chemical substances in quantitative terms. To achieve this they use a variety of models which are refined and adapted using theoretical and semi-empirical approaches that use purely chemical concepts that are not reducible to physics, as well as many physical techniques and concepts. Thinking along these lines chemistry is the science of the transformation of substances, including the transformation of substances that do not involve chemical reactions - after all the mathematics for describing chemical reactions isn't very different from that used for describing the grinding of solid particles or radioactive disintegration. This implies, inter alia, that metallurgy, usually distinguished from chemistry and

---

[295] Quotations from Primas (1991:163f). See also Primas (1983:308-17).
[296] Schummer (1997a, 1998a) refers to colloid and interface chemistry as solution theory and theory of phase equilibria. Note that chemical physics is not the same as physical chemistry.
[297] The *Zeitschrift für physikalische Chemie*, founded by Ostwald and van 't Hoff, started to appear in 1887. The American *Journal of Physical Chemistry* followed in 1896. Cf. Meyer (1889), Duhem (1899) and van't Hoff (1905). The latter defines physical chemistry as 'the science devoted to the introduction of physical knowledge into chemistry, with the aim of being useful to the latter.'

Block 5-2   Unit operations in chemical engineering.

> An industrial chemical process as well as a synthesis of chemical products on laboratory scale consists of a sequence of 'unit operations'. Examples of chemical operations are synthesis, oxidation, polymerisation. Physical operations like distillation, extraction, crystallising, melting can be used to divide mixed and pure materials. Mechanical operations such as filtration, centrifugation, grinding can be used to separate heterogeneous and homogeneous substances.
> Chemical operations are always followed by physical and/or mechanical operations to purify and concentrate the product ('down-stream processing'). Also note that for a chemical reaction to take place there has to be energy transfer and a mechanical pathway.

sometimes 'generalised' (together with polymer science) to material science would also be part of it, as too would so-called unit operations in chemical engineering (Block 5-2), irrespective of whether they are chemical, physical, or mechanical operations. And nuclear physics aimed at the production of materials might better be called nuclear chemistry.

Physical chemistry lends itself well to philosophical discussion about reductionism and realism. Examples are the discussion of interfaces [2.2], polywater [3.5], and capillary wetting phenomena [6.4]. In this section I focus on the relation of chemical thermodynamics and statistical mechanics.[298] I'll start with a few comments about temperature, in particular because the example 'temperature is mean kinetic (molecular) energy' has been used in hundreds of publications as an example of a straightforward reduction. Temperature also offers a clear example of how the path of reduction, if taken seriously, more or less automatically leads to eliminativism (Block 5-3).

The macroscopic concept of temperature was developed primarily in the context of thermometry (from Galileo onwards). Only in the past 100 years or so, were all kinds of theories added (and laws in which temperature occurs as a variable). The interpretation of temperature as mean kinetic molecular energy is only one of the laws or definitions in which temperature enters and it says something about the temperature of ideal gases, not about temperature 'in general'. Temperature is, in general, *not* the same as average mean molecular kinetic energy.[299] This might be so for gases that behave properly (i.e. 'perfect' gases of idealised 'billiard-ball'-molecules in random motion), but not for solids, plasma's or a vacuum. Even for ideal gases temperature cannot be identified with kinetic energy when different kinds of motion (translations, rotations, vibrations) are not in equilibrium, as in molecular beams. Further, for a dense gas at low temperature, quantum mechanics implies that the kinetic energy depends not only on the temperature. Hence the connection between temperature and mean kinetic energy is non-linear, its value depending on other characteristics of the system. Finally, even if temperature could be reduced, it doesn't yet follow that 'boiling point' can be reduced (Hooker 1981:497-500). There has also been talk of 'negative, infinite and hotter

---

[298] This section draws heavily on Sklar (1993) and Primas (1985b); see also Needham (1998).
[299] See Vemulapalli and Byerly (1999:28-32) and Sklar (1993, passim).

Block 5-3   Churchland's elimination of temperature.[301]

> *January 1985*
> Being warm is identical with having a certain mean level of microscopically embodied energies.
> Material-mode statements of identity can occasionally be made, of course; temperature = mean molecular kinetic energy.
>
> *March 1985*
> The common-sense notions of hot, warm, and cold are empirically incoherent. [They] are not just different in extension from the thermodynamic terms that displace them; nothing in nature *answers* to the collected laws of common-sense thermodynamics'.
> The irony is that *thermodynamic* temperature has turned out to enjoy no *uniform* essence in any case. A familiar property has fragmented again. Our taxonomies form, and dissolve, and reform, even as we watch.

than infinite temperatures' (Ehrlich 1982) - though this could be interpreted as an extension of the 'reduced' temperature concept at the microlevel, not as a problem for the reduction of macroscopic temperature to microscopic descriptions per se (Sklar 1993).

The question whether temperature can be reduced to statistical mechanics is a pseudo-problem – a debate that 'bespeaks an incredible ignorance of most elementary concepts of physics'.[300] Either temperature is to go completely (Block 5-3) or there has to be room for thermodynamic *and* statistical mechanical accounts as different 'perspectives' or illustrations of Bohr's principle of complementarity (Primas 1985b). The latter view allows for intertheoretic or interdiscourse relations of various sorts, without having to meet definitional characteristics of reduction, supervenience, or emergence.

There are many problems for those who advocate the reduction of (physical or chemical) thermodynamics, but all seem to be related to the macroscopic notion of equilibrium (which is *the* central notion of thermodynamics). The macroscopic concept of temperature only makes sense for systems in equilibrium. Operationally equilibrium means: no noticeable changes in macroscopic features. This implies that microscopically the status of the zeroth law of thermodynamics is unclear. This law says that there is a parameter which determines whether, before two (or more) systems come into contact, they will be in equilibrium or whether there will be an energy flow from one to the other. This parameter is the temperature.[302] But at the microlevel a system is never in equilibrium because of second order variation in statistical distribution of energy of the molecules. Microscopically there is no such thing as equilibrium. Temperature as defined by the zeroth law of thermodynamics is not reducible.

Coming now to the second law of thermodynamics, according to Sklar (1993), there is no satisfying explanation why systems strive to equilibrium *towards the*

---

[300] Primas (1991:164) commenting on Nagel's (1961) claim discussed in [2.2].
[301] Citations are from Churchland (1989:53,49,283f,285, cf. 31,60).
[302] Putting the emphasis slightly differently: 'The identification of heat - as the concept is used in thermodynamics - with molecular motion is wrong in principle' (Vemulapalli and Byerly 1999:29).

future, rather than in the direction of the past. Statistical thermodynamics is reversible; macroprocesses are irreversible. Sklar shows that various attempts to explain the time asymmetry fail: before the explanation starts, the asymmetry is presupposed. That is to say: there is no analogue of the second law of thermodynamics in statistical mechanics: a hundred years of research has not succeeded in reducing entropy, nor has it led to one microdescription of what entropy is. In thermodynamics entropy is additive and non-decreasing; Boltzmann's entropy displays only the first, Gibbs's entropy only the second.[303]

That there is no such thing as equilibrium at a microscopic level has vast consequences. For example, according to the micro account crystals cannot be perfect, because the 'positions' of atoms are average positions (they vibrate), whereas these average positions differ in different 'cells' (atoms are never in a homogeneous field of forces).

How one should deal with chemical reactions that do not develop towards thermodynamic equilibrium (like oscillating reactions)[304] is even less clear. Thermodynamics that takes on gravity, interfaces, or time is already 'underdeveloped' from a scientific point of view and raises many foundational issues, often of a paradoxical nature. For example, Gibbs' paradox[305] says that the entropy of mixing two substances (no matter how 'similar' they are) is $2R\ln 2$, but collapses to 0 when the substances are identical. This finds its base in the fact that mixtures can be separated, but pure substances cannot.[306] In physical chemistry one talks about solutions, mixtures, suspensions, colloids, alloys, etc. which are all defined as something being *distributed uniformly* in a *continuous medium*, but at the atomic level this is a contradictio in terminis. Aristotle already noted that mixing was a problem for any kind of atomism.[307] Later Duhem used a similar argument when opposing atomistic interpretations of (chemical) substances.[308] Further, as Vemulapalli and Byerly (1999:24) show, the chemical thermodynamics of multicomponent systems includes 'tricks' that are not part of thermodynamic theory. The behaviour of a system of N components is not reduced to the behaviour of N pure substances, but to that of N 'replicas' of the system. Equations used to estimate intermolecular energies (van der Waals, Lennard-Jones) 'cannot be deduced from any fundamental theory' (ibid).

Textbooks often misrepresent the contribution of Boltzmann and Gibbs to the development of statistical mechanics. Gibbs did not merely improve upon Boltzmann's formulation of statistical mechanics. He replaced Boltzmann's

---

[303] Sklar (1993:354-9), Callender (1999), Primas (1985b).
[304] Ostwald was the first to observe oscillating behaviour in the oxidation reaction of metals in acids. The phenomenon was first studied systematically by the Russian biochemist Belusov in 1951, but his paper was rejected by different chemical journals, because the accepted thermodynamic theory did not allow for systems oscillating between states of positive and negative entropy (Manzelli 1996). It was not until Prigogine opened the debate on chemical reactions out of thermodynamic equilibrium that the issue came to have some respectability.
[305] Denbigh and Redhead (1989) dissolve Gibbs' paradox by taking as central that entropy is an observable (though dispositional). For discussion see also Mosini (1995), Zemansky and Dittman (1981:362).
[306] R is the gas constant.
[307] Aristotle, *De generatione et corruptione* (*On Coming-to-be and Passing-away*), 328a10-15.
[308] Needham (1998); cf. main text to notes 17 and 18.

ALLEGED REDUCTION OF CHEMISTRY TO PHYSICS   127

reductionistic approach by taking the macroscopic laws and phenomena as basic and proposed abstract theoretical laws that were consistent with observations and empirically supported thermodynamic functions. That is to say: he proposed theories that could be tested empirically without assumptions about the microconstituents or microstructure of the system. Similarly, as already noted, Gibbs' phase rule is a law completely free from all hypothetical assumptions as to the molecular condition of the substances involved. Up to this moment tensions between macro- and micro-descriptions cause problems for students of interface chemistry, in which macroscopic parameters are used (to which Gibbs' theoretical considerations apply), but for which there is no plausible statistical mechanical interpretation available. This is the case in particular for three-phase systems containing solid-liquid, liquid-vapour, and solid-vapour interfaces.

Duhem conceived of generalised thermodynamics as the primary chemical theory and was dismissive of atomic theories.[309] Later writers, following Gibbs, assumed thermodynamics was a phenomenological theory, neutral with respect to atomic theories (Block 5-4). Thermodynamics gives a description of macroscopic phenomena using quantities like mass, volume, energy, temperature, which will remain true, whatever subsequent developments tell us about how these quantities might be modelled at a (sub)microscopic level.[310] When Perrin discovered Brownian motion this was generally seen as a victory for atomism (Nye 1972). But why not conclude that descriptions are needed at both levels (thermodynamic and statistical mechanics)? Perhaps a description in terms in molecules and in terms of substances should be seen as complimentary.

After a book-length account of the relations between thermodynamics and statistical physics, Sklar (1993:372-3) concludes:

> Even when we already have the statistical-mechanical kinetic equations at our disposal, we generally cannot hope to find the general solutions of it. Instead, the search for solutions is guided by the use of what we know, from the macroscopic thermodynamic-hydrodynamic level, about the structure of the solution we are seeking. The very structure of the solution found is guided by our antecedent knowledge at the empirical macroscopic level of the kind of solution we must look for. It is this already existing macroscopic account that is used in framing that deeper picture.

Or, to put it more vividly, if we ask for a statistical explanation of why my ice cream melted, we get a story using theorems that tell us what happens over a time span of less than one second, that presuppose infinitely many particles of zero-density and which offer no way to drop these idealisations.

The general point that it is not clear how reduction can work without making assumptions appealing to macrophysical concepts, shows up in all proposals for reduction, whether it is the reduction of chemical thermodynamics to statistical mechanics just reviewed, whether it is colour (Saunders and van Brakel 1997a),

---

[309] Needham (1996b). On the significance of the *essential* macroscopic character of thermodynamics in relation to chemistry see also Baracca (1996).
[310] For further discussion see Primas (1982, 1985b), Needham (1999a, 2000).

Block 5-4 Physical chemists on the relation between thermodynamics and statistical mechanics.[311]

Bumstead 1928	Nothing is assumed as to the mechanical nature of the systems considered, except that they are mechanical and obey Lagrange's or Hamilton's equations.
Zemansky 1957	If the molecular theory is changed or even discarded, the concept of pressure will still remain and will still mean the same thing to all normal human beings.
Baierlein 1971	The conceptual bases of [thermodynamics and statistical mechanics] are sufficiently different that one can expect only a close *correspondence* between the theoretical quantities, not a true equality or identity.
Denbigh 1981	Thermodynamics is independent of the fine structure of matter.
Eu 1992	The statistical assumption required for the Boltzmann equation is non-mechanical and thus distinctive and not deducible from the mechanical laws of classical or quantum dynamics.

schizophrenia (van Brakel 1996a), temperature (see above), or molecular chemistry (to which I now turn).

5.3 quantum chemistry

When presented with the question how to relate (classical) chemistry to quantum mechanics, various general stances are possible:

- Quantum mechanics is the true and complete theory of nature. All chemical phenomena (and all other natural phenomena) can be explained, in principle, in terms of the laws of quantum mechanics. There is room for conceptual autonomy of chemistry for pragmatic reasons, including crucially important heuristic considerations. But if all is said and done (even if we were never to reach that stage), all there is, is quantum mechanical descriptions, predictions, and explanations.
- Quantum mechanics is to chemistry, what mathematics is to economics. It is an abstract system, chemically not interpreted; 'a tool for handling experimental information about a given system' (Löwdin 1967:11).
- Quantum mechanics and the wave function in particular, is a model relative to molecular chemistry: 'we must continually make selections among many complementary models – from plastic balls and sticks to wave functions' (Carroll 1998:xxiii).
- Quantum chemistry is on the same ground as any other theory. In some cases quantum chemistry is of use to gain knowledge and understanding. In other cases the theory of electrovalences or thermodynamics is of more use.

These views are independent of one's view of the status of quantum mechanics proper. One might take the view that quantum mechanics as it is described in text

---

[311] Bumstead (1928:xxiv-xxv; *cf.* Gibbs 1902, ch. XIV), Zemansky and Dittman (1981:5; cf. Zemansky 1957:3), Baierlein (1971:473), Denbigh (1981:4), Eu (1992:3).

Block 5-5 Famous physicists about the 'standard' Kopenhagen interpretation of quantum mechanics.[313]

Bohr	The quantum postulate implies a renunciation as regards the causal space-time co-ordination of atomic processes.
Schrödinger	Bohr's approach to atomic problems is really remarkable. He is completely convinced that any understanding in the usual sense of the word is impossible.
Heisenberg	The idea of an objective real world whose smallest parts exist objectively in the same sense as stones or trees exist, independently of whether or not we observe them is impossible.
Einstein	I am, in fact, firmly convinced that the essentially statistical character of contemporary quantum theory is solely to be ascribed to the fact that this [theory] operates with an incomplete description of physical systems.
von Neumann	The present system of quantum mechanics would have to be objectively false, in order that another description of the elementary processes than the statistical one be possible.
Born	In quantum theory it is the principle of causality, or more accurately that of determinism, which must be dropped and replaced by something else.
Feynman	I think I can safely say that nobody understands quantum mechanics.
Gell-Mann	*Quantum mechanics* [is] that mysterious, confusing discipline, which none of us really understands but which we know how to use.

books, is waiting to be interpreted with reference to an 'underlying' deterministic world it is about, using a theory about hidden variables - 'Bohm mechanics' for example.[312] This stance would change one's view of the physical world completely, from being indeterministic on the 'standard' Kopenhagen interpretation of quantum mechanics (Block 5-5) to being deterministic according to 'Bohm mechanics'. But it would leave open each of the stances (or any other stance) about the relation of quantum mechanics to chemistry.

Since Dirac the received view has been that in principle, chemistry can be reduced to quantum mechanics. Nevertheless, all detailed investigations show that this cannot be done, or at least that the example fails to fit standard theories of reduction. In particular classical notions such as molecular structure or chemical bond are not easily reconciled with quantum mechanics (Block 5-7).[314] In so far as a solution of the Schrödinger wave equation for stationary states is possible, the

---

[312] Bohm mechanics is an empirically adequate revision of the formalism of quantum mechanics which presupposes deterministic particle trajectories - though particles that are (probably) unobservable in principle. For the difference between 'Bohr' and 'Bohm' mechanics and why Bohr 'won' see Cushing (1994).
[313] Most quotations I borrow from Cushing (1994).
[314] Pre-1990 publications problematising the in principle reduction of chemistry to quantum mechanics include, from a philosophy of science perspective, Bogaard (1978), Lévy (1979a, 1979b), Liegener and Del Re (1987a, 1987b), and from within quantum chemistry: Claverie and Diner (1980), Del Re (1974), Hartmann (1965), Löwdin (1967), Paolini (1981), Primas (1983, 1985a), Woolley (1978, 1986). For references to many earlier publications of Primas and Woolley see Claverie and Diner (1980) and Woolley (1986).

Block 5-6   Feynman telling his physics students about quantum chemistry.[316]

> *how to put six electrons into 'a benzene ring with six electrons removed'*
> Next put a second electron in. And now we make the most ridiculous approximation that you can think of - *that what one electron does is not affected by what the other is doing*. Of course they really will interact; they repel each other through the Coulomb force, and furthermore when they are both at the same site, they must have considerably different energy than twice the energy for one being there.
> In the case of benzene, the principal reason for the inconsistency is our assumption that the electrons are independent - the theory we started with is really not legitimate. Nevertheless, it has some shadow of the truth because its results seem to be going in the right direction. With such equations plus some empirical rules - including various exceptions - the organic chemist makes his way through the morass of complicated things he chooses to study.
> Don't forget that the reason a physicist can really calculate from first principles is that he chooses only simple problems. He never solves a problem with 42, or even 6 electrons in it.

results cannot be understood in terms of the size and shape of an individual atom or molecule. As Woolley (1978:1076) says [5.5]:

> In quantum theory an atom or molecule in a stationary state has no extension in space or time, so that it makes no sense to talk about the size or shape of an atom or molecule in such a state.

All phenomena in physics and chemistry are time-dependent of course, and 'stationary state' is an idealisation. Further, the Schrödinger equation can only be solved analytically for an isolated two particle system. In the crudest approximate solutions for bigger systems, all interactions between electrons are neglected (Block 5-6).[315] As a matter of course quantum chemistry is restricted to considering isolated systems, ignoring interatomic and intermolecular interactions.

Approximations to the wave equation cannot be taken as idealisations which, as they get better, converge to 'the truth'. Such approximations 'solve' the manybody problem but without saying how they do it (Woolley 1978:1075). As Hendry (1998:136) argues at length 'such features of the models as *are* invoked in explanations are *not* shared by the exact treatments'. Hence, different models will capture different features of the 'same' molecule for the purposes of different inquiries. In setting up a (time independent) Schrödinger equation a quantum chemist is selective. No other particles than nuclei and electrons are considered. Nuclear and electronic motions are separated, and it is assumed that the nucleus has a fixed position. The Born-Oppenheimer approximation typically places the nuclei at the empirically known equilibrium geometry, grounding quantum chemistry in the classical chemical molecular paradigm; i.e. the picture of a semi-rigid

---

[315] Hartree (1957:17): 'The approximation that has been most widely used is one in which each electron is regarded as being in a stationary state in the field of the nucleus and of the other electrons.'
[316] Feynman (1965:15.7-15.10).

ALLEGED REDUCTION OF CHEMISTRY TO PHYSICS 131

framework of atoms connected by bonds that rotate and translate in ordinary space as time elapses.[317]

The calculation deals with electrons and nuclei. Entities such as orbitals or atoms are not part of the system on which calculations are made (Block 5-9). In terms of quantum mechanics there may be *bonding*, but there are no *bonds* – neither in the sense of discrete electron pairs localised between pairs of atoms, nor in the sense of localised orbitals in spacetime. Further, when it is taken into account that two systems interact, it is assumed that the composite system can be represented as a product of the wave functions of the separate systems.

In quantum chemistry one can calculate bonding energies, without ever knowing what a bond is. A chemical bond is an entity with a length, an orientation, a dissociation energy, a contribution to the enthalpy of atomisation, an electric dipole moment, and other properties. But it does not 'appear' in the Schrödinger equation, nor does it emerge from it. Calculations from the Schrödinger equation, to obtain sufficient accuracy, describe an atom in terms of sums of contributions from numerous, literally billions, of electron configurations. Yet the atom itself cannot be said to possess any of the particular configurations that are used in the calculations. Particular electron configurations underlying the current notion of chemical bond are derived from spectral observations, not from quantum mechanics.

Moreover, what chemically speaking is considered to be one type of molecule is typically a quantum mechanical presupposition of different stationary states (of the same system. For example, ammonia, a (macroscopic) gas with the simple chemical formula NH_3 is, according to quantum mechanics, a superposition of two stationary states (the ground state and the first excited state). This only gives probability distributions for the position of the nitrogen and hydrogen atoms (Amann 1996) - a problem at the level of atoms similar to that of resonance structures at the level of chemical bonds. Different isomers cannot be picked out drawing on the quantum mechanical formalism alone.[318]

Lacking support for a reductionist programme is, as it were, built into quantum chemistry. A reduction of the Nagelian sort [2.2] is made difficult in principle because the Schrödinger equation has the peculiarity that a solution of the equation consists of indefinitely many eigenfunctions. Choosing the terms that are included in the Hamiltonian operator (for an isolated molecule) is based on judgements of what is relevant to the chemical system at hand.[319] A complete Hamiltonian would have to include all nuclear and electronic kinetic energy opera-

---

[317] According to Del Re (1998:94) we should speak of the Born-Oppenheimer *theorem*, formulated as follows: 'in the frame of quantum mechanics the nuclei of a molecule can be treated as classical particles practically at rest as far as electronic motions are concerned'.

[318] Any closed system corresponding to a given number of electrons and nuclei must share the same Hamiltonian, but, says Woolley (1991:26): the nature of this relationship has never been elucidated. See for examples and discussion also Amann (1991, 1996), Claverie and Diner (1980), Del Re (1998), Woolley (1991).

[319] The Hamiltonian is the energy matrix of coefficients used to (completely) describe a quantum mechanical system. The eigenvalues of an observable represent the various different possible outcomes of measuring the corresponding observable. Eigenvalues are the 'characteristic' values (energies) of the Hamiltonian.

Block 5-7 Chemical notions that are not amenable to rigorous quantum mechanical treatment (according to some authors).[320]

acidity, aromaticity, basicity, chemical bond, chemical potential, chemical reaction, chirality, donor and acceptor character, electronic configurations, electron orbitals or shells, electronegativity, functional group, hyperconjugation, internuclear distances, knot types in polymers (as in DNA), molecular structure and stability, moments of inertia, relative energy of s and p orbitals, resonance, size and shape of individual atoms, substituent effect, valence

tors, electrostatic interactions between all charged particles, and interactions for all magnetic moments due to spin and orbital motions of nuclei and electrons. In making the choice what terms to include, information is used that is not provided by quantum mechanics. In the words of Hendry (1998:137):

> We have not compared the approximate molecular models with what quantum mechanics says about molecules *simpliciter*, only with what quantum mechanics says about molecules *if we pretend they are isolated systems of point-mass nuclei and electrons that are subject only to Coulomb interactions*. This description is *itself* an idealised one: there *are* no isolated molecules; electrons and nuclei experience interactions that are non-Coulombic; nuclei are not without internal structure; and the Hamiltonian should be relativistic.

When doing calculations on a carbon monoxide molecule, it is split conceptually into two subsystems each described in isolation. The molecule is considered a system consisting of a rotator and a two-dimensional oscillator. There is a Hamiltonian for the rotator, for the oscillator, and there is a formula to 'add up' the two results to obtain the Hamiltonian of the whole system. But as Hüttemann (1998) and others have pointed out, although there is no doubt that this method works (in the case of carbon monoxide), the fundamental laws (the Hamiltonians) apply to idealised systems that are not part of the physical world: 'model Hamiltonians fit only highly fictionalised objects' (Cartwright 1983:136). At best the Hamiltonians are relevant only to isolated molecules, of which there are none in the real world. At the very best the idealisations can only be approached in carefully prepared experiments.

With the introduction of the fourth quantum number and Pauli's Exclusion Principle,[321] the coupling between electrons with anti-parallel spins in the homopolar bond came to be seen as an essentially quantum phenomenon. London 'considered spin to be the constitutive characteristic of quantum chemistry' (Gavroglu 1995:49) and in 1935 van Vleck and Sherman (1935:173) wrote in a long review on the quantum theory of valence (emphasis added):[322]

> The Pauli exclusion principle is the cornerstone of the entire science of

---

[320] See, for example, Primas (1983), Scerri (1991a), Amann and Gans (1989), Hoffmann (1998).
[321] Formulations of the Pauli exclusion principle include: No two electrons can occupy the same spin-orbital. No two electrons in the same system can have the same values for all their quantum numbers. In a system of identical fermions, no two particles can occupy the same state.
[322] Misquotation is easy here. French and Ladyman (1998:69) only quote the first clause of the citation ascribing to van Vleck and Sherman a blunt reductionistic position.

chemistry, for without it one would not have an understanding of the Mendeleev periodic table and the like. The reader will recall that it [the Pauli exclusion principle], *together with a knowledge of the relative order of binding of the various electronic orbits*, yields the characteristic structure of the periodic table.

This shows the enormous importance of Pauli's exclusion principle, but says little about the relation of the principle to quantum mechanics, whereas its philosophical significance has been neglected.[323] Though 'spin' appears as a natural consequence of *relativistic* quantum mechanics, it must be added on as an extra effect in nonrelativistic treatments. Spin has no classical counterpart. The introduction of Pauli's quantum number for electron spin is, from the point of view of the Schrödinger equation, completely ad hoc, whereas in terms of the more general Pauli antisymmetric exclusion principle, 'we lose the ability of identifying electrons in the atom and of assigning them to particular orbitals' (Scerri 1999b:300). The view that each electron is in a stationary state is at the basis of much of the current approach to chemistry teaching. However, it is only the system as a whole that possesses stationary states (Hartree 1957:14). Most formulations of quantum mechanics do not refer to individual particles at all.

Quantum mechanically speaking, a hydrogen molecule doesn't consist of two hydrogen atoms, each atom consisting of one proton and one electron. The two electrons are part of one system consisting of two protons and two electrons. Hence the two electrons lose their identity. But we also want to say that the two electrons in the hydrogen molecule cannot be in the same state (the Pauli principle). We must be able to refer to the two electrons if we are to formulate the antisymmetry requirement. This is puzzling (Gibbins 1987:117). In order to use quantum mechanics certain electrons are *selected* – those that turn out to be useful for remodelling the concepts of classical chemistry, but it is unclear what that could mean quantum mechanically speaking.

5.4 'ab initio' methods

The most theoretical approaches in quantum chemistry are referred to as 'ab initio' methods. Such methods consist, essentially, of three steps: write down the Hamiltonian for the system; select some mathematical functional form as the trial wave function; minimise the energy of the system with respect to variations in parameters in the wave function. Structural features of the molecule are explained by calculating the effect of changes in nuclear configuration on electronic energy.

Typically, in ab initio calculations relativistic mass effects are ignored, Hamiltonians are written for isolated molecules, nuclear and electronic motions are treated separately (the adiabatic approximation), the nuclei are given fixed positions (to remove nuclear kinetic energy operators) – the clamped nuclei or Born-Oppenheimer approximation, and all magnetic interactions are ignored (except in

---

[323] An early exception is Margenau (1944). See further Hall (1968), Scerri (1995).

special cases where one is interested in spin coupling).[324] That is to say, quoting from a 'standard' text book (Lowe 1993:351):[325]

> We are in effect electing to seek an upper bound for the energy of an idealised non-existent system – a nonrelativistic system with clamped nuclei and no magnetic moments.

The most common method is the Hartree-Fock method.[326] It consists of an iterative procedure which calculates the average configuration of all the electrons in the atom. It is also referred to with acronyms such as SCF-LCAO-MO. SCF stands for 'self-consistent field'. The method is described as self-consistent calculation since the energy of each electron is minimised with respect to every single other electron in the atom. AO stands for atomic orbital and MO for molecular orbital. Various models are available to construct molecular orbitals out of atomic orbitals. A molecular orbital is usually defined as a linear combination of atomic orbitals (an LCAO-MO), these atomic orbitals being located on the nuclei.

For atomic calculations the Hartree-Fock ab initio method replaces a multi-electron wave function with a product of one-electron orbitals, corresponding to a system of non-interacting electrons. Each electron is assumed to move in the field produced by the nucleus and a hypothetical charge cloud due to the other electrons. Analogously, a molecular orbital is a wave function which is a function of the coordinates of only one electron. Most current methods expand orbitals into a set of basis functions. The number of atomic (or molecular) orbitals equals the number of basis functions. A minimal basis set consists of one basis function for each inner shell and each valence atomic orbital.

A simple ab initio approach to the wave function takes into account the interactions between electrons in an average way.[327] It has some instantaneous electron correlation built into it, because it satisfies the antisymmetry requirement of the Pauli principle. The model can be improved by using one-electron spin-orbitals instead of one-electron spatial orbitals.[328] This model, taking into account spin explicitly (and not just the Pauli principle) still uses the approximation of taking the wave function as an antisymmetrised product of one-electron spin-orbitals and

---

[324] In semi-empirical models (such as the Hückel molecular orbital method) a molecular geometry is taken as given; it is assumed that the wave function is a product of one-electron functions and the Hamiltonian a sum of one-electron operators. A basis set is chosen based on experience. Interelectronic effects are taken into account via empirical parameters.

[325] As van Vleck and Sherman (1935:198) already remarked an atomic wave function cannot really be factored into one electron functions and there is always the distortion of the central symmetry due to the fields of adjacent atoms.

[326] For the difference with the valence bond method see note 90.

[327] For each electron the wave function is calculated in the field of both the nucleus and the remaining electrons. This process is repeated for each electron in turn until all resultant wave functions, and consequently the total (approximate) eigenfunction that is their product, are negligibly altered in consecutive iterations. Under these conditions the energy of the atomic system converges to a finite value.

[328] A spin-orbital is the product of a one-electron space orbital and a one-electron spin function. Slater incorporated exchange through the specific variation of the spin orbitals in a determinantal wave function. Slater-type orbitals are often used as the basis functions in atomic calculations. In a single-configuration description (i.e. with one Slater determinant) different sets of N orbitals can lead to the same overall N-electron Slater determinant wave function. Hence, any selection of orbitals (fitting a fixed Slater determinant) gives rise to artificially distinguished electrons.

ALLEGED REDUCTION OF CHEMISTRY TO PHYSICS 135

Block 5-8   Text book authors on the choice of basis sets.[329]

> WILSON
> The art of selecting a basis set is based on previous experience in treating similar systems using basis sets of comparable quality.
>
> LOWE
> One of the places where human decision can effect the outcome of variational calculation is in the choice of basis. Some insight into the ways this choice effects the ultimate results is necessary if one is to make a wise choice of basis, or recognise which calculated results are physically real and which are artifacts of basis choice.
> Problems with basis set adequacy are difficult to overcome completely. Fortunately, with a certain amount of experience, insight, and caution, it is nevertheless possible to carry out variational calculations and interpret their results to obtain reliable and useful information.
>
> LEVINE
> The use of an adequate basis set is an essential requirement for success of the calculation. In principle, this basis set should be complete. In practice, one is limited to a basis set of finite size; one hopes that a good choice of basis functions will give a good approximation to a complete set.

uses a finite (and hence incomplete) basis set of orbital functions, representing the orbitals as linear combinations of a complete set of basis functions.

Different choices of basis set will produce different wave functions and energies. Usually, basis functions are chosen because they are amenable to rigorous calculation not because they represent anything chemically significant. In general accuracy improves by including more effects in the basis set, but only if one chooses 'wisely' (Block 5-8). If one chooses 'wisely', the electronic energy will decrease with each basis set improvement. The limiting value is called the Hartree-Fock energy. However, this is not yet the 'true' energy of the system because the independence of electrons is still assumed (making use of a single determinantal wave function). The difference between the Hartree-Fock and 'exact' energy is called the correlation energy.[330] The disagreement between 'Hartree-Fock' and 'exact' grows progressively larger for larger atoms. Even at this stage 'exact' is still used in quotation marks.[331]

One method to account better for the interelectronic interaction is to express the wave function as a linear combination of functions corresponding to various electron configurations.[332] The number of possible configuration functions with the proper symmetry increases rapidly as the number of electrons and the number

---

[329] Wilson (1984:169) quoted in Scerri (1993:5), Lowe (1993:229,231), Levine (1991:461,291).
[330] The correlation energy has been defined as 'the difference between the exact nonrelativistic energy and the Hartree-Fock energy' (Levine 1991:318).
[331] Cf. the definition of 'exact' in note b to Table 11-2 in Lowe (1993:371): '"Exact" equals experimental with relativistic correction but without correction for Lamb shift.'
[332] This is called 'configurational mixing', 'superposition of configuration', or 'configuration interaction'. All refer to representing an atomic state not by a function representing a single configuration, but by the superposition of a set of functions, each of which has the form of the wave function for a single configuration.

of basis functions increase. One must therefore decide which type of configuration functions are likely to make the largest contribution to the wave function.[333]

The predictive accuracy of the Hartree-Fock method - note that it is checked against 'classical' data - is of the order of 0.5% to 1% of the total atomic energy. This may seem a reasonable figure, but the absolute value of 1% is of the same order of magnitude as the energy of a typical chemical bond (Scerri 1993:55). Most practising scientists are not very concerned about this because they are used to comparing their models and approximations with already known experimental data. But from the fact that the experimental value can be increasingly better approached, it doesn't follow that, theoretically speaking, the path followed will converge. Moreover, if no lower *and* upper bounds, or some other criterion for convergence or truncation can be obtained independently of experimental data, then these methods are not truly ab initio (Scerri 1991a, 1993, 1994).

The hybridity of the conceptual framework of quantum chemistry is well illustrated by the notion of electron orbitals and hybridisation (Block 5-9). Some will say reference to orbitals is not sanctioned by quantum mechanics; at that level they are ontologically redundant. Basis functions are not localised atomic orbitals. There are no atomic orbitals in a molecule. Others will say that an orbital is a mathematical function turning up in the standard approach that starts from the wave equation. In principle the same calculations could be carried out using matrix mechanics, but then there would be other mathematical entities corresponding to orbitals (Ogilvie 1994).

Hybridisation seems even more artificial. On the one hand it undermines microreductionistic pictures of even the simplest molecules. To deduce the tetravalence of carbon in the molecular orbital approach from 'underlying interactions' the hybridisation of orbitals has to be postulated.[334] On the other hand, hybridisation is a principle that does not belong to 'proper' quantum mechanics. Hybridisation seems more a case of an interaction, not of the components of the system, but of discourses at different levels of description.

The use of the expression ab initio easily leads to misunderstandings between philosophers of science and quantum chemists. Scerri (1992a) had pointed out that in the work of Schaefer et al on the bond angle of methylene, extrapolations are made from calculations performed on other molecules. Schaefer (1992) replied:[335]

> Those who have read the 1970 paper will join me in wondering what it is about the work [on the methylene molecule] that was not *ab initio*. The only remote possibilities that come to mind are the use of a) basis sets designed to give lowest possible energies for the carbon and hydrogen atoms, and b) configuration interaction methods proven to give accurate bond distances for very small diatomic molecules.

---

[333] Configurational mixing denies the atom having a fixed configuration (cf. the debate about resonance structures). But note that conventional theory of atomic spectra allows configurational mixing as well (Scerri 1991a).
[334] Also note that in metal carbonyls, carbon forms six bonds (yielding a 'molecule' with a plus two or minus two charge).
[335] To which Scerri (1992b) replied: 'the mere fact that a theory might yield a true prediction does not necessarily imply the truth of the theory itself'.

Block 5-9 Views on orbitals and hybridisation.[336]

Ogilvie	The qualitative ('hand waving') explanations of molecular structure and reactions based on orbitals and such ilk are not science (i.e. are nonsense) and should consequently be completely discarded.
Edmiston	Orbitals do have much meaning. Without these orbitals [atomic orbitals and molecular orbitals] the *understanding* of chemistry is not possible.
Simons	I agree with some of what Ogilvie says, and I concur, in particular with his view that orbitals are not real. Experimental data and the molecules, atoms, and substances we subject to scrutiny constitute the 'real' things in chemistry. *All* quantities derived from our data through the application of models, theories, and their associated equations are not 'real'.
Pauling	I read with much interest the paper by Ogilvie, which contains errors and logical flaws. The subheading of the Ogilvie paper, 'There Are No Such Things as Orbitals!,' is based on a misconception of the meaning of the word 'thing'. Ogilvie has surely shown that orbitals are objects of thought. Several of Ogilvie's other arguments seem to show that his understanding of quantum mechanics is at best superficial.
Scerri	Quantum mechanics shows categorically that the assignment of electrons to orbitals cannot strictly be maintained in many-electron systems. The assignment of electrons to particular orbitals is an approximation.
Hartree	The concept of an electron 'orbit' in an atom, as a continuum of successively observable positions like the orbit of the moon or of a planet, is not a significant concept.
Matteson	Hybridisation is not something that atoms do or have done to them. It is purely a mental process gone through by the chemist, who wants to group atomic orbitals according to their symmetry properties so he can talk about one localised bond and ignore the rest. Hybridisation does not change the shape of the electron distribution in any atom.
Ogilvie	Despite the fact that many authors of textbooks of general chemistry have written that CH_4 has a tetrahedral structure because of sp^3 hybridisation, there neither exists now nor has ever existed any quantitative theoretical or experimental justification of such a statement.
Carroll	Atomic orbitals [do *not*] actually hybridise. *We* hybridise the model [of molecular orbits] to produce a different model. Hybridisation is only a conceptual and a mathematical model that allows us to calculate (i.e., mathematically visualise) molecular parameters. Changing hybridisation [as in variable hybridisation] is simply modifying that original model to suit a current need.

I don't think there is any disagreement about facts. The question is whether the use of 'basic sets' and 'configuration methods' introduce background assumptions,

---

[336] Ogilvie (1990:288), Edmiston (1992), Simons (1991), Pauling (1992) – the latter three are comments on Ogilvie (1990); Scerri (1999b:394, cf. 1991a), Hartree (1957:4), Ogilvie (1994:181), Carroll (1998:30,44), Matteson (1974:5) quoted in Carroll (1998:30n114).

models, and intuitions that were not derived from quantum mechanical principles. As Woolley (1978:1078n16) points out 'ab initio' doesn't mean 'first principles'. The latter would mean description in terms of 'a Schrödinger equation in which *every* particle contributes to the kinetic and potential energy operators on equal footing.'

There is also the question of how to see the relation between the increasing accuracy of the ab initio methods and experimental (spectroscopic) data. At the time of the polywater episode [3.5], ab initio methods were quoted to determine molecular geometry to about 2%; dipole moments and force constants were systematically overestimated (10-25%) and dissociation energies were greatly underestimated (Allen and Kollman 1971a:461). Ab initio methods were of no help sorting out the rather 'open' issue of polywater. They could be made to confirm any hypothesis that was put forward.[337] Ab initio methods have improved since. The interpretative disagreement between Scerri and Schaefer is about the theoretical work of Bender and Schaefer (1970), who predicted a bond angle for methylene of 135.1°, convincing experimentalists that the methylene molecule was bent. As a consequence spectroscopic studies were reinterpreted to confirm the bent geometry predicted by the ab initio calculation. Schaefer (1986:1101) is absolutely right to claim this as a big success showing that theory is a 'full partner with experiment'. However, being a great success for theory and becoming a 'full partner with experiment', doesn't imply that the bond angle of methylene was derived *solely* from quantum mechanical first principles. For a start, there is still the matter of approximations. More theoretical work showed that 'we estimate that our theoretical bond angle of 134° is reliable to within 2°' - 134±2° seems to be the last word from theory.[338] I don't know whether even more accurate calculations have been reported in the meantime, but Schaefer (1986:1102) concludes the success story on methylene by saying (emphasis added):

> Fortunately, the equilibrium bond angle of triplet methylene is no longer a matter of debate. A *definitive experiment* [showed that] the equilibrium bond angle is 133.8° ± 0.1°.

As the theoretical models get better, it will happen more and more often that these results can challenge results from experimental chemistry or make predictions before experimental measurements are available, thus becoming a 'full partner with experiment'.[339] If this is so, it shouldn't surprise if ab initio methods are sometimes more precise and reliable than experiment for small molecules, especially for unstable systems. But the reliability of the choice of basis set and configuration interaction models that makes this success possible draws support from

---

[337] See quote of Everett, Haynes and McElroy on p. 96.
[338] McLaughlin, Bender, Schaefer (1972:357). See Scerri's (1993, 1994) detailed analysis of the case showing that the suggested error bars of ±2° are 'an extrapolation from the application of the method from one molecule to that of another', viz. comparison with theoretical and experimental values for bond angles in H_2O and NH_2 molecules and even then the value of 2 seems rather ad hoc (except that the later determined 'true' experimental value turns out to fall in this range).
[339] Perhaps the most impressive example today of 'getting the facts right before experiment' are the theoretical predictions of the ground and excited states of Si_2H_2.

trying out the models used on other molecules through trial and error guided by chemical 'experience' and experimental data.

My concern is not that ab initio methods do not provide analytic solutions to the 'full' Schrödinger equation. My concern is that approximate models and numerical methods are governed by what experimental data must be predicted, as well as a general chemical understanding of the system at hand. At the end of the 1970s, Bogaard (1978:351f) wrote:

> The minimisation of the system's potential energy can be accomplished in an indefinite number of ways, just as the mathematical techniques called upon can approximate proper wave functions in indefinitely many ways. Within this framework chemical results are left *under*determined by atomic physics. Only the particular chemical context can suggest what sort of 'superpositions' might be appropriate, and only experimental measurements can specify when enough mathematical terms have been added.

Of course the techniques of doing this have improved considerably in the past decades. By now there are even more sophisticated methods then the ones referred to above. However, this development is a continuation of the so called *aufbau* principle.[340] That's why I'm not impressed if it turns out that an earlier problem has been 'solved' by quantum mechanics. For example, Needham (1999b) referred to an example of Coulson (1961), who noted that on the basis of the Pauling principle one would expect oxygen to form covalent bonds with two hydrogen atoms, yielding a structure with an H-O-H bond angle of 90^0. Experimentally the value is 104.45^0 (which was then used to calculate the 'appropriate' amount of hybridisation). In response to this example Scerri (1999a:189) notes that 'the approximate solution to the time-independent Schrödinger equation for the water molecule does indeed yield the observed bond angle'. What this example shows is that as computers get bigger, more sophisticated approximation methods become feasible. But the general structure of how practice works remains the same. There are some classically observed data (which have been around for a long time) and with great effort it is possible to have the quantum mechanical formalism approximate these classically observed data better and better until it fits. Using the methods (models, approximations) that have worked, these methods are then extrapolated to similar molecules - similar according to chemical expertise. This may give the impression of ab initio calculations, but at every stage the experimental data steer the development of the models.[341]

Defenders of the Dirac school are quick to point out that for an isolated hydrogen molecule the theoretical calculations are exactly in line with experimental measurements. But it is instructive to see how many steps were necessary to work

---

[340] The 'aufbau principle', first used by Bohr, builds-up successive atoms by the addition of an extra electron to the previous atom. For discussion see Scerri (1991a).
[341] The following citation from a recent text book would seem to put the order of priority right (Munowitz 2000:R7.10): 'We begin with the facts of water's existence - most important, the molecule's near-tetrahedral bond angle of 104.5°. A tetrahedral geometry instantly suggests sp^3 hybridisation, a role for which oxygen is well suited.' And then, much later, come the quantum mechanical calculations.

enough refinements into an appropriate Hamiltonian before the appropriate approximations of the wave function produced the correct result for the electrostatic attraction gained by the two electrons being attracted by both nuclei.[342] Moreover, this particular molecule is atypical for a chemical molecule. The necessary refinements were very much inspired by physical considerations. If we turn to molecules of at least three atoms, the possibility of different structures arises (directions of bond angles etcetera) and Gell-Man's 'chemical questions being asked' will start to dominate the choice of auxiliary assumptions and approximation methods more and more.

Many theoretical chemists acknowledge or even stress the ineliminability of *chemical* arguments, knowledge, and experience in quantum chemistry (Block 5-10). But there is a constant pull towards the Dirac perspective. Whereas Coulson in the many editions of his influential book *Valence* took a nuanced view, in McWeeny's (19798-9) *Coulson's Valence* we find:

> The theory of valence is one aspect of a much bigger subject, for the laws of quantum mechanics allow us, in principle, to calculate not only the energy of an electronic system - and hence to predict the geometries and structures of molecules - but also all other properties of such systems.

Coulson's concern about the 'ideas, intuitions, and conclusions of the experimental chemist' [Block 5-10] is reduced to[343]

> Our 'understanding' of valence and of electronic properties must be in terms of simple physical models which can readily be visualised, and not in terms of numbers gushing from a computer!

The obligatory quote from Dirac follows. However throughout the book the actual support for the quantum mechanical approach is that what has been measured can, in some simple cases, be calculated.[344] The models quantum chemistry uses always have to be adapted to a specific situation and often contain approximations that are dictated more by computational possibilities than by chemical (or physical) intuitions. As Vancik (1999:254) puts it: 'Computer operations are *only optimisations* of molecular structures which have already been *a priori* conceived at the level of chemical principles'. One consequence is that different models may be used to capture different features of the system studied, without it making any sense which of the models is better in some absolute sense. Another consequence is (Carroll 1998:27f):[345]

---

[342] For the early calculations see Table II in van Vleck and Sherman (1935:188) quoting nine calculations of the ground state of H_2. At that time the best theoretical value was 4.70 electronvolt compared with the experimental value of 4.725±0.01; in the early 1930s the best value was 4 eV; in 1928 3.78 eV; in 1927 3 eV.
[343] Cf. Coulson (1961:9): 'To say that the electronic computer shows that $D(H-F) >> D(F-F)$ is not an explanation at all, but merely a confirmation of experiment. Any acceptable "explanation" must be in terms of repulsions between non-bonding electrons, dispersion forces between atomic "cores", and the like.'
[344] For example, McWeeny (1979:11): 'The charge density P can also be *calculated* from quantum mechanics, and generally speaking there is good agreement between observed and computed densities.'
[345] Or, much earlier, Schwab (1959:9): 'Wenn zwischen den klar erkannten Ansätzen, von denen

Block 5-10 Theoretical chemists on the relation of quantum mechanics and chemistry.[346]

COULSON 1961
It is not unfair to say that in practically the whole of theoretical chemistry, the form in which the mathematics is cast is suggested, almost inevitably, by experimental results. This is not surprising when we recognise how impossible is any exact solution of the wave equation for a molecule. Our approximations to an exact solution ought to reflect the ideas, intuitions, and conclusions of the experimental chemist.

PAULING 1993
I do not think that quantum mechanical calculations of molecular structure or crystal structure will ever make the sort of chemical arguments in my book [his 1939 *The Nature of the Chemical Bond*] obsolete.

SUTCLIFFE 1996
If one knows the Schrödinger equation for an isolated molecule, there is nothing in the mathematics that would lead one to expect molecule-like solutions among its eigenstates. If you know what you want to get out of the full wave function then you can usually get it by finding suitable approximations.

As the accuracy of the model is increased, its simplicity is decreased. We must choose the model that is sufficiently accurate for our computational purposes, yet simple enough that we have some understanding of what the model describes. Otherwise, the model is a black box, and we have no understanding of what it does, perhaps even no idea whether the answers it produces are physically reasonable.

Even if it were to make sense to talk about a reduction for the ideal model of an isolated molecule, it is not trivial to extend this to the reduction of a substance, let alone to a network representing the chemical affinities between different substances. Causally relevant factors are not only to be found in the relational properties of molecules, atoms, and smaller particles, but also in gradients of concentration or other parameters. And the latter may well be (partially) dependent on properties of macroscopic objects such as membranes. For example, in so-called 'ox-phos' reactions which depend essentially on the presence of a small membranous vesicle (a kind of bag with enzymes), the membrane's permeability is implicated in the reaction mechanisms (Allchin 1996).

French and Ladyman (1998:64) point out that 'an explanation of molecular structure may involve *a diverse array of models*, including, those whose equations are mathematically related to Schrödinger's equation' and argue that therefore it is correct to count these models as 'an application of quantum mechanics'. There is nothing wrong with the latter, except that there seems to be an irresistible tendency among many quantum chemists and philosophers to forget about the part of the quotation I emphasised and indulge rather in quoting Dirac.

---

wir ausgehen, und dem Rechnerergebnis, das wit wieder klar verstehen können, ein Chaos liegt, das nur die Maschine meistern kann, dann kann man wohl nicht mehr sagen, daß wir eine Gedankenkette vor unserem geistigen Auge haben, die wir überschauen'.

[346] Coulson (1961:113), Pauling (1993), Sutcliffe (1996). Cf. Löwdin (1967:11), Hartmann (1965), Hoffmann (1998).

5.5 holism of quantum mechanics and shape of molecules

Quantum mechanics predicts remarkable correlations in the behaviour of systems which may have previously interacted, but are now (apparently) physically and spatially separated. That is to say, in quantum mechanical systems there are correlations between different 'particles', e.g. electrons or photons, that make no classical sense. This was first highlighted in a famous thought experiment of Einstein, Podolsky and Rosen (1935). Experiments of Aspect, Dalibard and Roger (1982) very similar to the original 'EPR' thought experiment, and later variants on it, have confirmed that there is strong evidence that quantum mechanics violates some sort of principle of locality or separability. Quantum theory predicts (and experiments confirm) that when measurements are made on two particles that once interacted but now are separated by an unlimited distance, the results obtained in the measurement on one particle depend on the results obtained from measurement on the second particle. For example photons originating from the same source, following different paths, which are separated along this path in terms of their polarisation properties, display correlated properties, though no information exchange between them has taken place. That is, the 'profile' of two photons that have passed two different filters are correlated (i.e. are dependent on one another) though without interaction (i.e. information exchange).[347] Formulated for two particles that have previously interacted, the intuitions that quantum mechanics seem to violate can be formulated as follows (Healey 1991):

- Measurement of one particle should not depend on what measurement, is performed on the other – but it does.
- The 'real' state of the one particle cannot depend on the kind of measurement I carry out on the other particle – but it does.
- If two events occur there can be no causal connection between them if their 'spacetime' distance cannot be covered (i.e. if no light signal can reach the other in the time interval between the two measurements).

All three are a modern expression of the suggestion that 'actio in distans' is impossible. Quantum mechanics seems to suggest either that this *is* possible or that some very fundamental concepts or metaphysical assumptions have to be overhauled.[348]

The properties of quantum mechanical systems like photons or electrons are non-local in that they depend on what is happening 'elsewhere', without there being any causal interaction with 'elsewhere'. Such 'entangled' systems have properties that are not present in any subsystem. These holistic correlations are

---

[347] In the Aspect experiment two photons are sent through identically oriented but spatially separated polarisers. Each photon has a 50% chance to pass the respective polariser and a 50% chance to be absorbed. But either *both* photons pass or *both* fail to pass. The two photons show a perfectly correlated behaviour, despite being separated by an arbitrarily large distance.

[348] For a minority view saying that a 'common-cause account of EPR' can be given see Chang and Cartwright (1993:170): 'the widely perceived causal difficulties are created by the failure to be sufficiently realistic about quantum mechanics'. If probabilistic instead of deterministic causality is taken seriously, then there is no need for a causal process connecting the initial common cause and each of the two or more effects.

called EPR-correlations.³⁴⁹ EPR-correlations bring a 'hereditary' element into quantum mechanics. They entail that quantum mechanics is holistic in a very fundamental sense: 'everything' is 'EPR-correlated' with 'everything'. For two or more entangled systems, only the whole is in a pure state. A description of one of the systems does not completely specify the local observables of this system. Hence, strictly speaking only all quantum systems taken together are in a pure state. The ontology has to start from one global state of all quantum states taken together. Ultimately, the global quantum state of the world determines what there is. The 'material' world is one whole that is not constituted of parts.

Moving on to more fundamental physical theories such as quantum field theory will not change the picture (Esfeld 1999).³⁵⁰ In quantum field theory entities that are treated as physical systems in quantum mechanics (such as electrons) are conceived as properties of quantum fields. Again, because of entanglement only the whole of all quantum fields is in a pure state. It is only this whole that can serve as a 'supervenience' base of everything there is.

A consequence of the holism of quantum mechanics is that an object can only be defined in terms of its relations to its environment.³⁵¹ But 'the' environment, consisting of the rest of the universe, can never be given a precise description. As Bell describes it (1989:362):

> We are led, in the quest for more accuracy and completeness, to include more and more of the world in the wavy quantum mechanical 'system': the photographic plate that records the scintillations, the developing chemicals that produce the photographic image, the eye of the observer. But we cannot include the whole world in this wavy part. For the wave of the world is no more like the world we know than the extended wave of the single electron is like the tiny flash on the screen. We must always exclude part of the world from the wavy 'system', to be described in a 'classical' 'particulate' way, as involving definite events rather than just wavy possibilities.

So there is a big problem how to fit quantum mechanics and classical physics (and the rest) into one world. Even in the top journal *Physical Review*, we can find an author contemplating the thought that 'the idea that part of the world might be intrinsically classical is an obvious one' (Jones 1994). The author stresses that, contrary to what is often assumed, it is by no means clear that there is a simple combination of physical parameters that defines the classical limit of quantum mechanics. Moreover, the classical world is governed by gravity, which does not occur in standard quantum mechanics. Whether the quantum mechanical superposition principle holds at the macroscopic level is an open question. It is doubtful whether macroscopic systems can ever be considered as isolated quantum systems. There is no universal classical limit of quantum mechanics.³⁵²

---

³⁴⁹ Taking into account relativistic considerations it may be suggested that so called intrinsic properties such as charge and mass are relational as well.
³⁵⁰ On quantum field theory see Auyang (1995), Cao (1997).
³⁵¹ Amann (1990, 1993), Amann and Gans (1989), Primas (1985a, 1983:295).
³⁵² See for example, Primas (1983), Scheibe (1999), Chang and Cartwright (1993). As Primas (1990:61) points out the distinction between microphysics and macrophysics has become theoreti-

What has been said about objects in general applies to molecules in particular. In standard quantum chemical approaches a molecule is taken to be a set of nuclei and electrons, but 'in the physicist's description electrons and nuclei are correlated by Einstein-Podolsky-Rosen correlations so that neither electrons nor nuclei exist as individual objects' (Primas 1983:350). More specifically, each molecule is coupled to the environment and in particular to the radiation field, a system consisting of an infinite number of bosons. Physical systems are automatically coupled to their environment and are never closed in a strict sense (ibid:293):

> According to quantum mechanics the electrons of the moon are entangled with their radiation field. So without abstracting from the quantum structure of the radiation field, the moon cannot be an object.

Therefore *all* classical observables of molecular systems seem to be inherited from the environment in the sense in which a person's nationality is not intrinsic to the person, but inherited from their environment: 'a molecular object *adapts* its state to its environment' (ibid:304-6). According to Primas the crucial issue is not the approximations of quantum chemistry such as the Born-Oppenheimer description, but the breaking of the holistic symmetry of quantum mechanics by abstracting from the EPR-correlations. It is the Einstein-Podolsky-Rosen correlations that exclude any classical concept of object, shapes or other 'fixed' spatial structures such as presupposed in the notion of molecular structure (Block 5-11). When and where to suppress the EPR-correlations is not something that can be derived from quantum mechanics - more precisely: the only truly ab initio approach would not allow *any* EPR-correlation to be suppressed. But it is the decisions when to suppress them that, as it were, abstract objects out of the quantum mechanical formalism. Quantum mechanics describes the material world, in principle, as one whole. To separate out objects (including electrons, atoms and molecules) from this whole requires a justification which lies outside the principles of quantum mechanics. Only by forgetting about the EPR-correlations can one impose the concept of molecule on quantum mechanics. Therefore, quantum chemistry *borrows* the notion of molecular structure from classical chemistry.[353] As we saw in the previous section, molecular structure is the actualisation of a particular molecule from the many potentialities of a given set of nuclei and electrons, all 'present' in the same Schrödinger equation (Del Re 1998).

Amann (1993, 1996) has proposed that the notion of a Gestalt might be helpful to get a handle on the problem of a system having holistic correlations with its environment, which are 'severed' when observed - when the particular Gestalt emerges from the holistic relations.[354] On this view classical molecular observables can be generated by coupling a single molecule to its environment. But this is an interpretation that is clearly supplementary to the statistical formalism of

---

cally obsolete. Microscopic systems can have 'classical' properties and macrosystems can have 'quantal' properties.
[353] Primas (1983:292), Woolley (1978:1075).
[354] There are properties, like the charge of an elementary particle, that behave like classical properties, i.e. they have no holistic correlations with other particles, this makes quantum mechanics and classical physics even more difficult to reconcile (Amann 1996).

Block 5-11 Quantum mechanics, EPR-correlations, reification of objects.[355]

objects/molecules	EPR correlations
no objects	including all correlations
formless molecules	no correlations with environment
molecules with spatial structure	no correlations between nucleus and electrons
localised electron orbitals	no correlations between electrons

quantum mechanics. The quantum statistical description mixes all possible shapes and other molecular properties into one 'statistical pulp'. It is possible to bring such a system into a pure state, but then the state of the overall system changes; and this is only possible 'momentarily', because the position of a 'free' quantum 'particle' is smeared out in the course of time. Moreover there seems to be a bias in the choice of the term 'gestalt' – as an object of belief, but not 'real', because it cannot be found back in the mathematical equations.

Of course, the notion of structure has different meanings at different levels of description (Block 5-12). It is trivial that at whatever level of description we are interested in, there is structure. A reductionistic view will take for granted that structure at each level, if it is to make sense, will (eventually) be found back at the most fundamental level. Defenders of this view often seem to presuppose that the undeniable fact that quantum mechanical (mathematical) structures have been used to make excellent predictions of some (chemical) properties of molecules, is sufficient to prove that complete reduction is possible in principle. But molecular structure, composition, and bonding are intimately connected to notions of shape or orientation in space of objects - all of this has to be imported into quantum mechanics.

Woolley's paper 'Must a molecule have a shape?' (1978) has been referred to in many subsequent publications, though possibly due to its provocative title, it has not always been understood correctly.[356] He says for example:[357]

> The van 't Hoff notion of molecular structure is irreducible in classical chemistry - it is not derived from more primitive notions - we might expect a deductive account of the behaviour and properties of molecules according to quantum theory in which the *same* notion of molecular structure is a *derived concept*. The 'problem' of molecular structure arises because quantum chemistry has not achieved this result.

The ambiguity is in the term 'molecule'. For example, although Bunge (1982) is sceptical about the reduction of chemistry to physics, nonetheless it is concluded in García-Sucre and Bunge (1981) that 'quantum chemistry makes room for a concept of molecular shape, recently criticised by Woolley'. But what they mean by that is that the quantum system they call 'molecule', 'will extend over the entire space and its shape will be a closed surface infinitely far removed from the origin

---

[355] Adapted from Primas (1985a). See also Primas (1983:136-46,253,292-307,350).
[356] Amann (1992) chose the same title for his paper. For discussion on molecular shape see Amann (1992, 1996), Arteca and Mezey (1989), Claverie and Diner (1980), Del Re (1998), Löwdin (1991), Primas (1983), Ramsey (1997), Woolley (1978, 1986, 1991), Zeidler (2000). For discussion of the views of Woolley see Weininger (1964) and Ramsey (1997).
[357] Woolley (1986:204), summarising earlier work.

Block 5-12 Different notions of (molecular) structure.[360]

- Structure as a rough 'summary' of reaction possibilities, i.e. structure given by the (macroscopic) network of relations between substances and the chemical reactions into which they enter. This structure is a little potted theory in its own right and provides a model for the chemical reactions of macroscopic masses of the relevant compounds (their reactivity in various environments). When combined with different compounds, different parts of the structure may be relevant.
- Geometric structure at the level of molecules: spatial arrangement of atoms, crystal structures, structure of glasses, and so on.
- Valence structure, i.e. structure (topology) at the level of atoms and bonds between atoms: arrangement of electron configurations. At this level of structure structural and rotational isomers are introduced.
- Spatial arrangements at chiral centres in molecules of compounds exhibiting optical activity (introducing enantiomers).
- Mathematical structure of the quantum mechanical description for a particular system.

of the reference frame' and this system 'lacks a definite size'. That is to say the quantum mechanical 'molecule' is similar to the Stoic view according to which oxygen and hydrogen in water would occupy the same place.[358] But Woolley did not deny that a quantum chemical system 'will extend over the entire space and its shape will be a closed surface infinitely far removed from the origin of the reference frame'; and he would agree that it 'lacks a definite size' (as García-Sucre and Bunge say). Woolley's point was that the *classical* concept of molecule cannot be derived from quantum mechanics. The classical concept of molecule does not allow each molecule to 'extend over the entire space'.[359]

Ramsey (1997:244) has shown that shape cannot be reduced to quantum mechanics under almost any interpretation of reduction. The maximum that can be achieved is:

> *If* shape is conceived as a feature of some physical systems but is one which does not exist independently of measurement and time-scale considerations, then there is a sense in which shape is approximately ontologically reducible.

Ramsey suggests that microconstituents of substances are explanatory, but not intrinsic. For example, at very short time intervals there would be no interaction between nucleus and electrons. What we see depends on how we look: under some conditions of measurement, molecules do have a frame. This is missed if one only discusses the shape of isolated molecules, without taking into account that shape is dependent on how the molecule is picked out in measurement: 'different shapes appear depending on the experimental resolution and the correlative schema of approximation chosen to describe the experiment' (ibid:247). Shape is objective and

---

[358] See Needham (1996b, 1999c) on the Stoic theory of mixtures.
[359] Cf. the distinction Claverie and Diner (1980:60) make between quantum and classical structure. As Woolley (1986:203) points out he has never opposed molecular structure in the first sense. Moreover (ibid:204): 'I have never suggested that molecular structure should be abandoned.'
[360] See, for example, Schummer (1996a:252-289), Maccoll (1964:6).

real, but relative to the time scale of measurement (i.e. under particular conditions; how it has been prepared). Shape is not merely a concept.

There are always two options when confronted with a failed attempt at reduction. One can claim that this strengthens the autonomy of the domain that was going to be reduced - the pluralistic or promiscuous stance. But one can also claim that this shows that the 'old things' don't really exist – eliminative reductionism. Woolley (and perhaps Primas as well) may seem to draw eliminativist conclusions: there are no molecules; at least not in the sense 'molecule' is usually understood by chemists. For example, Ramsey (1997:234) says 'shape is an object of belief and not, *contra* Woolley and Primas, merely conceptual' and 'Woolley, Primas and others charge shape is only a "powerful and illuminating metaphor" (Woolley 1982:4)'. I think this presents a distorted picture of their views. In the case of Primas it overlooks his discussions of complementarity and realism (referred to elsewhere in Ramsey's paper).[361] In the case of Woolley it overlooks that an important motivation of this work is the observation that there is a 'real discontinuity' between the behaviour of small, isolated molecules and large molecules interacting with the environment; the former can be given experimental reality in rarefied gases or molecular beams (Woolley 1978:1074) – compared with 'classic' molecules these 'molecules' are 'a new state of matter' (Zeidler and Sobczynska 1995/96:529). Woolley (ibid) also says that every physical and chemical concept is only defined with respect to a certain class of experiments, so that it is perfectly reasonable for different sets of concepts, although mutually incompatible, to be applicable to different experimental situations – which is very similar to Ramsey's conclusions. Woolley's primary concern is that (1991:42,44):

> quantum chemistry beyond the Born-Oppenheimer approximation cannot claim to have a well-defined set of equations that determine the quantum states of molecules in general. The difficulty is not the complications of the equations; rather it is the lack of appropriate equations.

And he is looking for appropriate equations to describe experiments with isolated molecules.

What Primas and Woolley argue is that the concept of molecular shape (or bond) cannot be derived from quantum mechanics. That is to say, on most accounts of reduction, chemistry cannot be reduced to physics. But nothing follows as to one's stance on issues to do with concepts and their relation to what is 'objective and real'. Perhaps *both* quantum mechanical and classical chemical concepts are powerful and illuminating metaphors. Perhaps both refer (at least sometimes) to robust causal properties of the world. Why not leave room for quantum mechanics *and* molecular chemistry (*and* thermodynamics). We could be tolerant enough to leave equal *ontological* room for manifest water, water in terms of the thermodynamic theory of substances, the molecular structure of water ('constructed' out of spectroscopic measurements), the 'proper' quantum mechanical equations for an isolated water molecule, and experiments with isolated water

---

[361] On the issue of complementarity see Primas (1982), Claverie and Diner (1980:64f).

molecules which, depending on the measurement technique, show more or less of the 'classical' molecular structure.

The reason that worries about the reducibility of chemistry to quantum mechanics are easily dismissed as contentious is, I believe, due to the fact that microreductionism is presupposed as the goal of science and requires no further discussion. Even those who see problems for a smooth reduction of chemistry to quantum mechanics seem to adhere to this view. For example, Weininger (1984) gives a sympathetic hearing to views like those of Woolley (1978) and says the difference between such views and of those, for example of Bader (1990), is due to differing presuppositions. He doesn't however say what the differences are. Moreover Weininger says that Bader's approach 'is firmly grounded in standard quantum mechanics' (ibid:943). But Bader, for one, is very explicit about his presuppositions - note again the Kantian echo:[362]

> A scientific discipline starts with the classification of observations. It becomes exact, in the sense that predictions become possible, as soon as the classification represents the physics that underlies an observation.'

Moreover Bader accepts boundary conditions in his quantum mechanical definitions, the plausibility of which is justified with reference to 'the incorporation of chemical observations into the fabric of quantum mechanics', and ontological premises such as 'atoms are objects in real space'.[363] But if chemical observations are imported into quantum mechanics, as well as atoms as objects in real space, then the result is not solely the result of a derivation from quantum mechanical first principles. In the end, Bader says 'there is only one test of the validity of any theory and that is comparison with observation'.[364] Hence the situation is analogous to the debate on ab initio methods. In both cases an appeal is made to an inference to the best explanation: My theory (that of Schaefer, Bader, ...) is the best at saving the phenomena. My theory gives the most accurate and full description of the data. Hence it describes the world as it is.

Bader presents his theory as an alternative to the molecular orbital paradigm, because the latter would suggest that the chemist's talk of functional groups 'has no basis in theory'. In a range of publications he has argued that both atoms and functional groups are definable bounded regions of real space. But how atoms and functional groups are defined as 'quantum subsystems as regions in real space' remains difficult to assess (Block 5-13).[365] Bader admits that 'the total electron

---

[362] Bader, Popelier & Keith (1994:620). The original reads: 'Eine wissenschaftliche Disziplin beginnt mit der empirischen Klassifizierung von Beobachtungen. Sie wird exakt in dem Sinne, daß Vorhersagen möglich sind, sobald die Klassifizierung die Physik widerspiegelt, die einer Beobachtung zugrunde liegt.'
[363] Quotations from Bader (1990:3,5). Cf. also Eyring, Walter and Kimball (1944:1), quoted in Block 5-1: 'It is now well established that all matter is composed of a small number of particles'.
[364] Bader (1990:5; cf. 275). Also: 'The answer to the question whether or not these regions correspond to the atoms of chemistry should not be prejudiced by any existing orbital-based paradigm, but rather be determined solely by comparison of their predicted properties with the observed properties of matter' (Bader, Popelier & Keith 1994:622f).
[365] That 'the empirical classification of the observations of chemistry in terms of the properties assigned to functional groups is a consequence of and is predicted by physics' is, in nuce, to be found in 'Schwinger's formulation of physics'. Quotations from Bader, Popelier and Keith (1994:620-4).

ALLEGED REDUCTION OF CHEMISTRY TO PHYSICS   149

Block 5-13 Bader's 'derivation' of atoms and functional groups.[367]

> Two identical pieces of matter possess identical properties.
> This elementary fact should extend down to the atomic domain.
> If an atom is identical in two different systems or at different sites within a given system (e.g. in a solid), then it must contribute identical amounts to the total properties of the systems in which it occurs.
> Since the form of matter is determined by the distribution of charge throughout real space, two objects are identical only if they possess identical charge distributions.
> [The above] demands that an atom be a function of its form in real space, requiring that it be defined in terms of the charge distribution.
> Theory defines [atoms] through a partitioning of real space as determined by the topological properties of a molecular charge distribution, that is, by its form in real space.
> Atoms do indeed exist in molecules as separate definable pieces of real space.
> There is no question concerning the possibility of identifying atoms or functional groups with bounded regions of real space.
> There is no question concerning the possibility of obtaining a quantum mechanical definition of uniquely defined regions of real space.
> The theory demonstrates that the concepts of atoms and bonds may be rigorously defined and give physical expression in terms of the topological properties of the observable distribution of charge for a molecular system.

density distribution in a molecule shows no indication of discrete bonding or nonbonding electron pairs.'[366] But this merely raises the question:

> Where then to look for the Lewis model, a model which in the light of its ubiquitous and constant use throughout chemistry must be rooted in the physics governing a molecular system.

The answer is that: 'the Lewis model of the electron pair does find a more abstract but no less real physical expression in the topological properties of the Laplacian of the charge density.' This still rather careful formulation is then standardised into dogma, for example when Mainzer (1998:36,35) reports (emphasis added):[368]

> The topology of the measurable charge density *defines* the corresponding molecular structure. ... The gradient vector field *makes visible* the molecular graph with a set of lines linking certain pairs of nuclei in the charge distribution.

No doubt is left that reduction has been achieved. All assumptions and approximations are kept in Pandora's box. Another chemist might comment that Bader's approach uses intuitive ideas about molecules to introduce the concept of chemical bond into quantum chemistry, that his approach has little to say about energy and stability and their relevance for the existence of chemical substances, and that the

---

[366] Quotations in this paragraph are from Bader (1990:249,252,248). Most of them can already be found in Bader, Nguyen-Dang and Tal (1981).
[367] Citations in Block 5-13 are taken from Bader (1990:2,5) and Bader, Popelier and Keith (1994:622f); cf. Bader, Nguyen-Dang and Tal (1981:898)
[368] Mainzer gives only one specific reference to Bader (1990) a few paragraphs further, but it is clear that he is paraphrasing Bader's views; cf. Bader (1990:v).

three-dimensional aspect of molecules is virtually absent (Arteca and Mezey 1989).[369] But such criticism will be no more than an 'in-house' technical discussion and would never undermine the reductionistic programme.

---

[369] Cf. also Ogilvie (1994:182-3) and references given there, indicating that increased quality of the basis set may decrease localisation (of the electron density).

6 ceteris paribus

6.1 chemical laws and models

At the moment there is virtually consensus among philosophers of biology that biology is a science and that it has no laws in the sense of 'laws in physics'.[370] Here a law is, provisionally, understood as an exceptionless ('strict') counterfactual, supporting generalisation. There is considerable disagreement in what sense biology is different from physics, but consensus does exist that the difference is big. Typically in this discussion, chemistry (if mentioned) is lumped with physics,[371] though in the 'old days' it was lumped with biology and the rest of science (as sciences without 'proper' laws). Hence, it is not surprising that opinions have differed on how many chemical laws there are. Answers include: no laws, two laws, three, quite a few, very many.

If the number of laws is two these are the laws of definite proportions (ascribed to Proust and Richter) and of stoichiometry (of multiple proportions – Dalton's law), often associated with the development of the atomic theory in the early nineteenth century, but as Cassirer (1922:205) noted:[372]

> It is interesting that this law [of definite proportion] is at first conceived entirely independently of any conception of the constitution of matter, and in particular independently of the atomic hypothesis.

Christie (1994) shows that the laws of definite and multiple proportions have characters, which cannot be reconciled with philosophers' accounts of laws of nature. They are non-universal, and at least one of them is imprecise.[373] According to the first law any pure chemical compound is made up of its constituent elements in definite and invariant proportions (by mass). The law seems to draw a sharp distinction between compounds and solutions, but there is no such sharp boundary. In particular the law only *approximately* applies to network solids (crystals and polymers). The second law says that if two 'elementary particles' A and B unite to form more than one compound, then the invariant amounts of B which unite with the same mass of A, are in the ratio of two simple whole numbers. As Duhem said, the law is a truism because *any* ratio can be expressed as the ratio of whole numbers to within any desired precision - think of applying the law to hydrocarbons.[374]

---

[370] See for example Beatty (1997), Brandon (1997), and Sober (1997).
[371] For example, Rosenberg (1994) and McIntyre's (1997).
[372] The same point was made by Hegel concerning the law of multiple proportions (cf. note 43).
[373] Cf. the views of Caldin (1959) and MacDonald (1960) discussed in [1.4].
[374] No chemical analysis, no matter how refined, will ever be able to show the law of multiple proportions to be wrong (Duhem 1974:213, 1902:159-61). Cf. Polanyi (1958:40-43).

Often Mendeleev's law of periodicity is mentioned as the 'third' law of chemistry. It says that the properties of the elements are a periodic function of their atom numbers. There exists a periodicity in the properties of the elements governed by certain intervals within their sequence arranged according to their atomic numbers. Scerri and McIntyre (1997:223) say that it 'stands as an autonomous law of chemistry' and cannot be reduced, as Hettema and Kuipers (1988) argued, to a purely physical 'atomic theory'. If the elements within a group shared the same outer shell configuration, this could be argued, but this is neither necessary nor sufficient for inclusion of an element in any particular group of the table.[375] But exactly because of the reasons that count against it being reducible, it could be said that it is not a strict law. One might also wonder whether the order in the periodic table still holds if we look at elements under extreme conditions, for example at reaction conditions in vacuum at extremely high or low temperatures. The law expresses an appropriate trend among the properties of the elements and their compounds. If this law is nothing but a trend, perhaps it wouldn't be a law at all. This has been a subject of discussion since Mendeleev first proposed it. In Mendeleev's Table the atomic weights of tellurium and iodine are in reverse order to that indicated by their chemical properties. Mendeleev dismissed the anomaly assuming there must be something wrong with the determinations of the atomic weight of iodine. But Ostwald concluded from the same discrepancy that the periodic law was not a proper law, merely an imperfect empirical generalisation (van der Vet 1979b). Similar disputes arise when it's suggested that the periodic table is a theory.[376]

Other and perhaps better candidates for strict laws can be found in physical chemistry; for example (Duhem 1900:20):

> Under a given pressure and at a temperature below the transformation point relative to this pressure, only exothermic reactions can be produced in the system; at a temperature above the transformation point, only endothermic reactions can be produced.

Duhem also mentions the laws that govern conditions of equilibrium, either between different phases or concerning the equilibrium of chemical reactions.[377] All these laws have to take account of the dependence on temperature and pressure.[378]

As even the best examples of chemical laws seen to have exceptions, one of the suggestions has been that there are no laws in chemistry. Against this it has been argued that there are two kinds of laws in chemistry: first, statements that there are definite kinds of material; and second, statements concerning functional relations which express the properties of these substances.[379] The distinction runs

---

[375] See the discussion of Hundt's rule on p. 119.
[376] For example, Bogaard (1978:355n8) dismisses the suggestion that the periodic table is a theory because at best it is a universal descriptive.
[377] Duhem ascribes the laws mentioned to van't Hoff and Moutier, respectively. Cassirer (1922) mentions as major chemical laws Gibbs' phase rule, the law of definite proportions, and that of multiple proportions.
[378] For example: 'Under each pressure, equilibrium is produced at a certain temperature, being the transformation point relative to this pressure' (Duhem 1900:20).
[379] Caldin (1959), cf. Paneth (1962).

more or less parallel with the account of Duhem (1900). According to Laitko (1967) the notion of 'generality' in connection with laws should not be understood in terms of 'applicable to all objects', but as 'universal' even if it applies to a limited number of objects. Hence chemistry has millions of laws: each reaction equation provides a law and each chemical reaction proliferates into many laws, because of its sensitivity to temperature, pressure, catalysts, and other boundary conditions (which may interfere in crucial ways).[380] But others have argued that this choice trivialises the notion of law. Scerri (1999a) suggests that the periodic table might be the best example of a chemical law, because it says something about all elements.

If one applies 'strict' criteria, there are no chemical laws. That much is obvious. But are there *any* strict laws? The standard assumption of course is that there *are* strict laws in physics. But that assumption might well be mistaken. Pauling (1950:111) said:

> A physical law is a succinct description of the results of a number of experiments. It is not an inflexible, unchanging dogma. It describes only the experiments that have been carried out up to the time the law is stated. These basic laws of nature may, as a result of some new experiment not be exactly right next year.

In the traditional framework laws express comprehensive, strict, exceptionless connections governing the whole of nature. Even if such laws could be defended at the most fundamental level of physics, the great majority would not qualify. Hence there have been other proposals according to which a law of nature is a contingent regularity contained in a theory displaying maximum information content and simplicity. In one influential view the vague and epistemic notion of 'simplicity' is replaced by the more metaphysical notion of 'unification' (Kitcher 1989): the theory in which the law occurs should combine maximum empirical content with maximum degree of unification. In this case the fact that the law is ceteris paribus by adducing provisos that 'hedge' its regularity would be acceptable as long as it fulfilled an explanatory function meeting the virtues of maximum information content and maximum degree of unification. Hopefully the ceteris paribus clauses would eventually be integrated into a more comprehensive theory. If the notion of law is 'softened' in this way, there is no need anymore to argue 'in defence of psychological laws' (Carrier 1998), or chemical laws for that matter. It doesn't however substantially resolve the conflict of physics and chemistry in their priority dispute about matter versus different substances, because unification will always be bought at the cost of practically relevant content.

I have little to add to the recent flourishing of discussion on the use of models in chemistry. But as the terms will be used often in the rest of this chapter and the next, I'll make a few brief remarks, against the background of two large text blocks. In Block 6-1, I've listed a few contextualised occurrences of the term 'model' in chemical discourse. The ubiquity and vagueness of the term 'model' is

---

[380] Alternatively, the selection rules are laws. Given millions of colliding molecules selection rules determine (within limits) which complexes survive.

Block 6-1   Quotations illustrating the variety in the use of the term 'model' in chemistry.[382]

> The Lewis model of the electron pair is second only to the molecular structure hypothesis itself in providing a conceptual basis for much of present-day chemical thinking.
> The fundamental chemical process, bond making, requires for its description the most abstract of models, quantum mechanics, and its translation into the structural language (the universal model) of practising chemists.
> What we do explain with the model Hamiltonian, we could explain with the exact treatments.
> The use of model Hamiltonians (such as the hydrogen atom) depend on the prior specification of the molecular backbone, whose quantised motions give rise to the spectroscopic behaviour.
> Dendrimers can be used as synthetic models for enzymes and globular proteins.
> Models base on atoms have shown significantly weak points.
> In physical chemistry we find a pragmatic pluralism of models. For example, different acid-base concepts are not more or less right, but more or less convenient for particular situations.
> Brønsted-Lowry and Lewis acids illustrate the close connection in the philosophy of science between 'models' and 'natural kinds'.
> To write equations (models) for the specific heat for crystalline and amorphous solids, abstractions are made at as many as six levels.
> It is not true that in 1970 many scientists believed in the reality of the polywater model.
> Molecular modelling is a constitutive element of the practice of chemistry, but the concept of molecular structure seems to derive its meaning more from the way molecules are represented in models than from anything else.
> None of the existing models for liquid transport in porous media can explain the different 'types' of capillary rise of water in sand.
> Many models and analogies in scaling-up and scaling-down procedures in chemical engineering employ the method of dimensional analysis.

obvious. Though it would be possible to map these particular uses on proposed definitions of various types of models (Block 6-2),[381] still any order thus brought about will be artificial to a greater or lesser extent, because the type of models are themselves ideal types (= models) introduced by philosophers of science, whereas the actual use of the term 'model' is vague and isn't even clearly delineated from terms like 'law' or 'theory'. For example, Mendeleev's periodic system has been called a taxonomy, a law, a model, and a theory.

Below I will use the word 'model' innocently and at the same time as multifacetted (Block 6-2), except for one restriction. Broadly speaking my use is the 'old' notion of model in the philosophy of science, of the sort discussed in Hesse

---

[381] Block 6-2 is based on Achinstein (1968), Aronson, Harré and Way (1994), Black (1962), Day (1990), Hadorn (1997), Hesse (1966), Hüttemann (1997), Redhead (1980), van Fraassen (1989), Cartwright (1983), Lakoff (1987), Burger (1976) and other sources.
[382] Publications on chemical models include: Bhushan and Rosenfeld (1995), Francoeur (1997), Hartmann (1965), Hofmann (1990), Hoffmann, Minkin and Carpenter (1996), Mainzer (1998), Mulckhuyse (1961), Ramsey (1994), Röhler (1962), Scerri (1991a), Trindle (1984), Zhdanov (1963); and more recently: Del Re (2000), Francoeur (2000), Hendry (1998), Laszlo (2000), Ramberg (2000), Tomasi (1999), Trindle (1999), Zeidler (2000).

(1966) and exemplified by the imagined dialogue she presents between the 'realist' Campbell and the 'instrumentalist' Duhem. This notion is distinct from the formal notion of model that has become influential in the so called semantic approach in the philosophy of science (van Fraassen 1980:64):

> To present a theory is to specify a family of structures, its *models*; and secondly, to specify certain parts of those models (the *empirical substructures*) as candidates for the direct representation of observable phenomena. The theory is empirically adequate if it has some model such that all the appearances are isomorphic to empirical substructures of that model.

The weakness of the model-theoretic view of theories is the requirement that the observable substructure prescribed by the theory should match the structure of reality. But it says little or nothing about *how* these models are hooked up to the empirical world. Empirical substructures are abstract objects, 'abstracted' from observations or 'data models'. There is only talk of isomorphy between abstract objects, not between 'real' systems and empirical substructures. If we aim for empirical adequacy only, we can write better phenomenological laws than those a theory can produce. Model theory can give meanings to expressions of a language, but only given a metalanguage which is taken as the unquestioned background. That is to say: the semantics of an observation language is taken as given without even bothering to describe it in any detail.[383]

6.2 ceteris paribus laws

That there are no strict laws in the special sciences is a point often made - the special sciences including all sciences *except* physics (Fodor 1974). That there are no strict laws at all is perhaps less plausible, but it would be wrong to say that the conception of laws as exceptionless is 'virtually unchallenged'.[384] Here it will be suggested that *all* laws are ceteris paribus.[385] The boiling point of water is 100 degrees centigrade, but not if it is sea water, not if it is boiled on top of the Mount Everest or contained in small capillaries, or if heavy water (deuterium oxide) is used, *and so on*. And as we have seen in [3.3] the law that water is H_2O is equally imprecise and ceteris paribus. There is no way that one could write out the ceteris paribus or proviso conditions in full.[386] There exists a strong intuition that ceteris paribus clauses work differently when interpreting physical data as distinct from interpreting chemical (or psychological) data, but this intuition is rarely argued for.

---

[383] See Downes (1992), Peregrin (1997) and references given there.
[384] As for example, Horgan and Tienson (1990) do.
[385] For various types of argument in connection with the 'strictness' of physical laws see Cartwright (1983, 1994, 1995, 1997, 1999), Crane and Mellor (1990), Dupré (1993:159-67), Earman and Roberts (1999), Klee (1992), Hempel (1988), Lange (1993), Morreau (1999), Pietroski and Rey (1995), Putnam (1987:11,37), Scheibe (1991), van Fraassen (1989, 1992), Van Gulick (1992).
[386] The term 'proviso' stems from Hempel (1988). For discussion see Giere (1988), Mundy (1990), Lange (1993). Related terms include: ceteris absentibus, ceteris presentibus, hedged laws, other things being equal, other things being right. The issue of ceteris paribus or proviso conditions is related to that of the frame problem in artificial intelligence and the problem of complete description (van Brakel 1992b).

Block 6-2   Model types; 'S' and 'B' stand for 'object, system or process'.

scale model	a material S designed to reproduce as faithfully as possible the structure and appearance (or web of relationships) of a real or imaginary B (also: analogue model)
iconic/image model	representation of a system of purported entities in terms of another system of entities
simulation model	S simulates the behaviour of some B; S reproduces the input-output behaviour of B
simulacrum	model as a theory of a phenomenon; some properties will be 'genuine' properties of the objects modelled, but others will be merely properties of convenience
mathematical model	a) mapping which correlates the formulae of one system with those of another; b) using mathematical analysis for dealing with any number of variables to characterise phenomena in any domain; c) formal analogy between a physical theory and some mathematical structure
statistical model	fitting equations to data
model (logic)	interpretation or 'realisation' of a formal axiom system (turning the abstract calculus into meaningful sentences)
model (model theory)	mathematical structure called model of a given theory if the theory is entirely true with respect to this model alone
theoretical model	a theoretical (formal or linguistic) description of B in terms of the (rather well known) structure of S, but not 'literally' constructing a (material, conceptual, formal) model - ranging from heuristic fiction to ontological commitment
structural model	set of assumptions attributing an inner structure, composition or mechanism which manifests itself in other properties, proposed in the framework of some more basic theory
floating model	theoretical model disconnected from a fundamental theory
tinker toy model	mathematical simplification of a theory (also: impoverishment model)
hierarchical model	representation of natural kinds and relations between them; the type-hierarchy generates salience and similarity
ideal type	ideal forms of social phenomena (such as bureaucracy or capitalism) - there are no instantiations of the ideal type in actual social reality, but they are empirically possible
cultural model	cultural knowledge organised in sequences of prototypical events hierarchically related to other cultural models - also: mental model, cognitive model, schema, or script
cognitive model	a gestalt (complex structured whole) using propositional and image-schematic structuring and metaphoric and metonymic mappings
conceptual archetype	systematic repertoire of ideas by means of which a thinker describes, by analogical extension, some domain to which those ideas do not immediately apply (also: root metaphor)

Of course, physics provides models in which strict laws apply to closed systems, but in the real world there are no closed systems and in applying models to concrete situations, many additional assumptions have to be made with holistic indeterminacies of many sorts creeping in; as much in the application of quantum mechanics to molecular chemistry, as when economic models are applied to real people. Moreover, applying physical (and other laws) to concrete cases requires an appeal to initial and boundary conditions. As fixing these requires measurement, prediction (or retrodiction) is never strict, even before we enter the realm of quantum mechanics or quantum field theory. Then there is the problem that initial and boundary conditions are not freely choosable (Sklar 1990, Wilson 1990). More generally, any application of theory requires that the location is specified in pretheoretical terms (van Fraassen 1992). Further, in cosmology one assumes that the physical laws, at least the most fundamental ones, are not dependent on the global structure of the universe (Balashov 1994). But it is by no means self-evident that this is warranted. If one appeals to physical laws as intrinsic properties of the universe, there is no a priori reason why these properties would not vary during its evolution.

Cartwright has argued at length and given many examples illustrating how all theoretical laws of physics, including the most fundamental ones are ceteris paribus (Block 6-3). The strictness only applies when referring to models. Models deliberately construct falsifications of 'reality' so that the theory can deal with them. Moreover, there is more than one way to model a given situation, which may yield incompatible theoretical treatments. There are bits of the world that scientific models fit, in particular bits of the world inside laboratories. So the 'laws of nature' are not empty; they can be true, though nothing is said about their scope. This scope is not universal, because such 'model laws' only apply under very special (simplified) laboratory situations or 'thought experiments'. There is no fact of the matter about what a system can do (Cartwright 1997:78):

> What it does depends on its setting, and the kinds of settings necessary for it to produce systematic and predictable results are exceptional.

Neither down-ward, nor cross-wise reduction is needed: 'nature is governed in different domains by different systems of laws not necessarily related to each other in any systematic or uniform way: by a patchwork of laws' (Cartwright 1994:288f).

According to Cartwright we have ceteris paribus laws all the way down. What basic science should aim for is not to discover laws but to find out what stable capacities or natures are associated with features and structures in their domains, and how these capacities behave in complex settings. What happens in the world is the outcome of a range of (partly unknown) capacities that act in conjunction in complex ways, allowing at best probabilistic predictions. A law or theory will describe one or a few of these capacities only. The basic metaphysical intuition of Cartwright is that the real world is complex or messy whereas strict laws are (and should be) simple. In a terminology that is reminiscent of Feyerabend she says:[387]

---

[387] Cartwright (1999:12). And similarly for her motivation that (ibid:18): 'the great challenge that now faces philosophy of science [is] to develop methodologies, not for life in the laboratory where

Block 6-3   Cartwright on the ceteris paribus character of theoretical laws.[388]

> Only phenomenological laws can be tested by comparison with observations. Such laws do not explain anything; they *describe*. Examples of such laws are the law of Boyle-Gay-Lussac for ideal gases, Fick's equation for diffusion, or the law for radioactive decay. Because such laws cannot be derived from more comprehensive 'explanatory' theories (without adding assumptions that are not part of the comprehensive theory), the latter 'theoretical' theories and laws are only useful in the sense that they organise and classify knowledge in an elegant and efficient way, but that is all there is to it - no scientific realist conclusions are allowed.
>
> The central equations of fundamental physical theories such as quantum mechanics are stated in such abstract terms that they say very little about the specific situations they should apply to. Any application requires the addition of boundary conditions which uses knowledge that is not part of the theory. Fundamental laws do not govern objects in reality; they govern only objects in models. We build our devices to fit our well-understood models, for then we will know what to expect of the devices. Predictive closure only obtains in highly restricted circumstances,
>
> Any theory abstracts from 'ordinary' contexts; hence it cannot be *true* for real (actual) systems. The more comprehensive an empirical theory is, the less it can be true (metaphorically speaking) for concrete everyday situations or systems. This even applies to laws such as Newton's $F = ma$ ('the total force acting on an object is equal to the product of its mass and its acceleration'). To apply this law to a mechanical system a model is needed that gives an idealised description of the system. The 'F' is a theoretical term which is only partly interpreted by its place in the calculus of Newtonian mechanics. It is only via models that it can be connected to concrete systems.

> My investigations into how basic science works when it makes essential contributions to predicting and rebuilding the world suggest that even our best theories are severely limited in their scope: they apply only in situations that resemble their models, and in just the right way, where what constitutes a model is delineated by the theory itself.
>
> Physics in its various branches works in pockets, primarily inside walls: the walls of a laboratory or the casing of a common battery or deep in a large thermos, walls within which the conditions can be arranged *just so*, to fit the well-confirmed and well-established models of the theory, that is, the models that have proved to be dependable and can be relied on to stay that way.

Even the inference to the universal generality of quantum mechanics does not work. Its successes are impressive, but they merely refer to *carefully controlled* model situations (as in the Stern-Gerlach experiment). Further, many experiments that are of central relevance to quantum mechanics are understood in terms of calculations which draw on theories from classical physics. More fundamentally: no experiments on quantum systems are possible without using instruments that are

---

conditions can be set as one likes, but methodologies for life in the messy world that we inevitably inhabit.'
[388] Cartwright (1983, 1994, 1995, 1997, 1999).

described in classical terms, whereas quantum mechanics and classical physics cannot be united. Hence, as Hendry (1999:128) puts it [5.5]:

> It is one thing to have a quantum-mechanical account of the spin states of a silver atom, quite another to have a quantum-mechanical account of the whole Stern-Gerlach apparatus: the universality of quantum mechanics demands the latter.

Quantum mechanics is ceteris paribus in the sense that it presents a universal dynamic equation that applies in principle to every physical system, but this is only so *provided* no observation causes state changes between different times. Moreover there is no physical description of isolation that prohibits interaction. The notion of a sufficiently 'isolated' system is essentially contextual. Quantum physics works only in very specific kinds of situations that fit the very restricted set of models it can provide; it can be said to be true, but not universal.

In denying the existence of strict laws I don't mean to favour replacing them with capacities as the true necessary structurings of the world, as Cartwright does. I sympathise with Cartwright's critics when they argue that capacities too fail to escape the problem of context dependency (Morrison 1995) - though it is not altogether clear that Cartwright insists either on context independence or universality of capacities, as Morrison says. Cartwright (1995:180; cf. 1997) replies to Morris that capacities only operate if they are appropriately triggered, if they are within their domain or regime of application, and if they are not prevented by a physical interaction. Hence, she would seem to agree that capacities too are context dependent and this is consistent with her view that we live in a dappled world. Neither a law, nor a capacity is universal in the sense that it 'holds everywhere and governs all domains' (Cartwright 1994:281). But that means that all talk of capacities is itself a form of modelling; it doesn't warrant a metaphysically realist reading. If the capacities are hidden powers that only appear when *appropriately* triggered, it is not clear what they add to saying that laws are ceteris paribus. If a law tells us only what happens if all interfering factors are absent (i.e. never), how then can it be useful in situations where these factors are present? At this point it may be psychologically helpful to invoke capacities or natures, but it doesn't really change the epistemological condition. This doesn't replace realism by relativism; it merely restricts realism to situated cases. As Cartwright says: '*When* you can spray them, they exist.'[389]

Moreover, as I will show in the case studies that follow, laws and capacities are not only ceteris paribus all the way down, but all the way up as well. The dichotomy Cartwright seems to uphold between fundamental and phenomenological laws is not warranted. Also phenomenological laws are ceteris paribus. If there is *always* a ceteris paribus condition connected to a law, than laws do not refer to the empirical world, but to models of the world, in which, by stipulation, *every* ceteris paribus condition has been eliminated. It would be more appropriate to say that those statements called 'natural laws' are more like normative laws or norms.

---

[389] A more cautious variant of Hacking's (1983:23) famous 'If you can spray them, then they are real.

> Block 6-4   Protoscience.[391]
>
> All experimental science has a technical or instrumental basis that should not be identified with 'measuring experience', but with setting technical goals and their successful realisation. Against that background protoscience can be considered to be the study of the *normative* criteria for the use of experimental technology in science. It distinguishes itself from 'received' philosophy of science by focusing on the praxis in which new concepts and laws are not 'discovered', but 'constructed'. That is to say the commitment is explicitly *against* scientific realism. The methodological constructivism of 'protoscience' is a blend of instrumentalism and pragmatic realism. The historically grown scientific language is investigated both at the object level (vocabulary concerning operations, tools, instruments and technical terms) and the meta-level (the discourse *about* the object level). The aim of science is seen *not* as the description of nature, but as the theoretical-instrumental support of 'poietic practices', i.e. practices that aim at the production of material goods.
>
> There are similarities between protoscience and the operationalism of Bridgman (1927) and Dingle (1949a), the making-worlds view of Goodman (1978) and 'articulation' in the sense of Latour (1999), but it is different from all of them in starting explicitly from the daily life world - the *lebensweltliche Basis* (cf. Block 1-2) and poietic practices.

From a different background this view is also reached by applying the approach known as protoscience to chemistry.

6.3   protochemistry

Protochemistry aims at a reconstruction of chemical scientific knowledge.[390] It is a theory that is methodologically prior to chemistry and it makes a distinction between 'ordinary' chemical language and 'reflective' chemical terms which are part of a meta-language. The prime example of such a reflective term is 'stuff' (*hyle*, chemical substance). Only *after* the chemist has distinguished 'chemical stuff-properties' from other properties, does it make sense to introduce talk of atoms and molecules (Janich 1994a). When the protoscience approach (Block 6-4) is applied to chemistry, the difference with other sciences shows at once: the characteristics of the proto-concepts of chemistry are involved in practices such as cooking, brewing, dye-making, metallurgy, making medicinal stuffs, and so forth: colours, tastes, hardness and viscosity, melting and boiling, homogeneity and purity, toxicity, inflammability, and similar properties of materials. First reflection draws attention to the peculiarities of the notions 'stuff' or 'substance' - against the background of the practices just indicated, the 'stuff producing and processing practices'. This involves simple observations such as: if two arbitrary cut parts of a thing display the same 'essential' material properties then it is 'essentially' uniform.

On this view the most basic laws are those that state the existence of particular pure stuffs and their properties, that is the reproducible identification and

---

[390] Publications on protochemistry include: Gutman and Hanekamp (1996), Hanekamp (1997, 1998), Janich (1992, 1994a, 1994b, 1994c), Psarros (1994, 1995a, 1995b, 1998a, 1998b, 1999).
[391] See Lorenzen (1994) and Janich (1994a).

synthesis of substances and the reproducible measurement of their properties. These basic laws of protochemistry correspond to Cartwright's phenomenological laws or laws governing macroscopic phenomena where the manifest and scientific image are still intertwined. All further theorising remains dependent on the validity of these basic laws, which are tied to what it is technically possible to make.

More theoretical chemical laws, such as 'the big three' mentioned in [6.1], should be understood more as norms, not as 'natural laws'. For example, the 'laws' of constant and multiple proportions contain the undefined terms 'compound' and 'part' or 'element'. If one fills in those gaps one obtains a mix of macroscopic laws, operational definitions, and norms, which embed the basic laws into a theoretical model that is normatively applied to the design of experiments. For example, the law of definite proportions can be 'unpacked' as a mix of the following rules or norms (Psarros 1994):

- The sum of the weights of the compounds that enter into a (chemical) transformation is equal to that of the products that are formed.
- It must be possible to isolate the products of a transformation as pure chemical substances.
- A transformation that fulfils the previous two criteria is called a chemical reaction.
- Chemical reactions have to be carried out in such a way that its products are chemically pure substances with constant composition.

That scientific laws are normative in the sense of regulating scientific practice can also be illustrated with the phenomenon of 'closed theories' first noted by Schrödinger. An example would be classical hydrodynamics (F. Böhme 1980). The structure of this theory is determined by the empirical laws that characterise Newtonian fluids.[392] The theory sets the 'norm' for what is to be a Newtonian fluid. This normative truth will always apply to the structural relation between theory and empirical law, though this says nothing about whether there are actually any fluids that behave like Newtonian fluids.

The view of the advocates of protoscience that physical and chemical laws should be understood as norms is more extreme, but not very different from the view that all laws are ceteris paribus. Protochemistry is easily misunderstood as 'a form of logical positivism' or dismissed as 'abstractions endemic to systematic philosophy', which is 'distressingly remote from the practice of chemistry' (Ramsey 1998:413). And the view that all laws are ceteris paribus has met with equal incredulity. That the laws of physics are strict is part of the Myth of the Given (Block 6-5). For example, after having argued that ceteris paribus clauses are 'methodologically active in all domains' and being forced to the conclusion that 'the universe is an anomalous one', Klee (1992:402) continues:

> I trust that such a conclusion is manifestly false. There are perfectly good laws of nature in the physical universe.

---

[392] For which the parameter a in eq. (7.12) is equal to 1.

Block 6-5 Myth of the Given.[394]

> There is an ontologically given, categorically ready-made real world:
> - the world ultimately consists of certain kinds of entities, of the different properties of these entities, and of relations between these entities;
> - the world is independent of us and sliced into 'ready-made' entities and types of entities;
> - the changes in the world are governed exclusively by a closed set of strict laws.
>
> There is an irreplaceable, a priori privileged language for describing the world:
> - there is a transcendental logical and a priori privileged ideal scientific language;
> - there is a necessary or logical or ostensive connection between this language and the world;
> - the meaningfulness of all factual terms is based on causal interaction of a certain kind between the language user and the extra-linguistic world.
>
> Man can be engaged in non-conceptual but yet cognitive epistemic commerce with the world.

But he gives no reasons for this belief, nor any examples of such 'perfectly good laws of nature'. And Earman and Roberts (1999) in their dismissal of ceteris paribus talk despair:[393]

> We do not know how to begin to assess Cartwright's claim about context-specific factors that in principle elude theoretical treatment.

For them a 'ceteris paribus law' is an element of 'work in progress, an embryonic theory on its way to being developed to the point where it makes definite claims about the world'. However, it is instructive to see how their conviction that there are strict laws in physics works at the microlevel of their argumentation. When discussing the ceteris paribus character of thermodynamics, they say [5.2]:

> There are unresolved technical and conceptual problems in the reduction of thermodynamics to statistical mechanics (see Sklar 1993), but these problems do not affect the present discussion.

But at the end of that discussion they remind the reader:

> Note the important role played here by the availability of a microreduction: the reduction of thermodynamics to statistical mechanics is what makes it possible to give a clear sense to the crucial phrase, 'most of the intended applications'.

If there are always ceteris paribus conditions connected to theoretical laws, then theoretical laws do not refer to the empirical world, but to models of the world, in which, by stipulation, those conditions have been eliminated. What we have is phenomenological equations that normatively model phenomena vaguely described in common sense terms. These equations are more or less empirically adequate in the sense that they are more or less useful in guiding future actions

---

[393] The following quotations from Earman and Roberts (1999) are from p. 456, 464, 466, and 476n40.
[394] Adapted from Tuomela (1978).

that interest us. These laws are models of phenomena that are already severely regimented. More models at the theoretical level are added and the confirming experimental measurements are carried out in more and more artificial environments. There are *no* fundamental equations or theories that are *strictly* true of real world manifest situations. At best there are 'fundamental' theories that are true of their *conceptual* models, which give extremely accurate descriptions and predictions of the manifest properties of material realisations of its conceptual models. This explains the pull of an inference to the best explanation: if the model works *so well*, it would be a miracle if the model doesn't tell us how reality *is*. But this suggestion blurs the distinction between the impressiveness of some of these 'fundamental' theories - which is not under dispute - and the suggestion that they tell us how reality is – all of it.

I will now illustrate this intricate connection of 'basic laws', models, and the ubiquity of ceteris paribus conditions, using two detailed case studies; one concerning an anomaly in capillary liquid transport in porous media in the next section and one on the use of dimensionless numbers modelling heat, mass, and momentum transfer in chemical engineering in the next chapter. In both cases I will go into considerable scientific detail, as I did in the polywater case. Only thus is it possible to show the ubiquity of ceteris paribus assumptions and the presence of models at all levels of description. By taking case studies from more applied sciences, in this case chemical engineering science, it is possible to show how these ceteris paribus conditions and models are found in all relevant discourses: the manifest and the scientific, the micro and the macro, at the level of understanding and prediction, at the level of knowledge and the level of being able to do or make something.

6.4 the ubiquity of models and ceteris paribus conditions

The ubiquity of models and ceteris paribus conditions in chemistry and other sciences is easily overlooked if one only learns of the success stories that reach text book status - after Pandora's box is closed (Latour 1987, 1999). To appreciate where the modelling really goes on, one has to look for examples where the answers are not yet entrenched. I'll therefore consider a more open-ended example in some detail: the simple phenomenon of a wetting liquid advancing in a packed bed of rotund particles, which can be seen as a model of transport phenomena in porous media in which capillary forces are important. Knowledge of such transport phenomena in porous media is important for many applications in chemical engineering, as well as agricultural engineering, mining engineering, geo-hydrology, and other disciplines. In chemical engineering major applications include drying of porous or particulate materials and multi-phase transport phenomena in packed and fluidised bed reactors.

In Block 6-6 some equations and dimensionless numbers are listed that have been used in describing capillary liquid transport in porous media.[395] These equations say something about the velocity of the advancing liquid front, q or dx/dt, as

---

[395] See van Brakel (1975a) and Bear (1972) for details.

a function of the driving force, ψ, and the 'resistance' to flow, K, taking into account the geometrical properties of the porous medium. Note that some of these equations apply to capillary liquid transport in a cylindrical capillary; others to porous media in general. In its simplest form the generic equation (6.1) was first proposed in 1856 by Darcy; in 1907 Buckingham proposed a variant that can deal with capillary liquid transport.[396] Taken as the Darcy equation ψ is the pressure head and $x\cos\phi$ the elevation head. For capillary liquid transport K and ψ are dependent on u, the liquid content of the porous medium, and ψ is called the capillary potential; the latter is usually defined as a negative quantity. Hence mass transfer takes place from the lower capillary potential towards the higher potential.

In applied science and engineering the Washburn equation, eq. (6.9), is often used in the empirical form $x^2=ct$, with an empirically determined 'constant', c, neglecting the gravity term.[397] In high speed printing (very fast wetting), the Kozeny equation, eq. (6.4) is used, because the first two terms in eq. (6.4) cannot be neglected.[398] In some domains, when the Washburn 'doesn't work', alternative, purely empirical, correlations between x and t have been used.[399]

Usually the porespace in a porous medium is visualised as a more or less complicated assembly of isolated or interconnected capillaries. Such porespace models, used in describing transport phenomena in porous media, acknowledge the phenomenon of hysteresis. This includes the rain-drop effect and ink-bottle effect; i.e. the hysteresis between an advancing and receding contact angle (the angle which the liquid-vapour interface makes with the solid surface, θ) and the hysteresis effect of constrictions in the pore space. Still if capillary rise of a (wetting) liquid in a homogeneous irregular packing of rotund particles is considered, it appears that none of the existing porespace models is capable of explaining the observed phenomena.[400] For example, the capillary rise of glycol and of n-heptane, in the 'same' homogeneous packing of polystyrene spheres (same diameter of spheres, same porosity of packing), show strikingly different behaviours. When equilibrium has been reached, it can be observed that for glycol there is a sharp front between the completely saturated part of the porous medium and the dry part. In the case of heptane a wide saturation gradient is observed in the equilibrium situation.

---

[396] The Washburn equation was proposed in 1921; the Kozeny equation in 1927. Since these early days theoretically not much has changed. Consulting recent literature might even lead one to suggest that numerical methods have pushed aside the concerns of the early researchers.
[397] This equation is an integration of eq. (6.2).
[398] If these terms can be neglected eq. (6.4) reduces to eq. (6.3).
[399] For example, Schicketanz (1974) proposed $x=ct$, supporting it with reference to the Kozeny equation. But the data Schicketanz presents perfectly fit eq. (6.3) (van Brakel 1975b). No doubt in making this error, it plays a role that it is more cumbersome to calculate dx/dt and plot it against $1/x$, than to work with double logarithmic paper on which h and t are plotted directly. Using $x^2=ct$ or $x=ct$ for wetting phenomena is an extremely crude case of dimensional analysis [7.2], using the 'system constant' c as curve fitting parameter.
[400] See van Brakel (1975a). There has been no change in the models used in the past 25 years that addresses these problems, as recent articles in the *Journal of Colloid and Interface Science, Transport in Porous Media*, and *Water Resources Research* testify (Siebold et al 1997; Carmeliet, Descamps and Houvenaghel 1999; Marmur and Cohen 1997; Wang, Feyen and Elrick 1998).

CETERIS PARIBUS

Block 6-6 Equations used in describing capillary liquid transport in porous media. For symbols used see Block 6-7.

filtration equation (Darcy equation), capillary liquid transport (Buckingham equation):

$$q = \varepsilon \frac{dx}{dt} = -K \frac{d}{dx}(\psi + x\cos\phi) \tag{6.1}$$

wetting of porous medium in the absence of gravity:

$$\varepsilon \frac{dx}{dt} = \frac{K\psi_c}{x} \tag{6.2}$$

vertical capillary rise in porous medium:

$$\varepsilon \frac{dx}{dt} = \frac{K\psi_c}{x} - K \tag{6.3}$$

wetting of porous medium for $t \to 0$ (Kozeny equation):

$$\frac{d^2x}{dt^2} + \frac{1}{x}\left(\frac{dx}{dt}\right)^2 + \varepsilon \frac{g}{K} \frac{dx}{dt} - \frac{\psi_c g}{x} + g = 0 \tag{6.4}$$

modified Kozeny-Carman equation:

$$K = \frac{\rho g}{\eta} \frac{\delta}{\tau^2} \frac{\varepsilon^3}{(1-\varepsilon)^2 s^2} \tag{6.5}$$

pressure drop across curved surface (Laplace equation):

$$\psi_c = \frac{\sigma}{\rho g}\left(\frac{1}{\alpha} + \frac{1}{\gamma}\right) \tag{6.6}$$

maximum capillary rise in cylindrical capillary:

$$\psi_c = \frac{2\sigma\cos\theta}{\rho g d} \tag{6.7}$$

Hagen-Poiseuille equation:

$$q = \frac{\Delta P}{2\eta} \frac{d^2}{L} \tag{6.8}$$

capillary rise in cylindrical capillary (Washburn equation):

$$\frac{dx}{dt} = \frac{d^2}{2\eta} \frac{\psi_c - x\cos\phi}{x} \tag{6.9}$$

J-Leverett function (dimensionless):

$$\frac{\rho g \psi_c}{\sigma} \sqrt{\frac{K\eta}{\rho g \varepsilon}} \tag{6.10}$$

Kozeny function (dimensionless):

$$\frac{\Delta P}{\eta L} \frac{1}{qs^2} \frac{\varepsilon^3}{(1-\varepsilon)^2} \tag{6.11}$$

The saturation gradient for heptane is above the maximum capillary rise for glycol, suggesting that heptane wets polystyrene better than glycol (correcting for the difference in surface tension of glycol and heptane). In terms of the conventional porespace models the former suggests a model of equally sized cylindrical capillaries (predicting a sharp liquid front at equilibrium), whereas the latter result suggests a model of capillaries of varying size (predicting a saturation gradient at equilibrium). The two empirical observations therefore lead to ascribing inconsistent geometric pore space models to what is the 'same' homogeneous packing of polystyrene spheres.

The same anomaly is observed in the even more mundane experiment of capillary rise of water in a packing of sand particles. Confronted with such an anomaly, it would be an ostrich policy to observe, as Hofmann and Hofmann (1992) remark in a paper on the Darcy equation, that 'hydrologists in the field are forced to consider a multitude of models as simultaneously applicable to a given system'. One observes capillary rise in a porous medium (say water in sand), for the 'same' liquid (water) rising in the 'same' porous medium (same type of sand, same porosity, same average pore size) on different occasions. It turns out that there are vast differences in the velocity with which the liquid front advances and the extent to which it partially or fully saturates the pore space. Then it is not good enough to use post facto different models to describe the observed phenomena. None of the existing models for liquid transport in porous media can explain the variety of phenomena observed. Neither dimensional analysis, eqs. (6.10-6.11), nor the various empirical or semi-theoretical equations (Darcy, Washburn, Kozeny) are of any help. That some of these equations are derived from fundamental equations makes no difference, because the 'fundamental' equations are applied under unspecifiable boundary conditions. In a case like this, dimensional analysis is of even less use, because it may 'cover up' how the phenomena have been modelled in an inappropriate way. And it has the extra uncertainty that it can't do much with shape factors such as the tortuosity and constrictivity (Block 6-8), whereas the possibility of taking into account the pseudo-dimensionless contact angle θ is usually overlooked.

Why do all existing porespace models fail? The suggested 'phenomenal' explanation is that what is not the same (for unknown reasons like vvariations in surface treatment or contamination) is the contact angle. The proposed 'theoretical' explanation consists of two parts.[401] First, depending on the absolute value of the contact angle, the mechanism of capillary rise may be fundamentally different. Second, the contact angle may change *during* capillary rise because of 'film jumps'. The most fundamental mistake of all porespace models is the assumption of the presence of a large number of *isolated* menisci in the 'pores' of the porous medium. All models fail, because they all reduce the two variables that describe the curvature of the liquid-vapour surface to one 'average'. The liquid-vapour interface consists of *one* continuous meniscus, consisting of concave parts separated by anticlastic parts - anticlastic meaning that the two main radii of curvature of the liquid-vapour interface are of opposite sign – cf. eq. (6.6). For a packing of spherical particles the anticlastic parts or α-menisci lie between the adjacent parts of two spheres. The concave parts or γ-menisci are contained by three or more spheres.[402] The mechanism of capillary rise is governed by a continuous merging

---

[401] See van Brakel and Heertjes (1975, 1975/76, 1977, 1978).
[402] In 'α-menisci' and 'γ-menisci', 'menisci' is short for 'parts of the meniscus'.

Block 6-7   List of variables and symbols used in chapters 6 and 7.

dimensional variable	symbol	dimension
real numbers	a, b, c	[-]
heat capacity	C	[H/MT]
diameter (cf. Block 7-3)	d	[L]
molecular diffusion coefficient	D	[L^2/t]
gravitational acceleration	g	[L/t^2]
heat transfer coefficient	h	[H/tL^{2}T]
mass transfer coefficient	k	[L/t]
permeability of porous medium	K	[L/t]
length	L	[L]
mass	m	[M]
pressure difference	ΔP	[M/Lt^2]
overall liquid velocity	q	[L/t]
specific surface of particle or pore space	s	[L^2/L^3]
time	t	[t]
Cartesian coordinates	x, y, z	[L]
concentration	u	[M/M]
velocity	v	[L/t]
radii of curvature	α, γ	[L]
viscosity	η	[M/Lt]
porosity	ε	[L^3/L^3]
angle with vertical	ϕ	[L/L]
thermal conductivity	λ	[H/tLT]
capillary potential, pressure height	ψ	[L]
maximum value of ψ	ψ_c	[L]
dimensionless number variable	Π	[−]
cake resistance	Θ	[t/L]
density	ρ	[M/L^3]
surface tension	σ	[M/t^2]
angle of contact	θ	[L/L]
angular velocity	Ω	[t^{-1}]
shape factor	ξ	[L^3/L^3]

dimensions: [L] length; [t] time; [M] mass; [T] temperature; [H] heat.

and generation of α- and γ-menisci in the porespace, consisting of 'cavities' connected by 'windows'. Continuity of transport is sustained by two intersupporting mechanisms. Because the dependency of the curvature of α- and γ-menisci on θ, the contact angle, differs, at higher values of θ, the contribution of the γ-menisci becomes more significant, and thus fewer α-menisci are necessary to maintain continuity of transport. Therefore, above a certain value of θ, a saturation gradient will be observed.

One reason for overlooking that a meniscus has two radii of curvature, is that the mathematics of describing curved surfaces are extremely complex. For a long time the best that could be achieved was approximate calculations of the pendular 'saddle-like' surface of a liquid drop held by capillary forces around the contact point of two equal 'ideal' spheres. Calculations for menisci between three spheres are next to impossible. Even more impossible would be to introduce the full three dimensionality of the pore space, the fact that it is random, that the particles are non-spherical and the solid surface will not be smooth.

Block 6-8   Shape factors.

quantity	symbol	dimension	example
particle shape factor	ξ	L^3/L^3	Block 7-3
tortuosity of 'pores'	τ	L/L	eqs. (6.5, 6.13)
constrictivity of pipe or 'pores'	δ	?	eq. (6.13)
relative roughness (of solid surface)		L/L	Block 7-5
ratio of length and diameter of pipe		L/d	Block 7-5

The merging menisci model describes wetting phenomena in porous media better than any other model. But is it also a true picture of reality? In this particular case it might be possible in principle to observe the merging of α- and γ-menisci 'directly'. In principle it would be possible to 'observe' whether the mechanism of merging menisci is true, *provided* the technique of micro computer tomography is further developed and we can trust the standardisation methods underlying the transformation of two dimensional data to three dimensional pictures of interfaces. Then we would have shown that we can *make* instruments that would confirm that our model describes *exactly* what happens in the model.

Now that I have given an account (a model) explaining the apparent anomalies of capillary rise of various liquids in a packed bed of rotund particles, everything has its place again and text books can simply state that it was 'overlooked' for about a century that a liquid-vapour surface has, in general two radii of curvature, not one. But then I would have closed Pandora's box again. I wouldn't have told you about all kinds of other weird anomalies observed in capillary liquid transport that *cannot* be explained in terms of α- and γ-menisci [3.5]. And why am I so confident that no chemical reactions or physico-chemical interactions are taking place between the glycol (or heptane) and the polystyrene that may influence any of the three surface tensions involved (represented in the equations by the one σ, together with θ)?[403]

Or why shouldn't we invoke irreversible thermodynamics to tackle the anomalies? For example, in so called continuum theories of transport phenomena in porous media,[404] the Buckingham equation, eq. (6.1), is often rewritten as follows:

$$q = -D_u \frac{du}{dx} \qquad (6.12)$$

in which D_u, is called the moisture diffusivity. This is a bad idea, in particular if one is used to thinking of D as referring to molecular diffusion. Because empirical data show D strongly depends on the 'concentration' u, Luikov proposed to generalise eq. (6.12) to[405]

---

[403] Cf. the 'solution' of the polywater anomaly [3.5].
[404] In the continuum approach 'the actual multiphase porous medium is replaced by a fictitious continuum: a structureless substance, to any point of which we can assign kinematic and dynamic variables and parameters that are continuous functions of the spatial coordinates of the point and of the time' (Bear 1972:24). This theory is still being used; see for example Carmeliet, Descamps and Houvenaghel (1999:86-7).
[405] See van Brakel (1975a:212f) and references given there.

$$q = -D_u \frac{du}{dx} - \tau_{rm} \frac{dq}{dt} \tag{6.13}$$

in which τ_{rm} is called the relaxation time or period of propagation of moisture transport in capillary-porous media. Eq. (6.13) was proposed on the analogy with a similar relaxation time for heat transfer known from irreversible thermodynamics. However, this is nonsense and a case where Luikov's theory of similarity[406] is a clear example (to paraphrase Wittgenstein) of where formal approaches (such as irreversible thermodynamics) take off, scientific common sense goes on holiday. The driving force is not du/dx, but $d\psi/dx$. It is a case of hydrodynamic transport (driven by a capillary potential), not molecular diffusion. The only difference between eq. (6.12) and eq. (6.13) is that the latter contains two and the former one fudge factor. No wonder eq. (6.13) fits the data better.

What options are open to the reader of this text? No literature exists that gives as much as a provisional answer as to which of the stories to believe, except for the rhetorical suggestion of this author that the story about α- and γ-menisci is real stuff and invoking irreversible thermodynamics is humbug. Neither the phenomena, nor the phenomenal explanation, nor the theoretical explanation can be found in any text book. Some physical chemists will dismiss the whole story out of hand, suggesting that the alleged phenomena are an artefact of working with 'contaminated' materials. One day the capillary liquid transport anomaly I presented will be resolved (by accepting a new theory or model) or dissolved (for example dismissed as 'noise' due to contamination of the system), as happened in the polywater case. Pandora's box will again have been closed and contain all sorts of ceteris paribus conditions which to a sceptic will still be dubious. But the experiments will have been standardised in such a way that they fit the models and theories so that if you fail to confirm the model in your experiments it is your fault and not that of the received view on how to deal with the case at hand.

---

[406] Luikov (1966:74): 'the study of scientific generalisation of the data of a single experiment'.

7 modelling in chemical engineering

7.1 similarity considerations and dimensionless numbers[407]

In many cases, the behaviour of operating units or devices studied in chemical engineering science can be predicted by test procedures using a conveniently sized scale model. The interpretation of test data from such scale-model tests and application to full-sized equipment depends upon similarity constraints (Block 7-1). As a simple example consider two lengths of smooth tubing, one of which is 1 cm in diameter and 100 cm long (the model), and the other 1 m in diameter and 100 m long (the prototype). The model and prototype are geometrically similar since any two counterpart linear dimensions are in the ratio 1/100. Because the point velocities at the midpoint and the velocities at all other geometrically counterpart points between the centre and the wall bear a constant ratio, the model and prototype are kinematically similar too. If well-developed turbulent flow exists throughout the entire length of both model and prototype, the point velocity is independent of axial position. If the ratio of kinetic energy per unit volume (inertial forces) to fluid stress (viscous forces) is constant and no other forces (for example, surface tension or gravitational forces) exist to a significant degree, the two models are dynamically similar as well. If two systems are geometrically, kinematically, and dynamically similar, all velocities and forces are in a constant relationship at counterpart positions. In general, model and prototype will display similar behaviour if, for the same boundary conditions, the relevant dimensionless groups (Block 7-2) such as the Reynolds number have the same value for model and prototype. Note that many ceteris paribus conditions enter such similarity considerations when applied to actual cases.

The concept of similarity extends to many characteristics besides geometry, for example, in aeronautical engineering it may be specified that the mass distribution in a model be similar to that in the prototype or that the ratio of stiffness of homologous cross sections of a prototype wing and a small-scale model must be constant. Often scale effects are difficult to eliminate. For example the surface of a material that is relatively smooth on the prototype scale may be 'rough' on the model scale.

In chemical engineering, in addition to dimensionless numbers for fluid flow, dimensionless numbers for heat and mass transfer are of crucial importance (the second group in (Block 7-2). A simple example illustrating their use is to consider roasting turkeys. If two geometrically similar turkeys at initial temperature T_0 are cooked at a given surface temperature T_1 to the same dimensionless temperature distribution, then the dimensionless time, or Fourier number, with d the

---

[407] A much shorter version of chapter 7 was published as van Brakel (2000c).

Block 7-1   Similarity criteria.

> Geometric similarity exists when all counterpart length dimensions of the device bear a constant ratio.
> Kinematic similarity exists in a geometrically similar system of different size if all velocities of fluids at counterpart positions bear a constant ratio. Geometric similarity must exist for kinematic similarity to apply, otherwise counterpart positions would not exist.
> Dynamic similarity exists in two geometrically similar systems if all forces at counterpart positions bear a constant ratio.
> Thermal similarity exists if differences of temperature between particular points in one system bear a fixed ratio to difference of temperature between the corresponding points in the other system.
> Chemical similarity exists if the concentration of a reactant at any point in one system is in a fixed ratio to the concentration of this reactant at the corresponding points in the other system. [A precise formulation of chemical similarity would require considering the order of the chemical reaction(s) involved. In general thermal and chemical similarity require dynamic similarity.]

'characteristic' diameter of a turkey,[408] will be the same for both turkeys (if some 'reasonable' assumptions are fulfilled). This is the basis of roasting instructions for turkeys in cookery books and in the programming of modern ovens. Because of the temperature dependence of substance properties like density or viscosity, to achieve thermal similarity is more difficult than to achieve kinematic similarity. Comparison across different media is only possible if the physical characteristics of the two media have a similar temperature dependence.

The above account is the common way of introducing similarity considerations in Anglo-American literature. There exist 'alternative' similarity theories that are organised slightly differently. For example in the Kirpichev-Gukhman theory of similarity two phenomena are defined as similar if they are described by one and the same system of differential equations and have similar conditions of 'single-valuedness' for the geometrical properties of the system, all 'physical constants' that are essential for the phenomena under consideration, and the relevant initial and boundary conditions (Luikov 1966). This leads to slightly different definitions or role of some of the dimensionless numbers listed in Block 7-2. But the general approach is similar.

Historically, similarity considerations and dimensionless numbers used in modelling engineering applications are closely tied to the method of dimensional analysis. Using dimensional analysis, systems or devices can be reasoned about without explicit knowledge of the regularities that govern them (the 'laws' that allegedly apply to the system or device), requiring only knowledge of the relevant variables and their dimensional representation. Although dimensional analysis is now used much less in chemical engineering than a few decades ago, it is of (historical) interest because it involves a large number of intercalated models or ceteris paribus assumptions from the most 'fundamental' to the most 'applied'. Moreover, though the number of models for which numerical solutions of the (allegedly)

---

[408] For examples of characteristic length parameters see Block 7-3.

governing equations under the prevailing boundary conditions can be given has increased substantially, this doesn't mean that dimensionless numbers are disappearing from chemical engineering science and practice. In fact, new dimensionless numbers are still being added. For example, the (dimensionless) agitation cavitation number was introduced in the 1990s.[409] Let me give two more detailed examples from recent issues of *The Canadian Journal of Chemical Engineering*, of the use of dimensionless numbers and how they are tied in with model considerations.

Liu, Fryer, and Pain (1999) consider the 'influence of particle-specific gravity and particle shape on the averaged axial velocity of nearly neutrally buoyant particles in horizontal pipes'. This is a problem situated in the context of modelling two phase flow (such as hydraulic transport of particulate solids). Note that the quotation (actually the title of their article) already stipulates restriction to a model. They attempt to overcome limitations of earlier purely empirical work by setting-up a proper force-balance model for the system (taking into account both particle specific gravity and particle shape). However, the equation they derive between the average particle velocity and the modified particle Froude number still contains two constants which had to be determined by regression 'based on all the over 500 acquired experimental data'.[410] Moreover they can only provide approximate numerical solutions for spherical, cylindrical, disk and cubic particles, i.e. for models of actual particles, characterised using a shape factor (Block 6-8). Finally, they list several parameters that should be included in a fuller description. After having advocated that models for the curve fitting constants in their equations should be investigated, the last sentence of their article is:

> The study of single particles could be a first step towards investigations of multiple particle flows for industrial use.

Wang et al (1999) use a multiple linear regression technique to model the air-side performance of herringbone fin-and-tube heat exchangers in wet conditions. They present their results in terms of the Colburn j and Fanning friction factors.[411] The performance of the (highly streamlined) geometry is so complex that no single equation can describe the dependence of the heat transfer and friction on the Reynolds number, and the eight parameters characterising the geometry of the system. The curve-fitted correlation for the friction factor contains four constants, six dimensionless numbers, and five variables. The latter are powers of dimensionless numbers, all of which depend on the Reynolds number in complex ways: equations with up to four terms, containing natural logarithms in various places, several occurrences of the Reynolds number, as well as dimensionless numbers characterising the geometry of the system.

---

[409] The number is used to characterise multiple-impeller stirred tank reactors with boiling liquids.
[410] Liu, Fryer and Pain (1999:1087). They report that most (but not all) data fall within the 10% error boundary.
[411] See Block 7-2.for the definition of these numbers. 'The mean deviations of the proposed heat transfer and friction correlation are 5.75% and 8.27%, respectively' (Wang et al 1999:1229).

Block 7-2   Dimensionless numbers; for symbols see Block 6-7.

dimensionless numbers characterising flow of fluids		
Re Reynolds	ratio of inertia and viscous forces	$\rho dv/\eta$
Fr Froude	ratio of inertia and gravity forces	v^2/gd
f friction factor	dimensionless pressure drop	$d\Delta P/2L\rho v^2$
dimensionless numbers for heat and mass transfer		
Nu Nusselt	ratio of total and molecular heat transfer	hd/λ
Sh Sherwood	ratio of total and molecular heat/mass transfer	kd/D
Pr Prandtl	ratio of molecular and momentum heat transfer	$\eta C/\lambda$
Sc Schmidt	ratio of molecular and momentum mass transfer	$\eta/\rho D$
Fo Fourier (heat)	characterising, heat flux into a body	$\lambda t/\rho C d^2$
Fo Fourier (mass)	characterising mass flux into a body	Dt/d^2
dimensionless numbers written as products of other dimensionless numbers		
Ga Gallis	$FrRe^2$	
Le Lewis	$ScPr^{-1}$	
Pé Péclet	RePr, ReSc	
St Stanton	$ShRe^{-1}Sc^{-1}$ (mass), $NuRe^{-1}Pr^{-1}$ (heat)	
j_H Colburn j factor	$j_H = ShRe^{-1}Sc^{-0.33}$	
j_M Colburn j factor	$j_M = NuRe^{-1}Pr^{-0.33}$	
more dimensionless numbers		
Archimedes	inertia force, gravity force and viscous force	
Arrhenius group	activation energy and potential energy (determines rates of chemical reaction)	
Bingham number	yield stress and viscous stress	
Biot	boundary layer heat/mass flux at surface of body and heat/mass flux within body	
Bodenstein	mass transfer in beds of granular material	
Bond (Eötvös)	gravity and surface tension forces	
Brinkman	viscous heating and heat flow resulting from impressed temperature difference	
Capillary number	viscous and surface tension forces	
Deborah	duration of fluid memory and deformation process	
Derjaguin	thickness of coating and capillary length	
Euler	pressure force and inertia force	
Fluidisation number	fluidisation and minimum fluidisation velocity	
Grashof	accounts for change of density due to temperature or concentration gradient	
Gukhman	criterion for convective heat transfer in evaporation	
Kirpichev	modification of Biot number	
Knudsen	bulk and Knudsen diffusion in granular bed	
Kossovich	heat required for evaporation and heat used in raising temperature of body	
Lagrange	eddy mass transfer and molecular transfer	
Spalding function	temperature gradient at wall in dimensionless form	
Suratman	inertia force, surface tension force and viscous force	
Weber	inertia and surface tension forces	
Weissenberg	elastic and viscous forces	

In the sections that follow, I've restricted myself to the minimum number of variables (Block 6-7). This means that occasionally my equations look slightly different from those in practice, for example instead of the kinematic viscosity, I write ρ/η. This is innocent. Other simplifications are more problematic. Tricky variables I use include, D, the coefficient for molecular diffusion. It is not only dependent on the temperature, but, depending how it is defined, depends on the bulk density and the concentration of the components (which again can be defined in various ways). Also I exclude specifications of how to evaluate substance properties (different definitions of averages, 'bulk', 'at the film temperature'), various ways of defining the velocity ('average', 'bulk', 'film', 'at infinity', 'approach velocity'), or different ways of taking into account a temperature gradient. Note that each of these specifications exemplify the ubiquity of models and ceteris paribus conditions.

7.2 dimensional analysis

The method of dimensional analysis has its origin in the principle of similitude referred to by Newton.[412] Fourier (1822) was the first to apply the geometrical concept of dimension to physical quantities in his *Théorie analytique de la chaleur* :[413]

> It must now be remarked that every undetermined magnitude or constant has one *dimension* proper to itself, and that the terms of one and the same equation could not be compared, if they had not the same *exponent of dimensions*.

Fourier used the term 'dimension' in the sense it had already been used in the encyclopeadia of Diderot and d'Alembert.[414] He recognised the existence of dimensionless groups in his equations (Fourier 1822:158), but didn't see most of the consequences that were drawn out later. Maxwell in his *Treatise on Electricity and Magnetism* (1873, 1891) acknowledged the work of Fourier, expressed the requirement of dimensional homogeneity more precisely, used similarity considerations, and introduced the modern notation for dimensions, using capital letters in brackets (Block 6-7).[415] Lord Rayleigh made extensive use of the method of dimensional analysis in his *Theory of Sound* (1877, 1894):

> In the course of this work we have had frequent occasion to notice the importance of the conclusions that may be arrived at by the method of dimensions.

He contributed to many discussions on applying dimensional analysis to fluid flow and in 1915 got into a discussion in *Nature* on the principle of similitude.[416]

---

[412] Newton, *Principia*, II, proposition 32.
[413] Deposited with the *Académie des Sciences de Paris* in 1811, published 1822, quoted from the 1988 Gabay edition, p. 154f: 'Il faut maintenant remarquer que chaque grandeur indéterminée ou constante a une *dimension* qui lui est propre, et que les termes d'une même équation ne pourraient pas être comparés, s'ils n'avaient point le même *exposant de dimension*.'
[414] Dictionnaire des Sciences, volume IV, p. 1008.
[415] On the work of Maxwell in this connection see D'Agostino (2000:39-76).
[416] See Rayleigh (1892, 1904, 1915). The quotation is from Rayleigh (1945), vol. II, p. 429.

One of the first dimensionless numbers to gain widespread acceptance among physicists and engineers was the Reynolds number (in the 1880s).

Between 1890 and 1920 the 'method of dimension' or the 'procedure of dimensional analysis' was further developed. In its 'final' form it can be summarised as follows:[417]

Using previous experience decide on the general nature of the problem, including a judgement on which systems may be considered as similar 'in the relevant sense' (Block 7-1).

- Enumerate *all* dimensional variables that enter the 'fundamental' equations assumed to describe the phenomena (whether these equations are known or not, whether the boundary conditions to solve them can be specified or not).
- Enumerate *all* dimensional 'constants' (i.e. 'physical constants' such as density or the acceleration of gravity) that occur in these equations (but *not* universal constants like the Planck constant or the velocity of light).
- Select the most suitable fundamental magnitudes or units (such as [Length] and [Time]).
- Write the dimensions of all variables in terms of fundamental units – the latter number should be chosen to be as large as possible without introducing more dimensional 'constants'.
- Re-arrange all dimensional variables and 'constants' into dimensionless groups - choose dimensionless groups, $\Pi_1, \Pi_2, \Pi_3, ...$, such that variables that one is particularly interested in stand 'conspicuously by themselves' (Bridgman 1931:56).

The Π-theorem states that the number of dimensionless groups resulting from analysis will be equal to the number of variables n (obtained in step 2 and 3 of the 'procedure'), minus the number of dimensions p (corresponding to the number of fundamental units, specified in step 4). If two systems are similar in the relevant sense (i.e. in terms of the mechanisms described by the variables occurring in the Πs), then corresponding Πs must be the same in each case. In its most general form this can be written as:

$$\Phi (\Pi_1, \Pi_2, \Pi_3, ...) = 0 \qquad (7.1)$$

where $\Pi_1, \Pi_2, \Pi_3, ...$ represent the dimensionless groups of variables - hence called the 'Π-theorem'. Dimensional analysis does not indicate extraneous, omitted, or redundant variables (though it may offer some hints as to errors in reasoning). It only works if based on thorough familiarity with the relevant parameters.[418]

In general and often in practice there are so many Π's influencing a phenomenon that it is impossible to satisfy all the requirements of similarity at the same time - unless, contrary to one's aims, the systems being compared are completely

---

[417] Bridgman (1931:56), Focken (1953:124), Campbell (1920, 1924), Ellis (1966:143f), Sedov (1959), Isaacson and Isaacson (1975).

[418] Cf. the physicist Focken (1953:124f; *cf.* 108,110): 'an appreciation of the physical significance of the assumptions made [in dimensional analysis] is more important than elaborate formal treatment or mathematical refinement'; also Langhaar (1951:60).

identical. Hence choosing a limited number of Π's involves choosing a ceteris paribus model.

Buckingham (1914, 1915) popularised the method in the United States and his name is usually associated with the Π-theorem, though the theorem was already derived by Vaschy in 1892.[419] At the beginning of the 20th century, many physicists and engineers were already implicitly using it. The subject was more or less consolidated by Bridgman (1931 [1922]), in a series of five lectures presented at Harvard University.[420] When Buckingham first presented his work at the Spring Meeting of the *American Association of Mechanical Engineers* in 1915 one of the commentators praised his work in the following words:[421]

> The Π-theorem is closely analogous to thermodynamics and the phase rule. Thermodynamics affords certain rigid connecting links between seemingly isolated experimental results, while the phase rule tells us the number of degrees of freedom of a chemical system. The Π-theorem likewise affords rigid connecting links which not only serve as a check on the consistency of our results, but may greatly cut down the labour of experimenting. The fact that the paper contains nothing essentially new does not diminish its value. Gibbs' phase rule, too, was new only in form, not in substance, yet it served as the crystallising influence which caused an immense number of latent ideas to fall into line, and we may expect the Π-theorem to play a similar role.

Attempts have been made occasionally to apply dimensional analysis in social sciences, but with little success.[422] As Bridgman (1931:53) pointed out, dimensional analysis can only be applied reliably to systems whose 'fundamental' laws have been formulated. If there are as many dimensional variables as there are dimensional constants (the typical situation in social science), dimensional analysis has no information to give. There is by now an extensive literature, not reviewed here, addressing the question whether the Π-theorem is generalisable, for example to non-ratio scales.[423]

7.3 dimensional analysis in chemical engineering

Knowledge of combined mass, heat, and momentum transfer is crucial to chemical engineering. In studying these transport phenomena chemical engineering makes

---

[419] Other important contributions were made by Carvallo and Riabouchinsky. See for Buckingham's derivation his (1915:289-292) and for the priority disputes Monod-Herzen (1976:25,61), Macagno (1971), Bridgman (1931).
[420] For an alternative, more critical account see Campbell (1920:403-436); see also Ehrenfest-Afanassjewa (1915) for some subtle points. Extensive discussion took place in *The Philosophical Magazine*, in which Buckingham, Bridgman, Campbell (1924), as well as Ehrenfest-Afanassjewa (1926) participated. See Focken (1953:102-108) and Laymon (1991) for some of the issues and references.
[421] M.D. Hersey in Buckingham (1915:292-4)
[422] For the kind of pitfalls involved see the exchange between McGuire Pearson and Dobbert (1996) and Wormer (1986).
[423] For discussion see Krantz, Luce, Suppes and Tversky (1971:515-523).

extensive use of empirical correlations that are equations made up of dimensionless groups of variables raised to various powers.[424] Here is an example:[425]

$$\frac{kd}{D} = 2.0 + 0.6 \left(\frac{\rho v d}{\eta}\right)^{0.5} \left(\frac{\eta}{\rho D}\right)^{0.33} \tag{7.2}$$

There are different ways to obtain the form of correlations such as eq. (7.2): by dimensional analysis; by 'un-dimensionalising' the differential equations for heat, mass, and momentum transfer within the boundary layers, and by listing the transport mechanisms involved - if there are n mechanisms, they can be described by $n-1$ dimensionless numbers. Text books differ in the emphasis they put on these three methods and dimensional analysis in the narrow sense has been on the retreat in the past decades. However, the different methods draw on the same principles.

To see how an equation such as (7.2) is obtained using dimensional analysis, consider mass transfer between a bubble or a droplet of a fluid dispersed in another fluid - for example when a fluid with density ρ and viscosity η is flowing with average velocity v along a droplet with diameter d and mass transfer is taking place to (or from) the bulk of the fluid from (or to) the surface of the droplet by molecular diffusion of a reactant of which the diffusion coefficient in the fluid is D.[426]

Then it may be suggested that the partial mass transfer coefficient, k, is dependent solely on the parameters mentioned, or:

$$k = \Phi(v, \rho, \eta, d, D) \tag{7.3}$$

Assuming further that eq. (7.3) may be expressed as a power series with a sufficient number of constants, we can write:

$$k = c_1 v^d \rho^e d^f \eta^g D^h + c_2 v^i \rho^j d^k \eta^l D^m + ... \tag{7.4}$$

The constants in eq. (7.4) are dimensionless by definition; therefore, to be dimensionally consistent, each term in the series must have the same dimensions as the term on the left side of the equation. In order for the dimensions to be homogeneous, the following equations apply to the exponents (for the three dimensions length, time, and mass respectively):[427]

$$+1 - d + 3e - f + g - 2h = 0 \tag{7.5}$$

---

[424] Courses in transport phenomena in chemical engineering did not emerge until the 1950s. Erlier, the momentum transfer of fluid mechanics had been taught under civil or mechanical engineering, heat transfer under chemical or mechanical engineering and mass transfer or diffusion only in chemical engineering. Probably the first text book on 'integrated' transport phenomena for chemical engineers was used in the Netherlands in the 1950s. See acknowledgements in the book that 'made' Transport Phenomena a separate subject (Bird, Stewart, and Lightfoot 1960:x); see also Kramers (1989).
[425] For definitions of the dimensionless numbers see Block 7-2.
[426] Eqs. (7.3) and (7.4) represent step 2-4, eqs. (7.5-7) and (7.8) step 5 and eq. (7.9) step 6 in the 'procedure' of dimensional analysis.
[427] Since the dimensions of each term in the series are identical, only the dimensions of the first term need be considered for dimensional homogeneity.

$$-1 + d + g + h = 0 \tag{7.6}$$

$$-e - g = 0 \tag{7.7}$$

Since there are three equations and five unknowns we obtain

$$k = c_1 (v)^d (\rho)^{d-1+h} (d)^{d-1} (\eta)^{1-d-h} (D)^h \tag{7.8}$$

or, by rearranging the variables into dimensionless numbers:

$$\frac{kd}{D} = c \left(\frac{\rho v d}{\eta} \right)^d \left(\frac{\eta}{\rho D} \right)^{1-h} \tag{7.9}$$

or:

$$Sh = c Re^a Sc^b \tag{7.10}$$

with Sh the Sherwood number, Re the Reynolds number, and Sc the Schmidt number.

Equation (7.9) is in a form made up of several groups of variables, each group of which is in itself dimensionless. Such dimensionless numbers (the Π's in the Π-theorem) can often be read as a ratio of two transport mechanisms (Block 7-2). The equation is a restatement of eq. (7.3) with the necessary condition of dimensional consistency applied. The condition of dimensional consistency automatically brings into being a relationship among the exponents d, e, f, g and h in eq. (7.4) that must be satisfied in any dimensionally consistent equation. During the derivation, no conditions other than eq. (7.4) were applied to the exponents. Therefore, after dimensional analysis, the remaining exponents may be constant or variable, real or imaginary, positive or negative and may be functions of any of the dimensionless groups in the equation. Therefore it would be more correct to write them as variables (and c as well).

The constants c, a, and b in eq. (7.10) can be evaluated only from experimental data or may be 'borrowed' from related data using similarity considerations, For example eq. (7.2) written for mass transfer, uses the constants c, a, and b as determined for heat transfer:

$$\frac{hd}{\lambda} = 2.0 + 0.6 \left(\frac{\rho v d}{\eta} \right)^{0.5} \left(\frac{\eta C}{\lambda} \right)^{0.33} \tag{7.11}$$

or $Nu = 2.0 + 0.6 Re^{0.5} Pr^{0.33}$, with Nu the Nusselt number and Pr the Prandtl number.[428] Note that the Reynolds number occurs in both eqs. (7.2) and (7.11); the heat transfer coefficients h and λ in the Nusselt number correspond to the mass transfer coefficients k and D in the Sherwood number and similarly for the Prandtl and Schmidt number. Here it is assumed that there is an analogy between heat and mass transfer, assuming that the criterion for thermal similarity holds.[429]

---

[428] The number 2.0 is added because for Re = 0 (no flow), theoretical analysis yields Nu = 2.0.
[429] First proposed by Chilton and Colburn (1934): 'mass transfer factors can be estimated with a sufficient degree of certainty for design calculations, by analogy with heat transfer processes, not only for flow inside tubes, but also for flow across tubes and tube banks and flow over plane surfaces.'

Block 7-3   Characteristic length parameters used in Reynolds numbers.[430]

geometry	Reynolds number	characteristic length parameter
flow through a pipe	$\rho dv/\eta$	d, diameter of pipe
flow around sphere	$\rho dv/\eta$	d, diameter of sphere (or bubble)
flow of falling film	$\rho dv/\eta$	d, film thickness
packed beds	$\varepsilon \rho dv/\xi \eta$	d, diameter of equivalent sphere
agitated tank	$\rho \Omega d^2/\eta$	d, tank diameter; Ω, angular velocity
flow through annulus	$\rho dv(1-\kappa)/\eta$	d, diameter of inner cylinder; κ, ratio of the two cylinders

If the phenomenological equations for heat and mass transfer have the same mathematical form and the initial and boundary conditions are the same, then this may be the case. Because momentum transfer is vectorial, there can only be an analogy between all three transport phenomena if momentum transfer can be considered unidimensional (e.g. for momentum transfer in cylindrical tubes or along a flat plate).

Eq. (7.11), was obtained for a 'simple' geometry: a spherical droplet of diameter d surrounded by an 'infinitely' extended flowing fluid. Using 'characteristic' length parameters (Block 7-3), the approach is extended to more complex geometries. Although this text book example is almost half a century old, note that there always remain 'hidden' ceteris paribus conditions that may turn up in the future. Until the 1990s it was thought that the drag coefficient for rising and falling spheres would be the same, assuming that the same forces apply, but in opposite direction. It turned out not to be true for Newtonian liquids.

7.4   presuppositions of dimensional analysis

Many assumptions underlie the method of dimensional analysis and the use of dimensionless numbers. First, it is assumed that each quantity possesses a dimension proper to itself, which can be identified with classes of scales that are connected via linear transformations. This can be taken in a more realistic or a more conventionalistic way.[431] On the first reading dimensions are a unique and intrinsic property of a quantity; on the second reading the dimensions follow from the (arbitrary) choice of the units of measurement. The latter view has tended to dominate because two quantities may share the same dimensional formulae.[432]

It is commonly said that dimensions are always expressible as a product of powers and that the indices are always small integers or simple fractions, though the only reason that can be given is a 'felt need' to associate mathematical simplicity with physical reality. In practice, in chemical and other engineering disciplines, equations with non-integral exponents are common.[433] For example,

---

[430] Similar modifications apply to other dimensionless numbers containing a 'diameter'. An example is the 'particle modified Froude number' mentioned on p. 173.
[431] See the discussion in Johnson (1997:78-80).
[432] For example elastic modulus, energy density, pressure, resilience and stress all have dimensions $[ML^{-1}T^{-1}]$.
[433] This has caused some confusion in the theoretical literature (Dingle 1949b, Palacios 1964:45-6). The use of non-integral exponents has been common in dimensional analysis from the beginning. For an early example see Rayleigh (1904).

according to the Ostwald-de Waele model for non-Newtonian behaviour of liquids, the viscosity is expressed as follows:

$$\eta = c \left| \frac{dv_x}{dy} \right|^{a-1} \quad (7.12)$$

The parameters a and c have to be determined for each liquid; for 10% napalm in kerosene $c = 0.0893$ (lb$_f$sec^{a}ft^{-2}) and $a = 0.520$ (dimensionless).[434] Bird, Stewart and Lightfoot (1960:116; cf. 305) even go so far as to bring eq. (7.12) into the Reynolds number, defined as $Re_a = d^a v^{2-a} \rho / c$. They comment (emphasis added):

> *N.B.:* Whereas the foregoing analysis is *formally correct*, it should be kept in mind that the Ostwald-de Waele model of non-Newtonian behaviour is empirical and not followed exactly by real fluids.

Though this example has been quoted from a text book on transport phenomena of 1960, the power law for non-Newtonian fluids is still 'the best in town' and has not been replaced by more fundamental approaches.

A slightly different example is the 'filtration equation':[435]

$$q = -\frac{1}{\eta^a} \frac{\Delta P^b}{\Theta} \quad (7.13)$$

with q the fluid flow per unit area, ΔP the pressure difference, Θ the 'cake resistance' and the exponents a and b real numbers characterising 'the system'. The threat of 'broken' dimensions is easily removed in this case, by making the variables raised to powers of real numbers dimensionless by introducing 'reference' values (for some 'standard' or 'initial' situation):

$$q = -\frac{1}{(\eta/\eta_0)^a} \frac{(\Delta P / \Delta P_0)^b}{\Theta} \quad (7.14)$$

Any correlation between quantified parameters can be made dimensionally invariant, even dimensionally homogeneous, by inserting enough dimensional constants, as eqs. (7.13) and (7.14) show.

Again this example is taken from a 1960s text book of unit operations, and again the most up to date approaches to cake filtration still use power laws, though the powers have changed towards the 'constitutional' parameters underlying the cake resistance Θ. For example, Lee et al (2000) present a model of cake filtration that is only marginally different from models used 50 years before, though it tries to take into account the compactibility of the filter cake. Darcy's law and the Kozeny-Carman equation (Block 6-6) assume an isotropic, stationary porous matrix and Lee et al present a model in which the latter assumption is dropped. But they still assume 'that the gravity force and inertial terms can be ne-

---

[434] According to A.B. Metzner, *Advances in Chemical Engineering*, vol. I, New York: Academic Press, p. 103, quoted in Bird Stewart and Lightfoot (1960:13). Compare lb$_f$sec^{a}ft^{-2} with Williams (1892): 'if mass, length, and time are to be ultimately physical conceptions, we cannot give interpretations to fractional powers of M, L, and T' (quoted by Bridgman 1931:47).
[435] Cf. eqs. (6.2) and (6.5). For laminar flow a=b=1; for turbulent flow a=0.11 and b=0.55.

glected and that the solid particles are not deformable'. Further they assume that the porosity, ε, of the cake only depends on the compression pressure and that the compactibiltiy of the cake can be described by a 'solidosity' (1-ε) that is a smooth function of time and space. Then they can write the governing differential equations for the model. However to relate the model equations to experimental data, they still need what they call 'constitutive equations' for the permeability, K, and the porosity, ε, for which they use power laws similar to eq. (7.13). The final dimensionless equations contain five dimensionless numbers.

The assumption that each quantity possesses a dimension proper to itself is closely related to the assumption of dimensional invariance of qualitative laws. If an equation is dimension homogeneous it can be written using dimensionless numbers only, but does it *have to be* dimension homogeneous? Mathematically the Π-theorem is a consequence of the requirement that functions be 'invariant under similarities' (Whitney 1968:251) and in mathematics this can be given a precise definition. But it is less clear how similarity is to be understood for physical systems or chemical substances.[436] There are two ways at least to take mathematical equations in empirical sciences: either as a functional relation between empirical properties or as a relation between numbers. On the empirical interpretation of laws one would say that dimensional invariance is required because 'like can only be like'. The numerical interpretation of laws suggests that without dimensional invariance we could not know what the law says (Osborne 1978).[437]

A second assumption underlying dimensional analysis is the choice of fundamental units (dimensions) or primary magnitudes (quantities).[438] Block 6-7 is based on five 'fundamentals' or primary magnitudes. There is no solid justification for the choice – any number from 1 to 9 has been proposed in the context of dimensional analysis (Block 7-4); Some say that the choice is arbitrary or depends on the particular application.[439] Sometimes it is required that there is an operationally defined procedure of physical addition or concatenation for primary magnitudes.[440] If the latter condition does not apply, the status as primary quantity remains disputed, temperature being the prime example.[441] Secondary magnitudes are defined in terms of the primary magnitudes; often these definitions can also be considered as 'laws' (e.g. the definition of density as the ratio of mass and volume), undermining the distinction between analytic and synthetic statements (van Brakel 1984). However, there is no *necessary* connection between the

---

[436] There is an extensive literature on how to define the similarity of physical systems: Causey (1967), Krantz et al (1971:504-15), Bridgman (1931), Campbell (1920:412-20), Duncan (1953), Massey (1971).
[437] Note that two kinds of mappings occur in the latter version: properties into numbers (measurement) and numbers into numbers (the equation).
[438] Not to be confused with primary and secondary qualities.
[439] Bridgman (1922:23-4), Campbell (1928:255), Dingle (1942:322), Focken (1953).
[440] That is 'adding up' as in putting one length after another or putting two weights in the same pan of a balance. For length there exist formally equivalent empirically different concatenation operations (Ellis 1966, van Brakel and van der Peut 1979).
[441] Debates on the dimensions and other 'weird' properties of temperature have raged since the time of Mach. See Mach (1919), Tolman (1917), Buckingham (1914:357), Campbell (1920:396-402), Bridgman (1931), Ellis (1966), Palacios (1964).

Block 7-4   Number of fundamental or primary quantities or units.

Tolman	1917	5	length, time, mass, electric charge, entropy
Duncanson	1941	3-4	length, time, charge, mass?
Brown	1941	2	length, time
Guggenheim	1942	5	length, time, energy, charge, temperature
Huntley	1952	3	mass, length, time
Whyte	1954	0?	
Sedov	1959	3	mass, length, time
Post	1982	4	quantum of action (replacing mass), charge, length, time

units of derived and fundamental magnitudes (Campbell 1920) and the distinction between the two is not absolute; for example, density could be made a primary magnitude.[442] Finally, the number of fundamental units used in dimensional analysis is not necessarily the same as the minimum number of primary magnitudes to define all other physical magnitudes.

Attempts have been made to remove all dimensions so that only relations between numbers remain. For example, if the gravitational constant, the velocity of light and the viscosity of water are taken to be nondimensional constants, then all physical quantities will turn out to be nondimensional (Sedov 1959:7). Eddington attempted to provide a dimensionless physics, but did not find a satisfactory solution for the linear metric implicit in the physical interpretation of his formalism (Whyte 1954). But even if such a program would succeed, it would have little to say about the actual universe because *somebody* located *somewhere* has to apply the theory under initial and boundary conditions that cannot be reduced to pure numbers.

The Π-theorem implies that one gets the most information out of dimensional analysis if one lists the smallest number of quantities significant for the problem at hand and to have the largest number of acceptable fundamentals. Hence there have been proposals for legitimate means to *increase* the number of fundamentals (Huntley 1952), for example using vectorial quantities (in which L_x, L_y, and L_z for length and F_x, F_y, and F_z for force are used as separate dimensions).[443] For example, for the flow of liquid through tubes or the flow of electricity along wires the length and the cross-sectional area are 'independent' variables. Hence in dimensional analysis length and cross-section may be considered as independent fundamental quantities. One may also conceive of mass in two ways: related only to the inertial property of mass or to do merely with occupying space. And it can sometimes be useful to regard the number of atoms as having dimensions different from a pure number (Guggenheim 1942:496). But such choices of extra parameters are clearly guided by prior experience with the phenomena at hand.

---

[442] Cf. Duncanson (1941:448): 'there can be no such thing as the absolute dimensions of a quantity'.
[443] Probably the first to propose this was Williams (1892) who, for example, gave the dimension of density as $[MX^{-1}Y^{-1}Z^{-1}]$ and of curvature as $[Y^2X^{-1}]$ ([X], [Y], [Z] being the three dimensions of length).

Block 7-5 Assumptions made in proposing an empirical correlation between the Fanning friction factor, f, and the Reynolds number, Re, for flow in a cylindrical pipe.

- It is assumed that the solid is rigid and does not interact with the fluid.
- Pipes are assumed to be smooth; if surface roughness is taken into account, a plethora of proposals of how to model it pops up.
- Existing correlation charts for the dependence of f on Re only apply to long pipes with an 'established regime of flow'; otherwise the ratio of the length and the diameter of the tube enters the equations as a dimensionless number.
- In closed conduits at very high velocities or with rapidly varying pressures f depends on the Mach or Cauchy number (adding the acoustic velocity as a variable).
- In open channels gravity waves make f dependent on the Froude number.
- At very low velocities in shallow open troughs the Weber number might play up.
- Temperature differences between fluid and pipe wall may have an effect on the shear stresses.
- The usual correlations only apply to 'simple fluids', not to 'queer materials like greases, muds, cement slurries'.
- The behaviour of a system which is actually unstable cannot be completely predicted as this depends on random disturbances - chemical engineers have always known about chaos theory.

The Π-theorem can be formulated without reference to fundamental quantities as follows (Pankhurst 1971):

> The minimum number of independent dimensionless products to which the n different kinds of physical variables in a given problem can be reduced is $p-m$ where p is the number of equations needed to express the universal laws governing the system and to define the n quantities fully, and m is the number of additional kinds of physical quantities introduced in these equations.

However this simply means that disputes over the number of primary quantities move to disputes about the number of fundamental laws (Palacios 1964).

The use of dimensionless number correlations, obtained by dimensional analysis or otherwise, further rests on the assumption that implicit functions characterising the physical situation be complete, i.e. that all relevant dimensional constants and variables have been listed. As noted above, this involves a judgement of which parameters/mechanisms are *not* relevant.[444] For example, the friction factor which has been used for decades in numerous types of engineering applications 'of course could be affected by other dimensionless criteria' (Moody 1944:683). Its validity is therefore ceteris paribus. Such ceteris paribus considerations (Block 7-5) limit the choice of variables. This choice constitutes the 'real' model of the phenomena, the rest is mathematics.

---

[444] It is incorrect to say: 'choose those parameters on which the phenomenon depends', because, for example, atomic forces are 'relevant' to mechanical problems, but we don't have to take them into account in dimensional analysis.

Block 7-6 Some of the ceteris paribus conditions involved in deriving the Hagen-Poiseuille equation.

- The flow is laminar (i.e. the Reynolds number is less than about 2100).
- The density is constant ('incompressible flow').
- The flow is independent of time ('steady state').
- The fluid is Newtonian.
- End effects are neglected. If the section of pipe of interest includes the entrance region a correction must be applied.
- The fluid behaves as a continuum - this assumption is valid except if the molecular free path is comparable to or greater than the tube diameter.
- There is no slip at the wall (no tangential motion of the fluid relative to the wall) - for pure fluids this will be true if the previous assumption is met.
- The wall is impermeable (no motion perpendicular to the wall).

Of course ceteris paribus conditions are not restricted to dimensional analysis, but apply to the derivation of 'theoretical' equations as well. By taking a momentum balance over a section of a cylindrical tube one obtains the Hagen-Poiseuille equation:[445]

$$q = \frac{\Delta P}{2\eta} \frac{d^2}{L} \tag{7.5}$$

This equation is valid given how the model of a cylindrical tube has been specified in the theoretical context, but numerous ceteris paribus conditions emerge if it is to be used for real world cases (Block 7-6).[446] Moreover, the lists in Block 7-5 and Block 7-6 are not exhaustive and there are always as yet unknown provisos. Of course one can move up higher on the theoretical level and demand a solution of the full Navier-Stokes equation, and so on. But this doesn't make the ceteris paribus conditions go away.

Outside mathematics there are many kinds of numbers and it is not always easy to decide whether they are pure numbers, dimensionless numbers, dimensional numbers, or something else again (Block 7-7). A peculiar sort of non-dimensionality is that of standards like the standard meter: 'the length of the standard is not got by measurement but by *definition*, and so it is correctly represented dimensionally solely by unity' (Brown 1941:423).[447] Dimensionless quantities like angle are not, strictly speaking, dimensionless; they are just quantities measured on scales that are defined in ways which make them independent of the scale system.[448] If that 'restriction' is removed, the power of dimensional analysis might increase (Ellis 1966:145-151, Huntley 1952), but more often such dimen-

---

[445] Eq. (6.8) in Block 6-6.
[446] See Day (1990) for the history of the no-slip condition, which it took a while to get accepted. Day shows how the justification for the no-slip condition is holistically connected with the theory for which the model is constructed.
[447] As Wittgenstein (1953:§50) famously remarked: 'There is *one* thing of which one can say neither that it is one metre long, nor that it is not one metre long, and that is the standard metre in Paris.'
[448] The ambiguity can be resolved by distinguishing between 'unit' and 'dimension'; then the dimension of angle is [L/L].

Block 7-7 Variety of dimensionless numbers.

pure numbers	2, 4/3, e, π (?)
index of a power (natural numbers)	as in eq. (7.15)
index of a power (real numbers)	as in eqs. (7.2) and (7.14), pH (?)
non-integral exponents of magnitudes	as in eqs. (7.12) and (7.13)
dimensionless numbers so called	see Block 7-2
'artificial' dimensionless variables	as in eq. (7.14)
geometric numbers	angle, eccentricity ellipse, π (?)
shape factors	see Block 6-8
universal constants without dimensions	fine-structure constant, ratio of mass of proton and electron
material constants	ratio of specific heats
quantities measured on associative scales	temperature (?)
number of things	no dimension, but choice of unit is not arbitrary

sionless numbers are like the shape factors in chemical engineering (Block 6-8). For the latter there is no work to do in dimensional analysis. They occur in the list of relevant parameters when the problem is formulated and move straight to the resulting equation which states the relation between dimensionless numbers (Bridgman 1931:83). Moreover such shape factors only take on 'theoretical' values for idealised geometric models.[449] No matter how sophisticated the theoretical framework and no matter how powerful the numerical techniques to solve complicated sets of differential equations, they will always apply to strongly idealised initial and boundary conditions.

7.5   the model of dimensional analysis

A comparison with definitions of various types of models (Block 6-2) suggests not only that a whole variety of models are used in dimensional analysis, but that these model types themselves function as models in the sense of ideal types. Attempts to make definitive decisions as to what type of models are used to model flow in a pipe or in heat and mass transfer in a chemical reactor are spurious: similarities and differences crop up and disappear. Every S *and* B identified as figuring in arguments of the 'S is a model of B' relation becomes fluid on close examination, foregrounding a host of other models. Short of describing *particular* practices, only the most abstract notion of 'model' covers all uses (Block 7-8). The word 'model' is like Wittgenstein's example of 'game'. We know how to use the word, but it's spurious to ask for a definition guaranteeing 'correct' usage in every situation.

Each of the assumptions of dimensional analysis introduces its own models or ceteris paribus conditions. First, there are a number of background assumptions (models) that apply to all contexts where measurement takes place, in particular a

---

[449] Even less tractable are the 'modelless' empirical dimensionless numbers. For example, von Kármán proposed an equation for the Reynolds' stresses in which a parameter appears which has been described as 'a "universal constant" whose value is given as 0.40 by some investigators and as 0.36 by others' (Bird, Stewart and Lightfoot 1960:160).

Block 7-8 Most general characteristics of a model - the model of models.

- A model is a model *of* something; S is a model of B and the relation is asymmetric.
- A model is a representation of the real or imaginary B for which it stands - as with all representations, there are conventions for correct 'reading' of the model.
- A model serves a purpose, primarily its purpose is to 'read off' properties of B from the directly presented properties of S.
- A model appeals to notions such as isomorphy and structural similarity - other terms in this context: abstraction, idealisation, isolation, simplification, approximation, extrapolation, partiality.
- A model is *not* to be characterised primarily in terms of visual or imaginative metaphors, but in terms of things better and less known, controllable, or familiar - the 'same' thing may be B (unfamiliar) at one time, S (familiar) at another.

measurement theory (consisting of axioms and operational definitions) for each of the relevant quantities, the choice of fundamental units, and the requirement that all equations containing magnitudes have to be dimensionally invariant. The Π-theorem, which might be seen as the essence of dimensionless number modelling, draws on these background models and two more specific assumptions, which it shares with more theoretical approaches:

- First the ceteris paribus assumption or model that all relevant dimensional constants and variables have been listed. At a slightly more theoretical level this is the assumption of listing all the relevant phenomenological equations for the problem on hand.
- Second the idealisation or modelling of the initial and boundary conditions, which makes them suitable for mathematical treatment. For dimensional analysis in chemical engineering the most prototypical models are concerned with idealised geometries and the use of shape factors (Block 6-8 and Block 7-3). But there are also ceteris paribus conditions such as 'the inner wall of the vessel is chemically inert' or 'barring the presence of contaminating surfactants'. In addition there is the use of analogical modelling in the similarity considerations, as when the (approximate) analogy of heat, mass, and momentum transfer is exploited.

Note once more that dimensional analysis has no grip on shape factors: they are already dimensionless numbers. Neither are more theoretical approaches of much help to deal with real life initial and boundary conditions. Numerical solutions of phenomenological equations are only possible for the most simple geometries. In complex cases multiple regression scaling techniques are used which choose the relevant dimensionless numbers on the same basis as is advocated in 'pure' dimensional analysis. The increased power of these 'ab initio' approaches cum curve fitting techniques have 'made true' the earlier observation that 'the exponents [of dimensionless numbers] may be constant or variable, real or imaginary, positive or negative and may be functions of any of the dimensionless groups in the equation', as the example of Wang et al (1999) illustrates.[450]

In the past few decades significant advances have been made toward 'ab initio' design in chemical engineering. Complex sets of phenomenological equations can be numerically solved for relatively complex boundary conditions. But note that the design of equipment has gone hand in hand with finding solutions for the equations describing the heat, mass, and momentum transfer. For example, Satheesh, Chhabra and Eswaran (1999) have provided a numerical solution of the complete Navier-Stokes equations 'describing the steady flow of incompressible Newtonian fluids normal to an array of long cylinders', which they say 'is an idealisation of many industrially important processes'.[451] But note that the solutions of the equations are numerical, not analytic; they apply to a model or idealisation; which model is approached in some industrial processes by making practice more streamlined.

In contrast, for real world systems, even the most 'simple' sort of problems remain unsolvable. Maxey Flats (Kentucky, U.S.A.) is the world's largest commercial storage place for radioactive waste of low intensity. The management consortium said in 1963, when the site was opened, that it would take plutonium 24,000 years to get half an inch outside the borders of the site (i.e. into the soil outside the storage site). Less than ten years later the plutonium was already two miles away from the site. In retrospect, it turned out to be *impossible* to make any sensible models of the boundary conditions under which the phenomenological equations concerning the relevant diffusivities and permeabilities had to be applied (Shrader-Frechette 1997). If in 1963 the solution of the complete Navier-Stokes equation 'describing the steady flow of incompressible Newtonian fluids normal to an array of long cylinders' and similar ideal geometries would have been available, it would have made no difference for the prediction of possible leakage from the Maxey Flats site, because the absence of an appropriate model, not better 'ab initio' methods applied to simple (inappropriate) models, thwart description of the unmodelled world (Block 7-9).

In [6.2] I've argued, following the lead of Cartwright and others, that *all* laws and models are ceteris paribus and that the relation between a model and what it allegedly represents is symmetrical: fitting the model is a matter of mutual attunement of both model and reality. The example of dimensional analysis shows that if one moves further in the direction of 'real systems', the phenomenological equations take on the status of theories. Whether there are any Newtonian fluids in the real world is an open question and numerous other ceteris paribus conditions are introduced when applying the phenomenological equation to flow in a cylindrical tube (Block 7-5 and Block 7-6). Moreover, this cylindrical tube is itself an artefact. Because it has been made to narrow specifications, phenomenological equations can be applied to it under relatively simple boundary conditions. Already if one moves to flow through, say, a packing of sand particles or if capillary

---

[450] Discussed on p. 173.
[451] Such as flow in porous media, tubular heat exchangers, aerosol filters, flow through screens.

Block 7-9 Problems that may arise when reality hasn't already been streamlined to approximate the models to which the theoretical equations apply.[453]

- The combination of relevant processes is so unique and complex that no reproducible measurements can be carried out on scale models of the system.
- The system is so heterogeneous that it makes no sense to ascribe average values to the relevant parameters.
- The relevant phenomena or processes are dependent on their (unknown) history from which extrapolations are for the greater part arbitrary.[454]
- Little is known about the empirical adequacy of the models that are being used to make predictions.
- The predictions that are derived from the computer model of the system do not allow realistic testing.
- There is no consensus among scientists about the relevant parameters that are to be included in the model.

forces are present [6.4], modelling is only possible using geometric models that are strictly speaking meaningless (e.g. they contain discontinuities).[452]

Both the laws or dimensionless number correlations and the model situations to which they apply are models because both are subject to ceteris paribus conditions. Further these model situations are constructed by drawing on a plethora of other background models. What we have is a world of interrelated models, where no matter which model or description one picks out and tries to say what it is that is being modelled, what is being modelled is itself a model of something else. Instead of saying that the artefact S models the given B, it is better to say that S and B jointly make up B and S. The use of terms such as abstraction, idealisation, isolation, simplification, approximation, extrapolation, are *all* connected with the intuition that there is something real (i.e. a *part* of the real world) that is not partial, impoverished, and isolated, of which the model aims to be a good, relevant, appropriate likeness, albeit idealised and abstracted. However, because any description is already an abstraction, hence a model, a model is never a model of (part of) the world, but always a model description of a system (model) already described in some other way.

---

[452] For example: 'the fluid loses energy only during passage through the narrow channels and not while passing from one channel to the next through a junction' (Bear 1972:93).
[453] Adapted from Shrader-Frechette (1993, 1997).
[454] This is called hysteresis; an example is the mechanism of capillary liquid transport being dependent on the 'wetting' and 'drainage' history of the porous system.

8 conclusion

8.1 how to fit it all together

Examples such as the polywater episode show that many discourses contribute to the contestation whether polywater in the sense(s) intended, exists. Moreover, if manifest substances (such as water) and their manifest properties and relations are the base line, they are natural kinds, which, as Boyd (2000:66) puts it 'are features, not of the world outside our practice, but of the ways in which that practice engages with the rest of the world'. They are classifications that contribute to the formulation and identification of projectible hypotheses. For this projection to work it has to pick up causal affordances in the world; but there is no need either for strict laws in which the 'real' natural kinds figure, nor for the metaphysical suggestion that 'true' natural kinds 'have to cut nature at the joints'. Natural kinds that are practice dependent are no less real. But this realism allows promiscuity (Block 8-1). If there are three successful practices of cutting the joint between an acid and a base, it is not *necessary* to worry about which is the real one. I refer to this stance as a form of pragmatic or pluralistic realism or 'realism-with-a-small-r' (though nothing depends on the terminology).

If there are natural kinds successfully employed in different practices and there is no isomorphic mapping across discourses, as in a fully developed reductionism, then how do all these natural kinds or discourses fit together into one world? The dominant view is that everything fits together in the 'material' world of physics, whether reductionism works or not. But why is there such a strong intuition which leaps at once at every hint that this or that can be reduced to physics? I suggest there are three reasons:

- One would like to fit everything into one world.
- One prefers a scientific account over other accounts.
- Physics is the only science that claims to have something to say about everything: physical laws are universal; they apply to everything.

But there is a big difference between the possibly correct claim that physics has something to say about everything in the world of the (physical) spacetime manifold and the claim that physics has *everything* to say about every thing. The explanations that physicists provide are no doubt broad as well as deep, but their breadth has more to do with their touching the outer reaches of the universe (both the small and the big) than with their covering the same ground as more ordinary explanations.

If different discourses should be given equal standing in discussions and contestations about the 'same subject matter', then we need a different account of the nature of interdiscourse relations than the inherently asymmetrical intuitions of

Block 8-1 Dupré on the disorder of things.

> In his *The Disorder of Things* (1993) Dupré argues against determinism, reductionism, and essentialism from an empiricist perspective, advocating a form of pluralism or pragmatism characterised by:
> - A plea for promiscuous realism: natural kinds can overlap; this doesn't threaten their reality.
> - There are a great variety of causal efficacies or capacities that do not make up a closed system. There are genuinely causal entities at many different levels of organisation (drawing on Cartwright [5.2]).
> - There are no strict bridge laws because the individuals, events, systems at different levels coincide neither in practice nor in their idealisations
> - There are many different paths to knowledge. If there are many threads of partial order in the universe, the choice what to research is more open-ended.
> - No special weight should be placed on what things are made of (as distinct of their other 'form'-properties).

reduction, supervenience, and emergence. It would seem that there are three intuitions, motivations, or requirements that underlie contemporary discussions about interdiscourse relations (Kim 1990):
- All interdiscourse relations rest on an ultimate physical base (i.e. dependence).
- Somehow interdiscourse relations must be explained in terms of things indiscernible from the base domain being indiscernible from the supervening domain (i.e. covariance).
- Each supervening (or emerging) discourse should be autonomous (i.e. non-reducible).

Contemplating the intuitive meaning of 'dependence', 'covariance', and 'non-reducible', suggests that the task set is not an easy one (if coherent at all). The solution I propose is a form of anomalous monism, adapted from that introduced in the philosophy of mind by Davidson.[455] There are possible analogies with the views of philosophers such as Spinoza and Whitehead as well, though I won't pursue them here. Anomalous monism breaks the grip of the ontology - epistemology or object - property dichotomy, by reinterpreting the supervenience discourse. It takes a stance that says: there's no need for a *strict* separation between ontology and 'ideology', i.e. between things and properties;[456] no *strict* distinction between ontological, epistemological, and explanatory reduction and no *strict* distinction between ontological reduction and conceptual autonomy. The view presented has similarities to that of Dupré on 'the disorder of things' (Block 8-1), but Dupré takes Cartwright's capacities [6.2] rather than anomalous monism as a 'metaphysical base'.

The view proposed here differs from each of the proposals for interdiscourse relations discussed in [2.3] on the following points:

---

[455] See Davidson (1970, 1993); cf. van Brakel (1999b). To apply the idea to chemistry is briefly mentioned in van Brakel (2000a).
[456] The labelling 'ontology' (referents of names or values of variables) versus 'ideology' (referents of predicates) stems from Quine (1978:504).

CONCLUSION

- Interdiscourse relations (bridge laws, supervenience or emergence relations) are empirical ceteris paribus regularities, not metaphysical necessities - the only circumstances in which they might apply strictly, are model situations, completely isolated from the rest of the universe.
- Interdiscourse relations are symmetrical, leaving the autonomy of both sides intact (without preventing forms of explanatory interaction or extension *in both directions*, by borrowing, synthesis, criteria of overall coherence, and so on).[457]
- A strict separation between the two discourses that are connected by interdiscourse relations is usually not possible and primarily serves as a model imposed on a messy world.

Hence, interdiscourse relations between, say, macrochemistry and microchemistry or between microchemistry and quantum mechanics are best seen as symmetrical relations, locally valid under well described ceteris paribus boundary conditions - where the latter are governed by 'top-down' explanatory interests. The point is not a 'purely instrumentalist' or 'anti-realist' or 'constructive empiricist' one. I don't argue against the reality of atoms, or quarks, or the relevance of the mathematical structure of quantum mechanics for the reality of the physical world at some level of description. Nor do I argue against unification *per se*. There are many other paths to unification apart from reduction (or supervenience or eliminativism). For example, physical chemistry unifies, to a certain extent, physics and chemistry. All this can claim reality: charcoal, soot, diamond, and buckminsterfullerine; water and H_2O; temperature in thermodynamics and temperature in statistical mechanics and 'temperature' in quantum mechanics; neutrons and quarks; electrons and wave equations.

Most writers on the latter-day successors of reductionism (including advocates of 'radical' emergentism) propose, by implication, substance monism and explanatory pluralism. The only *things* there are, are physical things - all true descriptions and explanations concern the states and behaviour of things composed only of physical entities, but not all true descriptions or explanations are in, or are translatable into, the language of physics.[458] However, nothing forces us to this view. Scerri and McIntyre (1997:226) say:

> The question of the supervenience of chemistry on physics would seem to depend precisely on the empirical facts, and the conclusions which they support and not from more general philosophical musing about chemistry and physics.

But why bother to 'test' supervenience in the first place? Why not have physical and chemical and 'mixed' descriptions existing side by side, related by a variety of ceteris paribus interdiscourse relations, without any universal metaphysical glue,

---

[457] See Kitcher (1984b, 1989) for the notion 'extended explanation' and Needham (1999b) for 'overall coherence'.
[458] I am indebted here to Haldane (1996).

such as supervenience, to bind it all together? Why wouldn't the world be a very complicated and diverse place – Cartwright's 'patchwork' or 'dappled world'.[459]

Consider instead a form of anomalous monism applied to chemistry. On such a view causation is a relation between events, not between events *as* one thing or another. Events cause one another independently of how they are described, even independently of how they are identified. For example, the 'same' event can have a chemical and a neurophysiological description, or a moral and a physical description, or a macroscopic and a microscopic physical description, and so on. Of course, the only way to talk about causes giving explanations is under some description or other, but there is no need or ground for favouring one privileged description as more fundamental. The notion of event used here is a metaphilosophical term, but still close to the manifest image term 'event'. Justification of the proposal of anomalous monism (based on the primitive notion of 'event') is first grounded in philosophical discourse and this discourse is grounded, in the end, in the manifest image.

Most interpreters of Davidson's 'anomalous monism' have not read (or understood) what he says (van Brakel 1999b). The most common misreading is to assume that if (as Davidson says) 'psychological events are describable, taken one by one, in physical terms', it is entailed that 'presumably they can be explained in physical terms as well' (McIntyre 1999:380). The argument is easily transferable to the present discussion by substituting 'chemical' for 'psychological'. Though Davidson's definition of supervenience *does* imply that a change in psychological (or chemical) properties is always accompanied by a change in physical properties, it does *not* imply that the same physical properties change with the same psychological (or chemical) properties.[460] There is no reason therefore to suppose that the necessary change in physical properties of a particular event, can serve as an explanation, as McIntyre suggests.

Similarly, Davidson's anomalous monism has been accused of epiphenomenalism, because the presence of causal relevance in the scientific image (neurophysiology) may 'eclipse' the *appearance* of causal relevance in the manifest image (folk psychology).[461] Again the same reasoning might be applied to the relation of physics and chemistry: the physical cause would 'eclipse' the *appearance* of the chemical cause. Intuitively, the property from the 'lower', 'more general' level (physics) will mediate the causal relevance of the property from the 'higher', 'special science' level (chemistry), and not vice versa. If an event has a physical cause how could its occurrence be 'overdetermined' by having a chemical cause as well? The suggestion is that though chemical talk is of practical use, in the end, metaphysically or ontologically speaking, it is epiphenomenal. In the philosophy of mind epiphenomenalism has even more devastating consequences, because (Hornsby 1997:11)

---

[459] 'The dappled world is what, for the most part, comes naturally: regimented behaviour results from good engineering' (Cartwright 1999:1).
[460] Quoting Davidson (1993:7), substituting 'chemical' for 'mental' – though Davidson might consider chemistry as part of physics.
[461] See for example Pettit (1992), LePore and Loewer (1989).

It holds that nothing that we think (for instance) makes any genuine difference to what we say. It becomes impossible to see how we could have any view of *anything*.

However, moving back to chemistry and physics again, if you think P is one of the physical 'micro-realisations' of chemical event X, then why should you view P causes Q and X causes Q as competing hypotheses. The evidence you have may justify accepting both, because P and X each have their own identification. As Sober (1999:560) puts it: 'The reductionist claim that lower-level explanations are *always* better and the antireductionist claim that they are *always* worse are both mistaken.'

The overdetermination argument only works, if the concept of causation is associated with strict laws and strict definitions; if it is part of a closed, comprehensive system of laws, each of which is perfectly precise, explicit, and exception-free. Davidson endorses strict physical laws as one of the premises of his argument for anomalous monism, but as I argued in [6.2] we should give up these 'hidden' strict laws. At best there are a few strict laws plus prediction closure in some highly restricted circumstances. The idea too of closure of a system of physical laws is far from clear (Weingartner 1997). Moreover, if there is a threat of epiphenomenalism it doesn't only apply to the chemical or the mental; it applies to everything we apprehend in the world. A naturalised philosopher may take Quine or Dennett seriously if they speak of intentional or mental idiom as 'drama' or 'fiction', not something with firm, true ground. But a thorough naturalist (i.e. eliminativist) approach would require considering the whole of chemistry (including quantum chemistry) as fiction and drama (and most of physics as well). I suggest this undermines such a view on its own terms. Moreover, on the view of anomalous monism as understood here, the 'eclipse' argument doesn't even get started, because there are *neither* physical causes *nor* chemical causes. There are events causing other events and there can be all kinds of causal explanations of these events under various descriptions. Causally relevant properties are distinct from the causal relation itself. The everyday physical *and* the everyday chemical - where 'everyday' may refer to both the 'ordinary' everyday and the 'laboratory/theoretician's' everyday are equally part of a single (manifest) causal world view. There is not one preferred system of causal truths underlying all rational explanations.

A criticism of Davidson that is relevant in this connection is what one might refer to as the space-time ideology.[462] If events are so 'metaphysical' that they are *events* before they are *physical* or *mental* or *chemical* events, then there is a problem about how to identify these events. If there is no privileged vocabulary then the identification of an event will depend on how it is described and it seems a 'classical physics' assumption of Davidson to assume that these metaphysical events have a place in space-time. How otherwise could he be sure that every event falls under some (strict) physical law? Davidson's thesis about causation is

---

[462] Most forcefully argued in many papers by Hornsby (1997).

basically taken from a physicalistic discourse.[463] However, if we think in terms of world patches, instead of one closed space-time structure governed by strict laws, the space-time ideology disappears with Davidson's assumption of strict (physical) laws.

Anomalous monism is anti-reductionistic and ontologically pluralistic or even promiscuous in two senses. It has neither a need for vertical, downward (reductive) or upward (emergentist), metaphysical connections, nor a need for horizontal (cross-wise) metaphysical unification. What we have is a messy word and ceteris paribus symmetrical correlations between different patches and different discourses. In the terminology of Goodman's (1978) 'ways of worldmaking': Each discourse, each description is a version which is, if right, true for the world it is about. Different versions (world patches) can be connected using other versions (the successors of the reductionist's bridge laws), though some versions about 'the same patch' may be inconsistent relative to one another - hence Goodman's preference for talk of many worlds.

It might be thought that giving up strict laws takes the monism out of anomalous monism.[464] But this is not true, because as I argued above, a pure monism doesn't have to be purely physical. A monism of events as hypothesised bare particulars is consistent with a nomological pluralism according to which *all* laws are ceteris paribus. This opens up the assumption of the closedness of the physical world and the asymmetry of principles such as physical embedding or completeness;[465] and it leaves open the possibility that 'chance is a factor in the universe'.[466] The physical embedding of 'everything' can stay, of course. Even if there are no strict laws, physics can still claim that it has something to say about everything that occurs in time-space (as defined by physics). It still leaves room for the claim that two events somehow correlated, though not backed by *strict* physical laws, *must* be describable in physical terms too. But this claim would have to be argued separately and would require an independent definition of physics (no trivial matter).

There is a deep problem here in how we are to take various types of monisms. Often monism is held to imply an asymmetry of a reductionistic sort, as when one says materialism is a form of monism. But the anomalous monism I propose should be seen as *anomalous* in a variety of ways and should not be taken as a variant of materialism. Consider an analogy with Machean sense data. Sense data theories can be criticised if the sense data exemplify an interface between world and mind. But Mach held, at least in some of his publications, that both the mind and the world, as it were, emerge out of sense data.[467] Such sense data cannot be studied, but they make possible both physical and psychological things and explain, according to Mach, the neat correlation between say brightness as perceived

---

[463] Cf. Hornsby (1997:10).
[464] See for example Child (1994:220).
[465] Peacocke (1979), Owens (1992:118).
[466] van Brakel (1991b, 1993b).
[467] Mach (1919), van Brakel (1993c)- somewhat similar views are held by James, Whitehead, or even Ayer.

and brightness as measured physically.[468] The deep point is that it doesn't even matter much whether one tells the story in terms of sense data, or Whiteheadian events, or simple Tractarian objects, or whatever. The deep point is that *if* one appeals to metaphysics (which is what an appeal to the priority of the scientific image - read physics - would imply), then the fundamental building blocks of the world have to be neutral relative to the preoccupations of the various sciences and other human endeavours.

Anomalous monism is therefore to be understood as a variant of global token-token identity. We start with what scientists call a working hypothesis and philosophers might call a transcendental assumption: 'the world consists of events'. At this level it doesn't make sense to ask for criteria of identity, because such criteria require *descriptions* of the events. As Aristotle already said: there is no *thisness* without *suchness*. Sortals play an essential role in individuation. But, event, thing, individual, entity are not genuine sortals. Only proper sortals are capable of delimiting what falls in their extension in a definite way (and even then doubts can be raised about the definiteness).[469] How events are individuated is relative to the language used to describe them; the world does not come to us pre-sorted in a preferred way.[470] This doesn't mean that 'anything goes' - in fact making descriptions that 'work' and are 'right' is extremely difficult.[471] But it does mean that neither physics (nor chemistry), neither psychology (nor neurophysiology) can claim unique access to the preferred way of ordering and describing the world (of events, or objects for that matter) 'as it really is'. Events such as boiling water have coherent, intelligible persistence conditions, though the exact delineation of the event, both in everyday and chemical terms, as well as in physical terms is not possible *and not relevant*. Given the holistic character of quantum mechanics, any 'view from nowhere' delineation would be arbitrary and can only be motivated from considerations outside the domain of the fundamental theory.

There is no God's-Eye-point-of-view-meta-description that gives the only true identification of events. We can only ask about criteria of identification if something is in place - we can only quantify over something if there is already a domain of individuals. The request for individuation must stop somewhere. In a physical discourse events can be identified as physical events; in a chemical discourse events can be identified as chemical events. Similarly a system can have both quantum states and classical states (Cartwright 1994b:362). There's no contradiction here. Perhaps some ceteris paribus regularity can be found to correlate the two. That would be nice, but the use of quantum and classical discourse doesn't depend on it. Similarly, the event of a water wave in a swimming pool hitting a floating cork is not literally identical with any complex microscopic event.[472] More generally, it is not possible to say whether an event identified un-

---

[468] Actually this correlation isn't so neat as is generally assumed, also by Mach (Lockhead 1992, Saunders and van Brakel 1997a).
[469] See Quine (1960), Wiggins (1980), Lowe (1989).
[470] Carnap (1956), Lowe (1989), Quine (1993).
[471] Goodman (1978, 1984).
[472] The example is from Haugeland (1982). Cf. also Post (1987:196), who gives an argument against 'universal physical individuatability' and Putnam (1988) on computational plasticity.

der some physical description is exactly the same event (or not) as the one that is described under a chemical description (or a psychological, or a moral description, and so on).

Of course rough hints are possible, but they don't give exact descriptions, not even of the space-time boundaries of events - compare events happening to molecules under a physical or chemical description. As Ramsey (1997:237) argued in connection with the shape of molecules:

> The requirement that events at one level be naturally isolable at another is enormously implausible as intertheoretic compatibility condition.

Talking about one event and one other event is only by way of speaking, as there are no discourse-independent identification criteria of one or the other event. Any parcelling out of events bounded in time and space makes sense only given a particular discourse that is in place. It is not merely that at the lower level finer discriminations are made. It is that most (or all) of the structure at the higher level can only be recovered at the lower level by adding models, background assumptions, boundary conditions, and other ceteris paribus aspects to the fundamental laws. Chemical events, causally related to other events (chemical, physical, ...) are, as it were, too big and too rough-hewn to be picked out in the physical vocabulary.

It is not that there are first physical events and chemical events and then some physical events turn out to be identical with some chemical events. In ordinary language all events are EVENTS, no matter whether they are described in physical, social, chemical, or whatever terms. By assuming that each event that can be given a chemical description also has a physical description (although we cannot specify it or provide exact criteria of identification), some insight is gained in the *autonomy* of the chemical *and* the physical, while still keeping both in the same world. The purpose of anomalous monism is to make more sense of some well-known facts; nothing more, nothing less.

8.2   primacy of manifest over scientific image

In this last section I briefly return to the general question of the tension between the manifest and the scientific image in the sense of Sellars. In [2.1] I raised the question: where does chemistry fit in? Does it belong to the manifest or the scientific image? I suggested that Sellars, and most other philosophers and scientists would reply: the scientific image. My intention has been to raise some doubts about that apparent self-evidence, by showing that the pull towards unification tends to reduce the scientific image to Kant's 'proper science', i.e. physics. On such a view, strictly speaking, everything, including chemistry, is either eliminated or relegated to the manifest image. Instead I've argued in the previous section for a form of promiscuous realism or realism-with-a-small-r, which is pluralistic and pragmatic in the sense that there's room for the macroscopic, the microscopic, and the submicroscopic, for the scientific and the manifest, without the need to fit it all together in one rigid world, governed by strict laws, causal closure, rigid designators, and so on.

CONCLUSION   199

In previous chapters I've argued *not* that theories that are more 'fundamental' or more 'mathematical' are less good. I've argued that *if* the question of priority is pressed, it is the manifest substances, their properties and uses that form the methodological, epistemological, and ontological basis relative to which a discourse in terms of molecular structure or of quantum mechanics is constructed. As to the general relation of the scientific and the manifest image I want to go further. Sellars was absolutely right to raise the question of priority here and the answer has to be, the primacy of the manifest over the scientific image (within the constraints set by the regime of truth that produces the notions of manifest and scientific image). I will briefly indicate a variety of lines of argument that support this stance. Most of it doesn't bear directly on chemistry. Still the issue is directly relevant to the philosophy of chemistry, because of the perennial question of the relation of chemistry and physics. Only if the general claim for the primacy of the manifest image makes sense, can good sense be made of positions in the philosophy of chemistry such as protochemistry. If the primacy of the scientific image is taken for granted, one has already given in to the suggestion that the only way chemistry or any other science or human endeavour can become 'proper' is by being reduced to physics. Here then is a list of considerations in favour of the primacy of the manifest over the scientific image.

Attempts to provide a picture of science unified by one method have failed. Unity of method would require one unique set of epistemic virtues - the criteria that are used to decide what is a good scientific theory and how to assess new scientific knowledge. But if one looks at what epistemic virtues have been proposed by various scientists and philosophers of science (Block 8-2) one notes (van Brakel 1998b):

- Different philosophers emphasise different virtues.
- In different knowledge domains, different virtues come to the fore; there is not one domain (pure science, pure knowledge, pure philosophy) that sets some meta-epistemic standards that apply across the board.
- In each concrete case many virtues are relevant, some of which may conflict; due to the particularities of the situation, one virtue may completely overrule another.
- There is no single virtue whose application is straightforward; virtues are not like rules that can be applied mechanically to concrete cases - they are more like maxims.
- However one draws the dividing line between epistemic and pragmatic virtues, to claim truth, or empirical adequacy, or an economic rendering of our sensory input, or whatever, as the goal of science, such a goal is a value, not a scientific fact.

Epistemic virtues are grounded, in the end, in the manifest image. As Peirce noted, common sense plausibility judgements set the normative standards for scientific practice (van Brakel 1993b). We say science has done rather well, because it satisfies criteria that are not internal to or restricted to science or even to science-dominated cultures. Alternatively, if one claims a certain goal (or

Block 8-2 Epistemic or pragmatic virtues culled from writings of philosophers of science

empirical adequacy	'saving the phenomena', testability, falsifiability, refutability, predictive power
coherence	logical consistency, coherence (or 'naturalness'), consonance with other parts of science and (more controversially) consonance with broader world views (e.g. with metaphysical principles of natural order)
scope	generality, absence of ad hoc features (or 'naturalness'), causal specificity, accuracy, cumulativeness, fertility, fruitfulness, consilience, newness, unification
economic/pragmatic	simplicity, conservativeness, initial plausibility, practical use, economic description of phenomena, 'has to make sense', relevant, appropriate, informative, increasing our understanding
social	result of cooperation, open to criticisms from many sources, respect for all kinds of evidence, trust, taking into account the heterogeneity of the world
metaphysical (?)	eudaimonia, *summum bonum*, rightness, hope, aesthetic beauty, modesty

commitment to a particular method of inquiry) as *defining* science (as Quine might do), then judgements about the usefulness of this definition and the relevance of engaging in activities thus defined is a judgement within the realm of the manifest image. The most general ideas about what constitutes empirical inquiry are not a product of science or its development - although no doubt such ideas have been refined in science to fit more specific goals.

Even if there's something in the technical notion of bootstrapping, bootstrapping always draws on background hypotheses that are grounded in the manifest image. Bootstrapping may work locally, but only *given*, amongst many other things - a pretheoretic judgement about what counts as 'data'. If these pretheoretical judgements were completely wrong, bootstrapping would be powerless to repair the situation (Bealer 1992). There's no transcendental or self-correcting inductive method that systematically comes closer to what is right by the lights of all eternity. The only way out of it would be to appeal to exactly one best method of inquiry and exactly one best end of enquiry which gives THE answer. The only plausible answer of this sort would seem to be the view that *nothing* we currently believe exists (whether we tend to be empiricists, pragmatists, scientific realists, premodernists, or whatever), because *everything* supervenes in an epiphenomenal way on whatever the ultimate ontologically relevant referents are of the Theory of Everything.

Attempts to specify a reduction relation to fit all sciences into one world picture have failed. It isn't even possible to reduce chemistry to physics, or macrophysics to microphysics for that matter. Appeals to IDEAL PHYSICS, 'the best total causal account of the world,' or 'the language of completed total science' are either empty or a commitment to a value judgement not itself part of the ideal theory.[473] The unity and pluralism of the manifest traditions cannot fail, because they sustain everything. Cross-culturally only something like the manifest image is shared. People do not normally become tongue-tied, or experience serious rupture when they assimilate (new) science, because the manifest image (taken across time and place) is 'inherently' multi-faceted, multi-perspectival, open-ended, and so on. It can absorb anything to do with people. In contrast, the scientific image has unsurmountable difficulties trying to absorb the manifest image - when pressed it has to resort to eliminativism. The manifest image has no difficulty *in principle* absorbing anything, because all else *automatically* appeals to *some* of the certainties that are entrenched in the daily life world (again understood cross-culturally).

Even the unification of physics is not as straightforward as may appear from the science page of the newspaper. Current proposals for unification are little more than attempts to combine, without too many unrealistic assumptions, purely formal structures into one unified structure. Not only philosophers say this.[474] Even if we are prepared to accept the *idea* of the Theory of Everything, the possibility that *absolute* chance is a factor in the (this) universe cannot be excluded (van Brakel 1993b). This latter metaphysical speculation can be given a sound scientific backing. As Redhead (1991) says:

> It is generally agreed that the present state of the universe is the outcome of a series of randomly occurring symmetry breakings, which means that the TOE [Theory Of Everything] cannot itself predict many of the universe's essential features as we currently observe it - a very important limitation on what TOEs can be expected to deliver, even in principle.

Assume we accept that quantum mechanics forces us to the view that everything there is supervenes on chance events. But then we still need to decide on some initial and boundary conditions to apply the theory to the world. Moreover, the chance events described by quantum mechanics only exist relative to a higher order belief in the limit of chance (van Brakel 1991b). Therefore even for quantum mechanics there's no way to break out of dependence on the regularities perceived in the manifest image.[475]

Also there's no way of testing the predictions of quantum mechanics without appeal to macroscopic objects and colloquial language to describe experiments, as Bohr and Heisenberg stressed in different ways.[476] As Bell put it: 'we have this beautiful mathematics, and we don't know which part of the world it should be applied to'.[477] Think too of logic or mathematics: at the meta-meta-level we use common sense intuitions to 'prove' the relevance and validity of certain approaches.[478]

---

[473] Crane and Mellor (1990).
[474] See, for example, quotes of Feynman and other physicists in Gleick (1992) and Maudlin (1996).
[475] Cf. also Heisenberg (1942:246-58) who subdivides chemistry into 'heat', 'chemical laws', 'limits' [of distinguishing mechanical and chemical descriptions at the 'lowest' level of the *material world*], and 'chance'.
[476] Cf. the discussion between Bohr and Heisenberg recounted in Heisenberg (1972:129).
[477] Bell, in Bernstein (1991:52); cf. citation on p. 143. See, more generally, van Fraassen (1992).
[478] See quotations and references in van Brakel (1998a:79).

Then there is the suggestion that the content of the world and its dynamics is a closed system of natural kinds, causes or causal structures, and natural laws. As we have seen neither natural kinds [2.5] nor natural laws [6.2] live up to their universalistic expectations. Theoretical concepts in different discourses can divide the world differently. One and the same complex phenomena can be categorised in different ways by theories with different predicates or categories, and causal patterns discernible in one set of terms need not be perspicuously captured in another. One might respond: 'But surely some X-like entities are not X, because they are not micro-X'. But this is only so if we have already *decided* to use 'X' in that sense. Whales are fish if the latter means 'animals of the sea'. This is not to say that whales are not mammals. There's no a priori reason why whales couldn't be both mammals and fish. Only if we are already *inside* the scientific image is it wrong to say that whales are fish.

Of course, in the modern world, the manifest and scientific image cannot be neatly separated. [479] The manifest image is constantly modified under the impact of scientific developments. There is no denying that the scientific image has had enormous impact on the current manifest life forms. That doesn't however diminish the primacy of the latter. The fact that most of the posits of contemporary daily life in the western world have their origins in developments of science and its ocularcentric epistemology, doesn't change the fact that when their grounding status is disputed, adjudication will be governed by criteria that are not the product of science. Again, this is not to say that science has not produced all kinds of useful criteria of inquiry. Rather, it is to say that the judgement that these are good criteria is not itself a scientific judgement. Similarly, it is not denied that science is good at giving explanations. Rather it is to say that the judgement that one explanation is better than another is based on judgements grounded in the manifest image. Some kind of explanatory attitude is an essential part of any form of (human) life. This implies that not all explanations are equally valid - this is true whether or not a scientific image is around. One might say: 'Science could explain, for example, how explanation, communication, and normativity are possible among humans.' But how is the request for such an explanation ever finally justified? After all, there will always be alternative 'sciences' to offer explanations of whatever is considered relevant. To make a judgement with regard to these alternatives we are committed to making judgements on issues like 'deciding which features of science we value most,' 'rightness,' 'appropriateness to the circumstances,' and so on.

---

[479] Sellars would seem to agree that there is *no* distinctive scientific method (Sicha 1988).

references

Abbott, B. (1997) A note on the nature of 'water', *Mind*, 106: 311-319.
Abbott, B. (1999) Water = H$_2$O, *Mind*, 108: 145-148.
Abbri, F. (1996) Lavoisier and the foundations of chemistry, pp. 143-154 in *Philosophers in the Laboratory* (V. Mosini, ed.), Rome: Euroma.
Abramova, N.T., R.V. Garkovenko and E.F. Solonov (1963) Book review of M. Shakparonov *Chemistry and Philosophy* (1962) [in Russian], Voprosy filosofii, (8) 176-180.
Achinstein, P. (1968) *Concepts of Science: A Philosophical Analysis*, Baltimore: Johns Hopkins Press.
Ackerman, D. (1983) Natural kinds, concepts and propositional attitudes, *Midwest Studies of Philosophy*, 5: 469-486.
Adams, E. (1988) Stroll on surfaces, *Inquiry*, 31: 549-555.
Adickes, E. (1924) *Kant als Naturforscher*, Berlin: de Gruyter.
Akeroyd, (1986) A challenge to the followers of Lakatos, *The British Journal for the Philosophy of Science*, 37: 359-362.
Akeroyd, F.M. (1988) Research programmes and empirical results, *The British Journal for the Philosophy of Science*, 39: 51-58.
Akeroyd, F.M. (1990a) An oscillatory model of the growth of scientific knowledge, *The British Journal for the Philosophy of Science* 41: 407-14.
Akeroyd, F.M. (1990b) The challenge to Lakatos restated, *The British Journal for the Philosophy of Science*, 39: 437-439.
Akeroyd, F.M. (1993) Laudan's problem solving model, *The British Journal for the Philosophy of Science*, 44: 785-788.
Akeroyd, F.M. (1996) Poppers Beitrag für eine Philosophie der Chemie, pp. 67-76 in *Philosophie der Chemie: Bestandsaufnahme und Ausblick* (N. Psarros, K. Ruthenberg and J. Schummer, eds.), Würzburg: Königshausen & Neumann.
Akeroyd, F.M. (1997) Conceptual aspects of theory appraisal: Some biochemical examples, *Hyle*, 3: 95-102.
Aldrich, H.S., L.P. Gary, H.J. Lader, L.C. Cusachs and J.H. Corrington (1971) The structure of water-II, *Journal of Colloid and Interface Science*, 36: 536-542.
Alexander, P. (1985) *Ideas, Qualities and Corpuscles. Locke and Boyle on the External World*, Cambridge: Cambridge University Press.
Allchin, D. (1990) Paradigms, populations and problem-fields: approaches to disagreement, *Philosophy of Science Association*, 1: 53-66.
Allchin, D. (1992) How do you falsify a question?: crucial tests versus crucial demonstrations, *Philosophy of Science Association*, 1: 74-88.
Allchin, D. (1996) Cellular and theoretical chimeras: piecing together how cells process energy, *Studies in History and Philosophy of Science*, 27: 31-41.
Allchin, D. (1997) A twentieth-century phlogiston: constructing error and differentiating domains, *Perspectives on Science*, 5: 81-127.
Allen, L.C. (1971) An annotated bibliography for anomalous water, *Journal of Colloid and Interface Science*, 36: 554-561.
Allen, L.C. and Kollman, P.A. (1970) A theory of anomalous water, *Science*, 167: 1443-1454.
Allen, L.C. and Kollman, P.A. (1971a) Comparison of theoretical models for anomalous water, *Journal of Colloid and Interface Science*, 36: 461-468.
Allen, L.C. and Kollman, P.A. (1971b) What can theory say about the existence and properties of anomalous water, *Journal of Colloid and Interface Science*, 36: 469-482.
Amann, A (1991) Chirality: A superselection rule generated by the molecular environment?, *Journal of Mathematical Chemistry*, 6: 1-15.

Amann, A (1992) Must a molecule have a shape?, *South African Journal of Chemistry*, 45: 29-38.
Amann, A. (1993) The gestalt problem in quantum theory: generation of molecular shape by the environment, *Synthese*, 97: 125-156.
Amann, A. (1996) Individual interpretation of quantum mechanics and quantum theory of molecular shape, pp. 21-48 in *Philosophers in the Laboratory* (V. Mosini, ed.), Rome: Euroma.
Amann, A. and Gans, W. (1989) Theoretical chemistry en route to a theory of chemistry, *Angewandte Chemie (Int. Ed. Engl.)*, 28: 268-276.
AN SSSR (1954) *Der gegenwärtige Stand der Strukturtheorie in der organischen Chemie*, translated from the Russian (USSR Academy of Science, Department of Chemical Sciences) by G. Rudakoff and published by the Chemical Society of the DDR, Berlin: Akademie-Verlag (1956).
Arabatzis, T. (1998) How the electrons spend their leisure time: Philosophical reflections on a controversy between chemists and physicists, pp. 149-159 in P. Janich and N. Psarros (eds.), *The Autonomy of Chemistry: 3rd Erlenmeyer-Colloquy for the Philosophy of Chemistry*, Würzburg: Königshausen & Neumann.
Arbuzov, A.E. (1952) Butlerov's theory of chemical structure, *Bulletin of the Academy of Sciences of the USSR / Division of chemical science*, 1-11.
Armstrong, D.M. (1978a) *Nominalism and Realism*, Cambridge: Cambridge University Press.
Armstrong, D.M. (1978b) *A Theory of Universals*, Cambridge: Cambridge University Press.
Armstrong, D.M. (1983) *What Is a Law of Nature?*, Cambridge: Cambridge University Press.
Armstrong, D.M. (1989) *A Combinatorial Theory of Possibility*, Cambridge: Cambridge University Press.
Armstrong, D.M. (1999) A naturalist program: Epistemology and ontology, *Proceedings and Addresses of the American Philosophical Association*, 73: 77-89.
Aronson, J.L., R. Harré and E.C. Way (1994) *Realism Rescued: How Scientific Progress is Possible*, London: Duckworth.
Arteca, G.A. and Mezey, P.G. (1989) Molecular similarity and molecular shape changes along reaction paths: a topological analysis and consequences on the Hammond postulate, *Journal of Physical Chemistry*, 93: 4746-4751.
Aspect, A., J. Dalibard en G. Roger (1982) Experimental tests of Bell's inequalities using time-varying analysers, *Physical Review Letters*, 49: 1804-1807.
Au, T.K., A.L. Sidle and K.B. Rollins (1993) Developing an intuitive understanding of conservation and contamination: Invisible particles as a plausible mechanism, *Development Psychology*, 29: 286-299.
Aune, B. (1990) Sellars's two images of the world, *The Journal of Philosophy*, 87: 537-545.
Auyang, S.Y. (1995) *How Is Quantum Field Theory Possible?*, Oxford: Oxford University Press.
Averill, E. (1982) Essence and scientific discovery in Kripke and Putnam, *Philosophy and Phenomenological Research*, 43: 253-257.
Ayer, A.J. (1992) Reply to Hilary Putnam, pp. 454-465 in L.E. Hahn (ed.), *The Philosophy of A.J. Ayer*, La Salle: Open Court.
Ayers, M. (1991) *Locke, Volume II: Ontology*, London and New York: Routledge.
Bachelard, G. (1932) *Le pluralisme cohérent de la chimie moderne*, Paris: Vrin; second edition 1975.
Bachelard, G. (1934) *Le nouvel esprit scientifique*, Paris: Alcan; Paris: Presses Universitaires de France (reprint 1973),
Bachelard, G. (1942) *L'eau et les rêves: Essai sur l'imagination de la matière*, Paris: Corti.
Bachelard, G. (1949) *Le rationalisme appliqué*, Paris: Presses Universitaires de France.
Bachelard, G. (1971) Epistémologie de la chimie, in *Épistémologie*, Paris: Presses Universitaires de France.
Bader, R.F.W. (1990) *Atoms in Molecules: A Quantum Theory*, Oxford: Clarendon Press.

Bader, R.F.W., P.L.A. Popelier and T.A. Keith (1994) Die theoretische Definition einer funktionellen Gruppe und das Paradigma des Molekülorbitals, *Angewandte Chemie*, 106: 647-659.
Bader, R.F.W., T.T. Nguyen-Dang and Y. Tal (1981) A topological theory of molecular structure, *Reports of Progress in Physics*, 44: 893-948.
Baierlein, R. (1971) *Atoms and Information Theory: An Introduction to Statistical Mechanics*, San Francisco: Freeman.
Baird, D (2000) Encapsulating knowledge: the direct reading spectrometer, *Foundations of Chemistry*, 2: 5-46.
Balashov, Y. (1994) Uniformitarianism in cosmology: Background and philosophical implications of the steady-state theory, *Studies in History and Philosophy of Science*, 25: 933-958.
Balzer, W., C.-U. Moulines and J.D. Sneed (1987) The structure of Daltonian stoichiometry, *Erkenntnis* 26: 103-27.
Bantz, D.A. (1980) The structure of discovery: Evolution of structural accounts of chemical bonding, pp. 291-329 in *Scientific Discovery: Case Studies* (T. Nickles, ed.), Dordrecht: Reidel.
Baracca, A (1996) Chemistry's leading role in the scientific revolution at the turn of the century, pp. 61-80 in *Philosophers in the Laboratory* (V. Mosini, ed.), Rome: Euroma.
Barnett, D. (2000) Is water necessarily identical to H2O?, *Philosophical Studies*, 98: 99-112.
Bealer, G. (1992) The incoherence of empiricism, *Proceedings of the Aristotelian Society*, Vol. 66: 99-138.
Bear, J. (1972) *Dynamics of Fluids in Porous Media*, New York: Elsevier.
Beattie, J. (1993) Is Putnam inconsistent and parochial?, *Australasian Journal of Philosophy*, 71(3): 316-324.
Beatty, J. (1997) Why do biologists argue like they do?, *Philosophy of Science*, 64 (Proceedings): S432-S443.
Bechtel, W. (1988) Fermentation theory: empirical difficulties and guiding assumptions, pp. 163-180 in *Scrutinising Science* (A. Donovan, L. Laudan and R. Laudan, eds.), Dordrecht: Kluwer.
Bechtel, W. and R.C. Richardson (1992) Emergent phenomena and complex systems, pp. 257-288 in A. Beckermann, H. Flohr, and J. Kim (eds.) *Emergence or Reduction? Essays on the Prospects of Nonreductive Physicalism*, Berlin/New York: de Gruyter..
Beckermann, A. (1992) Supervenience, emergence, and reduction, pp. 94-118 in A. Beckermann, H. Flohr, and J. Kim (eds.) *Emergence or Reduction? Essays on the Prospects of Nonreductive Physicalism*, Berlin/New York: de Gruyter.
Beckermann, A., H. Flohr, and J. Kim (eds.) (1992) *Emergence or Reduction? Essays on the Prospects of Nonreductive Physicalism*, Berlin/New York: de Gruyter.
Bell, J.S. (1989) Six possible worlds of quantum mechanics, pp. 359-373 in S. Allén, *Possible Worlds in Humanities, Arts and Sciences. Proceedings of Nobel Symposium 65*, Berlin: Walter de Gruyter.
Bellu, E. (1973) The dialectical significance of chemistry in the works of Fr. Engels, *Revue roumaine des sciences sociales: série de philosophie et logique*, 17: 163-169.
Bellu, E. (1979) Conservation-transformation au niveau chimique, *Revue roumaine des sciences sociales: série de philosophie et logique*, 23: 559-569.
Benfey, O.T. (1963) Concepts of time in chemistry, *Journal of Chemical Education*, 40:574-577.
Bensaude, B. (1974) Histoires de la chimie, *Critique: Revue générale des publications françaises et étrangers*, 30: 790-799.
Bensaude-Vincent, B. (1994) La chimie: un statut toujours problématique dans la classification du savoir, *Revue de synthèse*, 115: 135-148.
Bensaude-Vincent, B. (1998) *Éloge du Mixte: Matériaux nouveaux et philosophie ancienne*, Paris: Hachette.
Bensaude-Vincent, B. and Abbri, F. (eds.) (1995) *Lavoisier in European Context: Negotiating a New Language for Chemistry*, Canton MA: Science History Publications.

Bensaude-Vincent, B. and I. Stengers (1993) *Histoire de la chimie*, Paris: La Découverte.
Bernal, J.D., Barnes, P., Cherry, I.A. and Finney, J.L. (1969) Letter to editor, *Nature*, 224: 393-394.
Bernstein, J. (1991) *Quantum Profiles,* Princeton: Princeton University Press.
Bernstein, R.J. (1966) Sellars' vision of man-in-the-universe, *The Review of Metaphysics*, 20: 113-143, 290-316.
Bhargava, R. (1992) *Individualism in Social Science: Forms and Limits of a Methodology*, Oxford: Clarendon Press.
Bhushan, N. and S. Rosenfeld (1995) Metaphorical models in chemistry, *Journal of Chemical Education*, 72: 578-582.
Bigelow, J. (1990) The world essence, *Dialogue*, 205-217.
Bigelow, J., B. Ellis and C. Lierse (1992) The world as one of a kind: Natural necessity and laws of nature, *The British Journal for the Philosophy of Science*, 43: 371-388.
Bird, R.B., W.E. Stewart and E.N. Lightfoot (1960) *Transport Phenomena*, New York and London: Wiley.
Black, M. (1962) Models and archetypes, pp. 219-233 in his *Models and Metaphors: Studies in Language and Philosophy*, Ithaca: Cornell University Press.
Blackburn, S. (1985) Supervenience revisited, in *Exercises in Analysis* (I. Hacking, ed.), Cambridge: Cambridge University Press.
Bocheński, I.M. (1963) *The Dogmatic Principles of Soviet Philosophy*, Dordrecht: Reidel.
Bogaard, P.A. (1981) The limitations of physics as a chemical reducing agent, *PSA* [*Philosophy of Science Association 1978*], 2: 345-356.
Bogdan, R.J. (ed.) (1984) *D.M. Armstrong*, Dordrecht: Reidel.
Boghossian, P. (1997) What the externalist can know a priori, *Proceedings of the Aristotelian Society*, 161-175.
Böhme, F. (1980) On the possibility of 'closed theories', *Studies in History and Philosophy of Science*, 11: 163-172.
Böhme, G. (1980) Aristotelische Wissenschaft - Aristoteles' Chemie: eine Stoffwechselchemie, pp. 101-120 in *Alternativen der Wissenschaft*, Frankfurt a/M: Suhrkamp.
Boyd, R. (1989) What realism implies and what it does not, *Dialectica*, 4: 5-29.
Boyd, R. (1991) Realism, anti-foundationalism and the enthusiasm for natural kinds, *Philosophical Studies*, 61: 127-148.
Boyd, R. (2000) Kinds as the "workmanship of men": Realism, constructivism, and natural kinds, pp. 52-89 in J. Nida-Rümelin (ed.), *Rationalität, Realismus, Revision*, Berlin: Walter de Gruyter.
Boyd, R., P. Gasper and J.D. Trout (eds.) (1991) *The Philosophy of Science*, Cambridge MA: MIT Press
Bradley, J. (1955) On the operational interpretation of classical chemistry, *The British Journal for the Philosophy of Science*, 6: 32-42.
Bradley, J. (1962) Discussion of professor Paneth's article, *The British Journal for the Philosophy of Science*, 13: 316-317.
Bradley, J. (1963) Discussion of professor F.A. Paneth's second article, *The British Journal for the Philosophy of Science*, 14: 39-40.
Brandon, R.N. (1997) Does biology have laws? The experimental evidence, *Philosophy of Science*, 64 (Proceedings): S444-S457.
Brandt, R. (1991) Kants Vorarbeiten zum Übergang von der "Metaphysik der Natur zur Physik": Probleme der Edition, pp. 1-27 in Forum für Philosophie Bad Homburg, *Übergang: Untersuchungen zum Spätwerk Immanuel Kants*, Frankfurt: Klostermann.
Bridgman, P.W. (1927) *The Logic of Modern Physics*, New York: Macmillan (1948).
Bridgman, P.W. (1931) *Dimensional Analysis*, Cambridge MA: Harvard University Press; quoted from 1963 reprint; 1st edition 1922.
Broad, C.D. (1925) *The Mind and its Place in Nature*, London: Kegan Paul, Trench, Trübner & Co.
Brown, G.B. (1941) A new treatment of the theory of dimensions, *Proceedings of the Physical Society*, 53: 418-432.

Brown, J. (1998) Natural kind terms and recognitional capacities, *Mind*, 107: 275-303.
Brummer, S.B., F.H. Cocks, G. Entine and J.I. Bradspeis (1971) Anomalous waterproperties and factors affecting its yield, *Journal of Colloid and Interface Science*, 36: 489-502.
Bruner, J. (1986) *Actual Minds, Possible Worlds*, Cambridge MA: Harvard University Press.
Brush, S.G. (1978) Why chemistry needs history - and how it can get some, *Journal of College Science Teaching*, May 1978.
Brush, S.G. (1999) Dynamics of theory change in chemistry: Part 1. The benzene problem 1865-1945, *Studies in History and Philosophy of Science*, 30: 21-79.
Buchwald, J.Z. (1992) Kinds and the wave theory of light, *Studies in History and Philosophy of Science*, 23: 39-74.
Buck, P. (1998) Chemical comprehension versus physical comprehension: The case of Julius Robert Mayer, pp. 119-132 in P. Janich and N. Psarros (eds.), *The Autonomy of Chemistry*, Würzburg: Königshausen & Neumann.
Buckingham. E. (1914) On physically similar systems; illustrations of the use of dimensional equations, *The Physical Review*, 4: 345-376.
Buckingham. E. (1915) Model experiments and the forms of empirical equations, *Transactions of the American Association of Mechanical Engineers*, 37: 263-296.
Budreiko, N.A. (1970) *Filosofskie voprosy khimii* [Philosophical Problems in Chemistry], Moscow: Syssh. Shkola.
Bumstead, H.A. (1928) Preface to Gibbs (1928), pp. i-xxvii.
Bunge, M. (1982) Is chemistry a branch of physics?" *Zeitschrift für allgemeine Wissenschaftstheorie*, 13: 209-223.
Bunge, M. (1985) *Treatise on Basic Philosophy*, Dordrecht: Reidel, vol. 7, pp. 219-230.
Burbidge, J.W. (1993) Chemistry and Hegel's logic, pp. 609-617 in *Hegel and Newtonianism* (M.J. Petry, ed.), Dordrecht: Kluwer.
Burbidge, J.W. (1996) *Real Process: How Logic and Chemistry Combine in Hegel's Philosophy of Nature*,
Burger, T. (1976) *Max Weber's Theory of Concept Formation: history, laws, and ideal types*, Durham NC: Duke University Press.
Bykov, G.V. (1962) The origin of the theory of chemical structure, *Journal of Chemical Education*, 39: 220-224.
Caldin, E.F. (1959) Theories and the development of chemistry *The British Journal for the Philosophy of Science*, 10: 209-222.
Caldin, E.F. (1961) *The Structure of Chemistry in Relation to the Philosophy of Science*, London and New York: Sheed and Ward.
Callaway, H.G. and J. van Brakel (1996) No need to speak the same language, *Dialectica*, 50: 253-297.
Callender, C. (1999) Reducing, thermodynamics to statistical mechanics: the case of entropy, *The Journal of Philosophy*, 96: 348-373.
Campbell, N.R. (1920) *Physics: The Elements*, Cambridge: Cambridge University Press; republished as *Foundations of Science: The Philosophy of Theory and Experiment*, New York: Dover (1957).
Campbell, N.R. (1924) Dimensional analysis, *The Philosophical Magazine* [6] 47:481-494.
Cao, T.Y. (1997) *Conceptual Developments of 20th Century Field Theories*, Cambridge: Cambridge University Press.
Carmeliet, J., Descamps, F. and Houvenaghel, G. (1999) A multiscale network model for simulating moisture transfer properties of porous media, *Transport in Porous Media*, 35: 67-88.
Carnap, R. (1956) *Meaning and Necessity*, Chicago: The University of Chicago Press, pp. 206-221.
Carnap, R. (1963) Reply to Beth, in *The Philosophy of Rudolf Carnap* (P.A. Schilpp, ed.), La Salle: Open Court.
Carrier, M. (1990) Kants Theorie der Materie und ihre Wirkung auf die zeitgenössische Chemie, *Kantstudien*, 81: 170-210.

Carrier, M. (1991) Kraft und Wirklichkeit. Kants späte Theorie der Materie, pp. 208-230 in Forum für Philosophie Bad Homburg, *Übergang: Untersuchungen zum Spätwerk Immanuel Kants*, Frankfurt: Klostermann.
Carrier, M. (1993) What is right with the miracle argument: establishing a taxonomy of natural kinds, *Studies in History and Philosophy of Science*, 24(3): 391-409.
Carrier, M. (1998) In defence of psychological laws, *International Studies in the Philosophy of Science*, 12: 217-232.
Carroll, F.A. (1998) *Perspectives on Structure and Mechanism in Organic Chemistry*, London: Brooks/Cole.
Cartwright, N. (1983) *How the Laws of Physics Lie*, Oxford: Clarendon Press.
Cartwright, N. (1994) Fundamentalism vs. the patchwork of laws, *Proceedings of the Aristotelian Society*, 94: 279-292.
Cartwright, N. (1995) The metaphysics of the disunified world, *PSA 1994*, 2: 357-364.
Cartwright, N. (1997) Where do laws of nature come from?, *Dialectica*, 51: 65-78.
Cartwright, N. (1999) *The Dappled World: A Study of Boundaries of Science*, , Cambridge: Cambridge University Press.
Cassam, Q. (1986) Science and essence, *Philosophy*, 61: 95-107.
Cassirer, E. (1953 [1923]) *Substance and Function* and *Einstein's Theory of Relativity* (Trans. by W.C. Swabey and M.C. Swabey), New York: Dover.
Causey, R. (1977) *Unity of Science*, Dordrecht: Reidel.
Causey, R.L. (1967) Derived measurement, dimensions, and dimensional analysis, *Philosophy of Science*, 36: 252-270..
Cerruti, L. (1998) Chemicals as instruments: a language game, *Hyle*, 4: 39-61.
Chang, H. and N. Cartwright (1993) Causality and realism in the EPR experiment, *Erkenntnis*, 38:169-190.
Child, W. (1994) *Causality, Interpretation and the Mind*, Oxford: Clarendon Press.
Chilton, T.H. and A.P. Colburn (1934) Mass transfer (absorption) coefficients: prediction from data on heat transfer and fluid friction, *Industrial and Engineering Chemistry*, 26: 1183-1187.
Chomsky, N. (1995) Language and nature, *Mind*, 104: 1-61.
Christie, M. (1994) Philosophers versus chemists concerning 'laws of nature', *Studies in History and Philosophy of Science*, 25: 613-630.
Churchland, P.M. (1979) *Scientific Realism and the Plasticity of Mind*, Cambridge: Cambridge University Press.
Churchland, P.M. (1989) *A Neurocomputational Perspective*, Cambridge MA: The MIT Press.
Churchman, C.W. and B.G. Buchanan (1969) On the design of inductive systems: Some philosophical problems, *The British Journal for the Philosophy of Science*, 20: 311-323.
Clark, A. (1989) The particulate instantiation of homogeneous pink, *Synthese*, 80: 277-304.
Clark, S.R.L. (1991) How many selves make me?, pp. 17-34 in *Human Beings* (D. Cockburn, ed.), Cambridge: Cambridge University Press.
Claverie, P. and S. Diner (1980) The concept of molecular structure in quantum theory: Interpretation problems, *Israel Journal of Chemistry*, 19: 54-81.
Clifford, J. [1975] Properties of water in capillaries and thin films, in F. Franks (*ed.*), *Water: A Comprehensive Treatise*, volume 5, Plenum, New York, pp. 75-132.
Cornforth, J.W. (1993) The trouble with synthesis, *Australian Journal of Chemistry*, 46: 157-170.
Coulson, C.A. (1961) *Valence*, Oxford: Oxford University Press; first edition 1952.
Coulson, C.A. (1970) Recent developments in valence theory, *Pure and Applied Chemistry*, 24: 257-287.
Crane, T. (1991) All the difference in the world, *The Philosophical Quarterly*, 41(162): 1-25.
Crane, T. and D.H. Mellor (1990) There is no question of physicalism, *Mind*, 99: 185-206.
Crosland, M. (1996) Changes in chemical concepts and language in the seventeenth century, *Science in Context*, 9(3): 225-240.
Currie, G. (1988) Realism in the social sciences: social kinds and social laws, pp. 205-227 in *Relativism and Realism in Science* (R. Nola, ed.), Dordrecht: Kluwer.

Currie, G. (1990) Supervenience, essentialism and aesthetic properties, *Philosophical Studies*, 58: 243-257.
Cushing, J.T. (1994) *Quantum Mechanics: Historical Contingency and the Copenhagen Hegemony*, Chicago: The University of Chicago Press.
D'Agostino, S. (2000) *A History of the Ideas of Theoretical Physics*, Dordrecht: Kluwer.
D'Amico, R. (1995) Is disease a natural kind?, *The Journal of Medicine and Philosophy*, 20: 551-569.
Daly, C. (1998) What are physical properties?, *Pacific Philosophical Quarterly*, 79: 196-217.
Davidson, D. (1970) Mental events, in his *Essays on Actions & Events*, Oxford: Clarendon Press (1980) 207-224.
Davidson, D. (1986) A nice derangement of epitaphs, in *Truth and Interpretation* (E. LePore, ed.), Oxford: Basil Blackwell, pp. 433-446.
Davidson, D. (1993) Thinking causes, in J. Heil and A. Mele (eds.), *Mental Causation*, Oxford: Clarendon Press, pp. 3-18.
Davies, M. and Humberstone, L. (1980) Two notions of necessity, *Philosophical Studies*, 38: 1-30.
Day, M.A. (1990) The no-slip condition of fluid dynamics, *Erkenntnis*, 33: 285-296.
De Sousa, R. (1984) The natural shiftiness of natural kinds, *Canadian Journal of Philosophy*, 14: 561-580.
Del Re, G. (1974) Current problems and perspectives in the MO-LCAO theory of molecules, pp. 95-136 in P.-O. Löwdin (ed.), *Advances in Quantum Chemistry*, vol. 8, New York: Academic Press.
Del Re, G. (1987) The historical perspective and the specificity of chemistry, *Epistemologia*, 10:231-240.
Del Re, G. (1996) The specificity of chemistry and the philosophy of science, pp. 11-20 in *Philosophers in the Laboratory* (V. Mosini, ed.), Rome: Euroma.
Del Re, G. (1998) Ontological status of molecular structure, *Hyle*, 4: 81-103.
Del Re, G. (2000) Models and analogies in science, *Hyle*, 6: 5-15.
Del Re, G., Barbier, C. and Vincent, C. (1981) Bond polarisation in the FeCO System: Semi-empirical MO-SCF (BMV) Calculations, *International Journal of Quantum Chemistry*, 19: 1057-1063.
Delhez, R. (1974) Bachelard et la chimie, pp. 121-133 in *Bachelard*, publication of the Centre Culturel Internationale de Cerisy-La-Salle.
Denbigh, K. (1981) *The Principles of Chemical Equilibrium*, 4th ed., Cambridge: Cambridge University Press.
Denbigh, K.G. and M.L.G. Redhead (1989) Gibbs' paradox and non-uniform convergence, *Synthese*, 81: 283-313.
Derjaguin, B.V. and N.V. Churaev (1971) Investigation of the properties of water II, *Journal of Colloid and Interface Science*, 36: 415-426.
Derjaguin, B.V. and N.V. Churaev (1973a) The question of "anomalous water" [in Russian], *Kolloidnyi Zhurnal*, 35: 814-815.
Derjaguin, B.V. and N.V. Churaev (1973b) Nature of "anomalous water", *Nature*, 244: 430-431.
DeVidi, D. and G. Solomon (1994) Geometric conventionalism and Carnap's principle of tolerance, *Studies in History and Philosophy of Science*, 25: 773-783.
Devitt, M. (1990) Meanings just ain't in the head, pp. 79-104 in *Meaning and Method: Essays in Honour of Hilary Putnam* (G. Boolos, ed.), Cambridge: Cambridge University Press.
Devitt, M. and Sterelny, K. (1987) *Language and Reality: An Introduction to the Philosophy of Language*, Oxford: Basil Blackwell.
Diamond, A.M. (1988) The polywater episode and the appraisal of theories, pp. 181-198 in *Scrutinising Science* (A. Donovan, L. Laudan and R. Laudan, eds.), Dordrecht: Kluwer.
Dieks, D. and de Regt, H.W. (1998) Reduction and understanding, *Foundations of Science*, 1: 45-59.
Dingle, H. (1942) On the dimensions of physical magnitudes (second paper), *Philosophical Magazine*, 33: 692-697.

Dingle, H. (1949a) The nature of scientific philosophy, *Proceedings of the Royal Society of Edinburgh*, (4) 409.
Dingle, H. (1949b) On the dimensions of physical magnitudes (Seventh paper: a paradox in dimensional theory), *Philosophical Magazine*, 40: 94-99.
Dirac, P.A.M. (1929) Quantum mechanics of many-electron systems, *Proceedings of the Royal Society of London*, A123: 714-733.
Dirac, P.A.M. (1939) The relation between mathematics and physics, *Proceedings of the Royal Society of Edinburgh*, 59 (1938/39) 122-129.
Dobrotin, R.B. and V.P. Barzakovsky (1963) Concerning the specificity of the chemical form of the motion of matte [in Russian]r, *Voprosy filosofii* (5) 100-105.
Donahoe, F.J. (1969) 'Anomalous' water, *Nature*, 224: 198.
Donnan, F.G. (1951) Ostwald memorial lecture, pp. 1-17 in F.G. Donnan et al., *Memorial Lectures delivered before the Chemical Society 1933-1942*, London: The Chemical Society.
Donovan, A., and others (1988) *The Chemical Revolution: Essays in Reinterpretation*, *Osiris*, 2nd series, volume 4.
Douven, I. and J. van Brakel (1995) Is scientific realism an empirical hypotheses, *Dialectica*, 49: 3-14.
Douven, I. and J. van Brakel (1998) Can the world help us in fixing the unique reference of natural kind terms?, *Journal for General Philosophy of Science*, 29: 59-70.
Downes, S.M. (1992) The importance of models in theorising: A deflationary semantic view, *PSA 1992*, 1: 142-153.
Drews, A. (1894) *Kants Naturphilosophie als Grundlage seines Systems*, Berlin: Mitscher & Röstell.
Duhem, P (1899) Une science nouvelle, la chimie-physique, *Revue philomatique de Bordeaux et sud-ouest*, Paris: Hermann, pp. 205-219, 260-280.
Duhem, P. (1900) L'œuvre de M.J.H. Van 'T Hoff: A propos d'un livre récent, *Revue des questions scientifiques*, 1: 1-27.
Duhem, P. (1902) *Le mixte et la combinaison chimique: Essai sur l'évolution d'une idée*, Paris: Naud; reprint Paris: Fayard (1985); originally published in *Revue de Philosophie*, 1 (1900) 69-99, 167-197, 331-357, 430-467, 730-745.
Duhem, P. (1974) *The Aim and Structure of Physical Theory*, New York: Atheneum; translation of the second edition of *La théorie physique: son objet, sa structure* (Paris, 1914).
Dummett, M. (1978) *Truth and Other Enigmas*, London: Duckworth.
Duncan, A. (1996) *Laws and Order in Eighteenth-Century Chemistry*, Oxford: Clarendon Press.
Duncan, W.J. (1953) *Physical Similarity and Dimensional Analysis*, London: Edward Arnold.
Duncanson, W.E. (1941) The dimensions of physical quantities, *Proceedings of the Physical Society*, 53: 432-448.
Dupré, J. (1986) Sex, gender, and essence, *Midwest Studies in Philosophy*, 11: 441-457.
Dupré, J. (1989) Wilkerson on natural kinds, *Philosophy*, 64: 248-251.
Dupré, J. (1993) *The Disorder of Things: Metaphysical Foundations of the Disunity of Science*, Cambridge MA: Harvard University Press.
Düring, I. (1944) *Aristotle's Chemical Treatise: Meteorologica Book IV*, Göteborgs Högskolas Årsskrift 1944:2; reprint New York: Garland (1980).
Dussort, H. (1956) Kant et la chimie, *Revue philosophique*, 81: 392-397.
Earley, J.E. (1981) Self-organization and agency: in chemistry and in process philosophy, *Process Studies*, 11(4): 242-258.
Earley, J.E. (1992) The nature of chemical existence in P.A. Bogaard and G. Treash (eds.), *Metaphysics as Foundation*, Albany: State University of New York Press.
Earley, J.E. (1998) Modes of chemical becoming, *Hyle*, 4: 105-115.
Earman, J. and Roberts, J. (1999) *Ceteris paribus*, there is no problem of provisos, *Synthese*, 118: 439-478.
Eddington, A. (1921) *Space, Time and Gravitation*, Cambridge: Cambridge University Press.

Editorial (1964) The role of the use of chemistry in construction of communism [in Russian], *Voprosy filosofii*, (1) 3-10.
Edmiston, C. (1992) The nature of the chemical bond - once more, *Journal of Chemical Education*, 69: 600.
Ehrenfest-Afanassjewa, T. (1915) Der Dimensionsbegriff und der analytischen Base physikalischer Gleichungen, *Mathematische Annalen*, 77: 259-276.
Ehrenfest-Afanassjewa, T. (1926) Dimensional analysis viewed from the standpoint of theory of similitudes, *The Philosophical Magazine* [7] 1: 257-272.
Ehrlich, P. (1982) Negative, infinite and hotter than infinite temperatures, *Synthese*, 50: 233-277.
Einstein, A., B. Podolsky and N. Rosen (1970 [1935]) Can quantum-mechanical description of physical reality be considered complete?, pp. 122-142 in S. Toulmin (ed.), *Physical Reality*, Evanston/London: Harper & Row.
Eisvogel, M. (1996) Wem nützt eine Philosophie der Chemie - den Chemikern oder den Philosophen, pp. 95-110 in *Philosophie der Chemie: Bestandsaufnahme und Ausblick* (N. Psarros, K. Ruthenberg and J. Schummer, eds.), Würzburg: Königshausen & Neumann.
Elder, C.L. (1994) Higher and lower essential natures, *American Philosophical Quarterly*, 31: 255-265.
Elder, C.L. (1995) A different kind of natural kind, *Australasian Journal of Philosophy*, 73: 516-547.
Elgin, C.Z. (1983) *With Reference to Reference*, Indianapolis: Hackett.
Ellis, B. (1957) A comparison of process and non-process theories in the physical sciences, *The British Journal for the Philosophy of Science*, 8: 45-56.
Ellis, B. (1966) *Basic Concepts of Measurement*, Cambridge: Univ. Press.
Endicott, R.P. (1998) Collapse of the new wave, *The Journal of Philosophy*, 95: 53-72.
Engels, F. (1940) *Dialectics of Nature*, New York: International Publishers.
Engels, F. (1979) *Friedrich Engels über die Dialektik der Naturwissenschaften*, texts collected and published by B.M. Kedrov, Köln: Pahl-Rugenstein (for Berlin: Dietz).
Ereshefsky, M. (1991) Species, higher taxa, and the units of evolution, *Philosophy of Science*, 58: 84-101.
Ereshefsky, M. (1995) Critical notice Dupré (1993), *Canadian Journal of Philosophy*, 25: 143-158.
Esfeld, M. (1999) Physicalism and ontological holism, *Metaphilosophy*, 30: 319-337.
Eu, B.C. (1992) *Kinetic Theory and Irreversible Thermodynamics*, New York: Wiley.
Everett, D.H., Haynes, J.M. and McElroy, P.J. (1969) Letter to editor, *Nature*, 224: 394.
Everett, D.H., J.M. Haynes and P.J. McElroy (1970) The story of anomalous water, *Science Progress* (Oxford), 59: 279-308.
Everett, D.H., J.M. Haynes and P.J. McElroy (1971), Colligative properties of anomalous water, *Nature*, 226: 1033-1037.
Eyring, M., Walter, J. and Kimball, G.E. (1944) *Quantum Chemistry*, New York: Wiley.
Falvey, K. and Owens, J. (1994) Externalism, self-knowledge, and scepticism, *The Philosophical Review*, 103(1): 107-137.
Farmer, J.D. (1990) A Rosetta stone for connectionism, *Physica D*, 42: 153-187.
Farrell, R. (1983) Metaphysical necessity and epistemic location, *Australasian Journal of Philosophy*, 61: 283-294.
Faustov, A.P. (1955) Concerning some statements of the periodical law [in Russian], *Voprosy filosofii*, (2) 168-178.
Fedyakin, N.N. [1962] Change in water structure on condensation in capillaries [in Russian], *Kolloidnyi Zhurnal*, 24: 497-501.
Feibleman, J.K. (1954) Theory of integrative levels, *The British Journal for the Philosophy of Science*, 5: 59-66.
Feynman, R.P., Leighton, R.B. and Sands, M. (1965) *The Feynman Lectures on Physics: Quantum Mechanics*, Reading, Mass.: Addison-Wesley.
Field, H. (1972) Tarski's theory of truth, *The Journal of Philosophy* 69: 347-75.

Flanagan, O. (1995) Deconstructing dreams: The spandrels of sleep, *The Journal of Philosophy*, 92: 5-27.
Focken, C.M. (1953) *Dimensional Methods and Their Applications*, London: Edward Arnold.
Fodor, J. A. (1974) Special sciences, *Synthese*, 28: 77-115; also in his *The Language of Thought*, Cambridge MA: Harvard University Press, 1975, pp. 9-25.
Forbes, G. (1985) *The Metaphysics of Modality*, Oxford: Clarendon Press.
Forbes, G. (1997) Essentialism, pp. 515-536 in B. Hale and C. Wright (eds.), *A Companion to the Philosophy of Language*, Oxford: Blackwell.
Förster, E. (1991) Die Idee des Übergangs. Überlegungen zum Elementarsystem der bewegenden Kräfte, pp. 28-48 in Forum für Philosophie Bad Homburg, *Übergang: Untersuchungen zum Spätwerk Immanuel Kants*, Frankfurt: Klostermann.
Förster, E. (1993) Introduction, pp. xv-lviii in Kant's *Opus postumum* (E. Förster, ed.), Cambridge: Cambridge University Press.
Forum für Philosophie Bad Homburg (1991) *Übergang: Untersuchungen zum Spätwerk Immanuel Kants*, Frankfurt: Klostermann.
Fourier, J.B. (1822) *Théorie analytique de la chaleur*, Paris: Didot; reprint Sceaux: Gabay (1988).
Fowkes, F.M., R.W. Lovejoy and J.S. Chow (1971) Anomalous water experiments with silica powders, *Journal of Colloid and Interface Science*, 36: 522-528.
Francoeur, E. (1997) The forgotten tool: The design and use of molecular models, *Social Studies of Science*, 27: 7-40.
Francoeur, E. (2000) Beyond dematerialization and inscription: Does the materiality of molecular models really matter?, *Hyle*, 6: 63-84.
Frank, P. (1946) *Foundations of Physics*, International Encyclopaedia of Unified Science, volume I, Number 2, Chicago: The University of Chicago Press.
Franks, F. (1981) *Polywater*, The MIT Press, Cambridge MA.
French, S. and Ladyman, J. (1998) Semantic perspective on idealisation in quantum mechanics, *Poznan Studies in the Philosophy of the Sciences and the Humanities*, 63: 51-73.
Frenking, G. (1998) Heretical thoughts of a theoretical chemist about the autonomy of chemistry as a science in the past and the present, pp. 103-108 in P. Janich and N. Psarros (eds.), *The Autonomy of Chemistry: 3rd Erlenmeyer-Colloquy for the Philosophy of Chemistry*, Würzburg: Königshausen & Neumann.
Freundlich, H. (1922) *Kapillarchemie: Eine Darstellung der Chemie der Kolloide und verwanter Gebiete*, Leipzig: Akademische Verlagsgesellschaft.
Fricke, M. (1976) The rejection of Avogadro's hypotheses, pp. 277-308 in Howson (1976).
Friedman, M. (1992) *Kant and the Exact Sciences*, Cambridge MA: Harvard University Press.
Fuchs, G. (1964) Erkenntnistheoretische Probleme in der modernen Chemie (Eine Auseinandersetzung mit Georg Maria Schwab), pp. 32-53 in *Philosophische Probleme der Chemie und ihrer Geschichte*, published by Technische Hochschule für Chemie "Carl Schorlemmer", Leuna-Merseburg.
Fuchs, G. (1965) Theorie und Empirie in der Quantenchemie, *Deutsche Zeitschrift für Philosophie*, 13 (Sonderheft): 338-344.
García-Sucre, M. and M. Bunge (1981) Geometry of a quantal system, *International Journal of Quantum Chemistry*, 19: 83-93.
Garfield, J.L. (1989) The myth of Jones and the mirror of nature: Reflections on introspection, *Philosophy and Phenomenological Research*, 1: 1-26.
Garkovenko, R.V. (1963) Qualitative diversity of chemical particles and some methodological problems of chemistry [in Russian], *Voprosy filosofii*, (5) 89-99.
Garkovenko, R.V. (1964) Some theoretical questions of the chemical ways of industry's development [in Russian], *Voprosy filosofii*, (8) 9-17.
Garkovenko, R.V. (1970) *Philosophical Problems in Modern Chemistry* [in Russian], Moscow.
Gascoigne, R.M. (1961) The basic concepts of modern chemistry, pp. 68-80 in G.H. Aybward (ed.) *Approach to Chemistry*, Sydney: University of New South Wales.
Gavroglu, K. (1995) *Fritz London: A Scientific Biography*, Cambridge: Cambridge University Press.

Gavroglu, K. (1996) Appropriating the atom at the end of the 19th century: Chemists and physicists at each other's throat, pp. 93-106 in *Philosophers in the Laboratory* (V. Mosini, ed.), Rome: Euroma.
Gavroglu, K. and A. Simoes (1994) The Americans, the Germans, and the beginning of quantum chemistry: The confluence of diverging traditions, *Historical Studies in the Physical and Biological Sciences*, 25: 47-110.
Gay, H. (1976) Radicals and types, *Studies in History and Philosophy of Science*, 7: 1-51.
Gay, H. (1978) The asymmetric carbon atom, *Studies in History and Philosophy of Science*, 9: 207-308.
Gee, M.L., T.W. Healy and L.R. White [1990] Hydrophobicity effects in the condensation of water films on quartz, *Journal of Colloid and Interface Science*, 140: 450-465.
Gell-Mann, M. (1994) *The Quark and the Jaguar: Adventures in the Simple and the Complex*, New York: Freeman.
Gey, E. and Parthey, H. (1984) Quantenchemie - eine interdisziplinäre Entwicklung, *Wissenschaft und Fortschritt*, 34(11): 282-285.
Ghiselin, M.T. (1987) Species concepts, individuality, and objectivity, *Biology and Philosophy*, 2: 127-143.
Gibbins, P. (1987) *Particles and Paradoxes. The Limits of Quantum Logic*, Cambridge: Cambridge University Press.
Gibbs, J.W. (1902) *Elementary Principles in Statistical Mechanics*, New Haven CT: Yale University Press.
Gibbs, J.W. (1928) *Collected Works*, vol. 1, New Haven CT: Yale University Press, London: Longmans.
Giere, R. (1988) Laws, theories, and generalisation, pp. 37-46 in *The Limitations of Deductivism* (A. Grünbaum and W.C. Salmon, eds.), Berkeley: University of California Press.
Giere, R.N. (ed.) (1992) *Cognitive Models of Science*, Minneapolis: University of Minnesota Press.
Gingold, M.P. (1973) L'eau dite anormale: revue générale, *Bulletin de la Société Chimique de France*, (5) 1629-1644.
Glas, E. (1999) The evolution of a scientific concept, *Journal for General Philosophy of Science*, 30: 37-58.
Gleick, J. (1992) *Genius: The Life and Science of Richard Feynman*, New York: Pantheon.
Goodman, N. (1978) *Ways of Worldmaking*, Indianapolis: Hackett.
Goodman, N. (1983) *Fact, Fiction, and Forecast*, Cambridge MA: Harvard University Press.
Goodman, N. (1984) *Of Mind and Other Matters*, Cambridge MA: Harvard University Press.
Goodman, N. and C.Z. Elgin (1988) *Reconceptions in Philosophy & Other Arts & Sciences*, Indianapolis: Hackett.
Graham, L.R. (1971) *Science and Philosophy in the Soviet Union*, London: Allen Lane.
Griffiths, P.E. (1996) Darwinism, process structuralism, and natural kinds, *Philosophy of Science*, 63: S1-S9.
Grimes, T.R. (1991) Supervenience, determination, and dependency, *Philosophical Studies* 62: 81-92.
Guggenheim, E.A. (1942) Units and dimensions, *Philosophical Magazine*, 33: 479.
Gutmann, M. and G. Hanekamp (1996) Abstraktion und Ideation - Zur Semantik chemischer und biologischer Grundbegriffe, *Journal for General Philosophy of Science*, 27: 29-53.
Gutting, G. (1987) Gaston Bachelard's philosophy of science, *International Studies in the Philosophy of Science*, 2: 55-71.
Haack, S. (1992) Extreme scholastic realism: Its relevance to philosophy of science today, *Transactions of the Charles S. Peirce Society*, 28: 19-50.
Haack, S. (1992) Extreme scholastic realism: Its relevance to philosophy of science today, *Transactions of the Charles S. Peirce Society*, 28: 19-50.

Haberditzl, W. and H. Laitko (1967) Reduziert sich die theoretische Chemie auf angewandte Quantenmechanik und Quantenphysik?, *Wissenschaftliche Zeitschrift der Humboldt-Universität Berlin, mathematisch-naturwissenschaftliche Reihe*, 16: 961-962.
Habermas, J. (1968) *Technik und Wissenschaft als >Ideologie<*, Frankfurt a/M: Suhrkamp.
Hacking, I. (1983) *Representing and Intervening*. Cambridge: Cambridge University Press.
Hacking, I. (1991) A tradition of natural kinds, *Philosophical Studies*, 61: 109-126, 149-154.
Hadorn, G.H. (1997) Webers idealtypus als Methode zur Bestimmung des Begriffsinhaltes theoretischer Begriffe in den Kulturwissenschaften, *Journal for General Philosophy of Science*, 28: 275-296.
Haldane, J.B.S. (1938) *The Marxist Philosophy and the Sciences*, London: Allen & Unwin.
Hall, P.J. (1986) The Pauli exclusion principle and the foundations of chemistry, *Synthese*, 68: 267-272.
Han, E. (1996) Die Notwendigkeit einer Philosophie der Chemie, pp. 59-66 in *Philosophie der Chemie: Bestandsaufnahme und Ausblick* (N. Psarros, K. Ruthenberg and J. Schummer, eds.), Würzburg: Königshausen & Neumann.
Hanekamp, G. (1997) *Protochemie: Vom Stoff zur Valenz*, Würzburg: Königshausen & Neumann
Hanekamp, G. (1998) Valence and the chemical bond: crossing the border between chemistry and physics, pp. 161-167 in P. Janich and N. Psarros (eds.), *The Autonomy of Chemistry: 3rd Erlenmeyer-Colloquy for the Philosophy of Chemistry*, Würzburg: Königshausen & Neumann.
Hanfling, O. (1984) Scientific realism and ordinary usage, *Philosophical Investigations*, 7(3): 187-205.
Hare, R.M. (1984) The inaugural address: Supervenience, *Proceedings of The Aristotelian Society*, Suppl. Vol. 58: 1-16.
Harré, R. and E.H. Madden (1975) *Causal Powers*, Oxford: Blackwell.
Hartmann, H. (1965) *Die Bedeutung Quantentheoretischer Modelle für die Chemie*, Wiesbaden: Franz Steiner.
Hartmann, M. (1948) *Die philosophische Grundlagen der Naturwissenschaften*, Jena: Fischer.
Hartree, D.R. (1957) *The Calculation of Atomic Structures*, New York: Wiley.
Haugeland, J. (1982) Weak supervenience, *American Philosophical Quarterly*, 19: 93-103.
Healey, R.A. (1991) Holism and nonseparability, *The Journal of Philosophy*, 88: 393-421.
Heber, G. (1964) Eine Bemerkung über die quantenmechanische Theorie der Valenz, *Deutsche Zeitschrift für Philosophie*, 12: 985-986.
Heinig, K. (1975) Immanuel Kant und die Chemie des 18. Jahrhunderts in der Darstellung der Wissenschaftsgeschichte, *Wissenschaftliche Zeitschrift der Humboldt-Universität zu Berlin, gesellschafts- und sprachwissenschaftliche Reihe*, 24 (2) 191-194.
Heisenberg, W. (1942) *Ordnung der Wirklichkeit*, quoted from his *Gesammelte Werke, Physik und Erkenntnis 1927-1955*, München/Zürich: Piper (1984).
Heisenberg, W. (1972) *Physics and Beyond: Encounters and Conversations*, New York: Harper & Row.
Hempel, C.G. (1965) *Aspects of Scientific Explanation.*, London: Macmillan.
Hempel, C.G. (1988) Provisos: A problem concerning the inferential function of scientific theories, pp. 19-36 in *The Limitations of Deductivism* (A. Grünbaum and W.C. Salmon, eds.), Berkeley: University of California Press.
Hendry, R.F. (1998) Models and approximations in quantum chemistry, pp. 123-142 in N. Shanks (ed.), *Idealisation in Contemporary Physics*, Poznan Studies in the Philosophy of the Sciences and the Humanities, vol. 63, Amsterdam/Atlanta: Rodopi.
Hendry, R.F. (1999) Molecular models and the question of physicalism, *Hyle*, 5: 117-134.

Hesse, M.B. (1966) *Models and Analogies in Science*, Notre Dame IN: University of Notre Dame Press.
Hettema, H. and T.A. Kuipers (1988) The periodic table - its formalisation, status, and relation to atomic theory, *Erkenntnis*, 28: 387-408.
Hiebert, E.N. (1996) Discipline identification in chemistry and physics, *Science in Context*, 9: 93-119.
Hirsch, E. (1982) *The Concept of Identity*, New York: Oxford University Press.
Hirsch, E. (1993) *Dividing Reality*, New York: Oxford University Press.
Hoffman, J. and G.S. Rosenkrantz (1994) *Substance Among other Categories*, Cambridge: Cambridge University Press.
Hoffmann, R. (1990) Molecular Beauty, *The Journal of Aesthetics and Art Criticism*, 48(3): 191-204.
Hoffmann, R. (1995) *The Same and Not the Same*, New York: Columbia University Press.
Hoffmann, R. (1998) Qualitative thinking in the age of modern computational chemistry-or what Lionel Salem knows, *Journal of Molecular Structure*, 424: 1-6.
Hoffmann, R. and P. Laszlo (1991) Representation in chemistry, *Angewandte Chemie Int. Ed. Engl.*, 30:1-16.
Hoffmann, R., Minkin, V.I. and B.K. Carpenter (1996) Ockham's razor and chemistry, *Bulletin de la Société Chimique de France*, 133: 117-130 [also: *Hyle*, 3: 3-28]
Hofmann, J.R. (1990) How the models of chemistry vie, *PSA 1990*, 1: 405-419.
Hofmann, J.R. and P.A. Hofmann (1992) Darcy's law and structural explanation in hydrology, *PSA 1992*, 1: 23-35.
Hooker, C.A. (1981) Towards a general theory of reduction, *Dialogue*, 20: 38-60, 201-235, 496-529.
Hoppe, H.G. (1969) *Kants Theorie der Physik: Eine Untersuchung über das Opus postumum von Kant*, Frankfurt a/M: Klostermann.
Horgan, T. (1982) Supervenience and microphysics, *Pacific Philosophical Quarterly*, 63: 29-43.
Horgan, T. and J. Tienson (1990) Soft laws, *Midwest Studies in Philosophy*, 15: 256-279.
Hornsby, J. (1997) *Simple Mindedness: In Defence of Naive Naturalism in the Philosophy of Mind*, Cambridge MA: Harvard University Press.
Howson, C. (ed.) (1976) *Method and Appraisal in the Physical Sciences*. Cambridge: Cambridge University Press.
Howson, C. and A. Franklin (1991) Maher, Mendeleev and Bayesianism, *Philosophy of Science*, 58: 574-585.
Hüffemann, A. (1998) Laws and dispositions, *Philosophy of Science*, 65: 121-135.
Hulswit, M. (1997) Peirce's teleological approach to natural classes, *Transactions of the Charles S. Peirce Society*, 33: 722-772.
Humphreys, P. (1996) Emergence, not supervenience, *Philosophy of Science*, 64 (Proceedings) S337-S345.
Hunsberger, I. M. (1954) Theoretical chemistry in Russia, *Journal of Chemical Education*, 31: 504-514.
Huntley, H.E. (1952) *Dimensional Analysis*, London: MacDonald.
Hurley, S.L. (1985) Supervenience and the possibility of coherence, *Mind*, 94: 501-525.
Hüttemann, A. (1997) *Idealisierungen und das Ziel der Physik: Eine Untersuchung zum Realismus, Empirismus und Konstruktivismus in der Wissenschaftstheorie*, Berlin: de Gruyter.
Ionidi, P.P. (1958) *The Philosophical Significance of the Periodic Law of Mendeleev* [in Russian], Moscow.
Isaacson, E. de St. Q. and M. de St. Q. Isaacson (1975) *Dimensional Methods in Engineering and Physics*, London: Edward Arnold.
Jackson, F. (1998)*From Metaphysics to Ethics: A Defence of Conceptual Analysis*, Oxford: Clarendon Press.
Janich, P. (1992) Chemie als Kulturleistung. Zum Selbstverständnis der Chemie im Spiegel der Kulturgeschichte", pp. 161-173 in *Chemie und Geisteswissenschaften* (J. Mittelstraß and G. Stock, eds.), Berlin: Akademie Verlag..

Janich, P. (1994a) Protochemie: Programm einer konstruktiven Chemiebegründung, *Journal for General Philosophy of Science*, 22: 71-87 [also in: *chimica didactica*, 21 (1995) 111-128].
Janich, P. (1994b) Probleme der Bestimmung von Grundbegriffen der Chemie, pp. 11-26 in *Philosophische Perspektiven der Chemie* (P. Janich, ed.), Mannheim: Wissenschaftsverlag.
Janich, P. (1998) Die Selbständigkeit der Chemie im Verhältnis zu ihren Nachbarwissenschaften: Ziele und Ergebnisse des 3. Erlenmeyer-Kolloquiums für Philosophie der Chemie, pp. 1-30 in P. Janich and N. Psarros (eds.), *The Autonomy of Chemistry: 3rd Erlenmeyer-Colloquy for the Philosophy of Chemistry*, Würzburg: Königshausen & Neumann.
Janich, P. (ed.) (1994c) *Philosophische Perspektiven der Chemie*, Mannheim: BI-Wissenschaftsverlag.
Janich, P. and N. Psarros (eds.) (1996) *Die Sprache der Chemie*, Würzburg: Königshausen & Neumann.
Janich, P. and N. Psarros (eds.) (1998) *The Autonomy of Chemistry: 3rd Erlenmeyer-Colloquy for the Philosophy of Chemistry*, Würzburg: Königshausen & Neumann.
JCIS (1971) Discussion (in part) of papers presented at the 44th National Colloid Symposium (Anomalous Water Symposium), *Journal of Colloid and Interface Science*, 36:562-566.
Johnson, D.M. (1990) Can abstraction be causes?, *Biology and Philosophy*, 5: 63-77.
Johnson, P. (1997) *The Constants of Nature. A Realist Account*, Aldershot: Ashgate.
Johnston, M. (1997) Manifest kinds, *The Journal of Philosophy*, 94: 564-583.
Jones, K.R.W. (1994) Exclusion of intrinsically classical domains and the problem of quasiclassical emergence, *Physical Review*, 50: 1062-1069.
Jordan, P. (1957 [1947]) *Das Bild der modernen Physik*, Frankfurt a/M: Ullstein.
Kamlah, A. (1984) A logical investigation of the phlogiston case, pp. 217-38 in *Reduction in Science* (W. Balzer et al., eds.), Dordrecht: Reidel.
Kant, I. (1993) *Opus postumum* (E. Förster, ed.), Cambridge: Cambridge University Press.
Karpinskaya (1966) Philosophical problems of molecular biology [in Russian], *Voprosy filosofii*(1) 65-74.
Kedrov, B.M. (1949) Dalton's atomic theory and its philosophical significance, *Philosophy and Phenomenological Research*, 9: 644-662.
Kedrov, B.M. (1956) *Spektralanalyse*, Berlin: VEB.
Kedrov, B.M. (1962) Über das Verhältnis der Bewegungsformen der Materie in der Natur, in: *Philosophische Probleme der modernen Naturwissenschaft*, Berlin: VEB.
Kedrov, B.M. (1965) The general course of the cognition of substance [in Russian], *Voprosy filosofii*, (4) 92-101.
Kedrov, B.M. (1969) *Tri aspekta atomistiki*, vol. I: *Paradoks Gibbsa*, vol. II: *Uchenie Dalt'ona*, vol. III: *Zakon Mendeleeva*, Moscow.
Kekulé, A. (1859) *Lehrbuch der anorganischen Chemie*, Erlangen.
Kemeny, J.G. and P. Oppenheim (1956) On reduction, *Philosophical Studies*, 7: 6-19.
Kent, W. (1958) Scientific naming, *Philosophy of Science*, 25: 185-194.
Kern, C.W. and M. Karplus (1972) The water molecule, in F. Franks (ed.), *Water. A Comprehensive Treatise vol. 1. The Physics and Physical Chemistry of Water*, New York: Plenum Press.
Khalidi, M.A. (1993) Carving nature at the joints, *Philosophy of Science*, 60: 100-113.
Kim, J. (1984) Concepts of supervenience, *Philosophy and Phenomenological Research*, 45: 153-176.
Kim, J. (1990) Supervenience as a philosophical concept, *Metaphilosophy*, 21: 1-27.
Kim, J. (1999) Supervenient properties and micro-based properties: A reply to Noordhof, *Proceedings of the Aristotelian Society*, 99: 115-118.
Kincaid, H. (1987) Supervenience doesn't entail reducibility, *Southern Journal of Philosophy*, 25: 343-356.
Kincaid, H. (1997) *Individualism and the Unity of Science: Essays on Reduction, Explanation, and the Special Sciences*, Lanham: Rowman & Littlefield.
Kitaygorodsky, A.I. (1966) The problem of theory in chemistry [in Russian], *Voprosy filosofii*, 20: 75-86.

Kitcher, P. (1978) Theories, theorists and theoretical change, *The Philosophical Review*, 87: 519-547.
Kitcher, P. (1984a) Species, *Philosophy of Science*, 51: 308-333.
Kitcher, P. (1984b) 1953 and all that: a tale of two sciences, *The Philosophical Review*, 93: 335-373.
Kitcher, P. (1989) Explanatory unification and the causal structure of the world, pp. 410-503 in *Minnesota Studies in the Philosophy of Science* (P. Kitcher and W. Salmon, eds.), vol. 13.
Kitcher, P. (1993) *The Advancement of Science: Science Without Legend, Objectivity Without Illusion*, Oxford: Oxford University Press.
Klee, R. (1992) Anomalous monism, *ceteris paribus* and psychological explanation, *The British Journal for the Philosophy of Science* 43: 389-403.
Klein, U. (1994a) Origin of the concept of chemical compound, *Science in Context*, 7: 163-204.
Klein, U. (1994b) Robert Boyle: Der Begründer der neuzeitlichen Chemie, *Philosophia Naturalis*, 31: 63-106.
Kluge, A.G. (1990) Species as historical individuals, *Biology and Philosophy*, 5: 417-431.
Klüver, J. and Müller, W. (1972) Wissenschaftstheorie und Wissenschaftsgeschichte: Die Entdeckung der Benzolformel, *Zeitschrift für allgemeine Wissenschaftstheorie*, 3: 243-266.
Körner, S. (1991) On Kant's conception of science and the *Critique of Practical Reason*, *Kantstudien*, 82: 173-178.
Kramers, H. (1989) Chemical engineering in the Netherlands 1935-1965, pp. 143-152 in *One Hundred Years of Chemical Engineering* (N.A. Peppas, ed.), Dordrecht: Kluwer.
Krantz, D.H., Luce, R.D., Suppes, P. and Tversky, A. (1971) *Foundations of Measurement*, vol. I, New York/London: Academic Press.
Kripke, S. (1971) Identity and necessity, pp. 135-164 in M.K. Munitz (ed.), *Identity and Individuation*, New York: New York University Press.
Kripke, S. (1980) *Naming and Necessity*, Oxford: Blackwell.
Kroon, F.W. (1985) Theoretical terms and the causal view of reference, *Australasian Journal of Philosophy*, 63: 143-166.
Kubbinga, H.H. (1984) Les première spécification dite 'moléculaire' de l'atomisme épicurien: Isaac Beeckman (1620) et Sébastien Basson (1621): Le concept d'individu substantiel' et d'espèce substantielle, *Revue d'Histoire des Sciences*, 37: 215-233.
Kubbinga, H.H. (1988) The first 'molecular' theory (1620): Isaac Beeckman (1588-1637), *Journal of Molecular Structure (Theochem)*, 181: 205-218.
Kuhn, T.S. (1952) Robert Boyle and structural chemistry in the seventeenth century, *Isis*, 43: 12-36.
Kuhn, T.S. (1962) *The Structure of Scientific Revolutions*, International Encyclopaedia of Unified Science, volume II, Number 2, Chicago: The University of Chicago Press; 2nd enlarged edition 1970.
Kuhn, T.S. (1991) The road since structure, *PSA1990*, 2: 3-13.
Kuhn, T.S. (1993) Afterwords, in *World Changes: Thomas Kuhn and the Nature of Science* (P. Horwich, ed.), Cambridge MA: The MIT Press, pp. 311-341.
Kultgen, J.H. (1958) Philosophic conceptions in Mendeleev's *Principles of Chemistry*, *Philosophy of Science*, 25: 177-184.
Kursanov, D.N., M.G. Gonikberg, B.M. Dbinin, M.I. Kabachnik, E.D. Kaverzneva, E.N. Prilezhaeva, N.D. Sokolov, and R. Kh. Freidlina (1950) K voprosu o sovremennom sostoianii teorii khimicheskogo stroeniia, *Uspekhi Khimii*, 19: 529-544; translated as: The present state of the chemical structural theory, *Journal of Chemical Education*, 29 (1952) 2-13.
Kuznetsov, V.I. (1963) New ideas about the discontinuous and continuous nature of the chemical organisation of substance [in Russian], *Voprosy filosofii*, (10) 61-70.
Kuznetsov, V.I. (1964) On the nature of the qualitative change in chemistry [in Russian], *Voprosy filosofii*, (11) 94-102.
Kuznetsov, V.I. and A.N. Shamin (1993) Butlerov's and Kekulé's contributions to the formation of structural organic chemistry, pp. 211-220 in J. Wotiz (ed.), *The*

Kekulé Riddle. A Challenge for Chemists and Psychologists, Clearwater (Fl): Cache River Press.
Kuznetsov, V.I. and A.A. Pechenkin (1972) Conceptual problems of chemistry: The theory of resonance [in Russian], *Voprosy filosofii*, (5) 75-86.
Laitko, H. (1965) Zum Verhältnis von Chemie und Physik", *Deutsche Zeitschrift für Philosophie*, 13 (Sonderheft): 330-338.
Laitko, H. (1967) Philosophische Fragen der Chemie. Einführung in die Problemsituation, pp. 107-137 in *Naturforschung und Weltbild. Eine Einführung in philosophische Probleme der modernen Naturwissenschaften* (M.Guntau and H. Wendt, eds.), Berlin: VEB.
Laitko, H. (1996) Chemie und Philosophie: Anmerkungen zur Entwicklung des Gebietes in der Geschichte der DDR, pp. 37-58 in *Philosophie der Chemie: Bestandsaufnahme und Ausblick* (N. Psarros, K. Ruthenberg and J. Schummer, eds.), Würzburg: Königshausen & Neumann.
Laitko, H. and W.-D. Sprung (1970) *Chemie und Weltanschauung: Standpunkte der marxistischen Philosophie zu einigen philosophischen Problemen der modernen Chemie*, Leipzig-Jena-Berlin: Urania; 2nd edition 1973.
Lakoff, G. (1987) *Women, Fire and Dangerous Things*, Chicago: The University of Chicago Press.
Lange, M. (1993) Natural laws and the problem of provisos, *Erkenntnis*, 38: 233-248.
Langhaar, H.L. (1951) *Dimensional Analysis and Theory of Models*, New York: Wiley
LaPorte, J. (1996) Chemical kind term reference and the discovery of essence, *Nous*, 30: 112-132.
LaPorte, J. (1998) Living water, *Mind*, 107: 450-455.
LaPorte, J. (2000) Rigidity and kind, *Philosophical Studies*, 97: 293-316.
Laszlo, P. (1993) *La parole des choses ou le langage de la chimie*, Paris: Hermann.
Laszlo, P. (1998) Chemical analysis as dematerialization, *Hyle*, 4: 29-38.
Laszlo, P. (2000) Playing with molecular models, *Hyle*, 6: 85-97.
Latour, B. (1987) *Science in Action*, Milton Keynes: Open University.
Latour, B. (1999) *Pandora's Hope: Essays on the Reality of Science Studies*, Cambridge MA: Harvard University Press.
Laudan, L. (1984) *Science and Values: The Aims of Science and their Role in Scientific Debate*, Berkeley: University of California Press.
Lauth, B. (1989) Reference problems in stoichiometry, *Erkenntnis* 30: 339-62.
Lauth, B. (1993) Physical constants and reference dynamics, *Journal for General Philosophy of Science*, 24: 63-86.
Lavoisier, A.L. (1789) *Traité élémentaire de chimie*, Paris: Cuchet; *Principles of Chemistry* (Edinburgh 1790).
Laymon, R. (1991) Idealisations and the reliability of dimensional analysis, pp. 146-180 in *Critical Perspectives on Nonacademic Science and Engineering* (P.T. Durbin, ed.), Bethlehem: Lehigh University Press.
Le Poidevin, R (1994) The chemistry of space, *Australian Journal of Philosophy*, 72: 77-88.
Lee, D.J., S.P. Ju, J.H. Kwon and F.M. Tiller (2000) Filtration of highly compactible filter cake: variable internal flow rate, *American Institute of Chemical Engineers Journal*, 46: 110-118.
Leibniz, G.W. (1981) *New Essays on Human Understanding* (P. Remnant and J. Bennett, eds.), , Cambridge: Cambridge University Press.
Leplin, J. (1988) Is essentialism unscientific?, *Philosophy of Science*, 55: 493-510.
LePore, E. and B. Loewer (1987) Mind matters, *The Journal of Philosophy*, 84: 630-642.
Lequan, M. (2000) *La Chimie selon Kant,* Paris: Presses Universitaires de France.
Levere, T. (1971) *Affinity and Matter*, Oxford: Oxford University Press.
Levine, I.N. (1991) *Quantum Chemistry*, Prentice-Hall International.
Levinson, J. (1991) Attribute, *Handbook of Metaphysics and Ontology* (H. Burkhardt and B. Smith, eds.), München: Philosophia, 1: 65-70.
Lévy, M. (1979a) Les relations entre chimie et physique et le problème de la réduction, *Epistemologia*, 2: 337-70
Lévy, M. (1979b) Le problème de la réduction de la chimie à la physique, Doctoral Dissertation Université Catholique de Louvain.

Lewis, D. (1983) New work for a theory of universals, *Australasian Journal of Philosophy*, 61: 343-377.
Lewis, E. (1996) Introduction to *Alexander of Aphrodisias: On Aristotle Meteorology 4*, London: Duckworth.
Li, C. (1993) Natural kinds: Direct reference, realism, and the impossibility of necessary a posteriori truth, *Review of Metaphysics*, 47(2): 261-276.
Liegener, C. and G. Del Re (1987a) The relation of chemistry to other fields of science: Atomism, reductionism, and inversion of reduction, *Epistemologia*, 10: 269-284.
Liegener, C. and G. Del Re (1987b) Chemistry vs. physics, the reduction myth and the unity of science, *Zeitschrift für allgemein Wissenschaftstheorie*, 18: 165-174.
Liegener, C.-M. (1994), Die Stellung der Chemie zur Physik - Symptome des Reduktionismus, pp. 95-100 in *Philosophische Perspektiven der Chemie* (P. Janich, ed.), Mannheim: Wissenschaftsverlag.
Lippencott, E., Cessac, G.L., Stromberg, R.R and Grant, W.H. (1971) Polywater - A search for alternative explanations, *Journal of Colloid and Interface Science*, 36: 443-460.
Lippincott, E.R., R.R. Stromberg, W.H. Grant, and G.L. Cessac [1969] Polywater, *Science*, 164: 1482.
Liu, S., P.J. Fryer and J.P. Pain (1999) Influence of particle-specific gravity and particle shape on the averaged axial velocity of nearly neutrally buoyant particles in horizontal pipes, *The Canadian Journal of Chemical Engineering*, 77: 1083-1089.
Lockhead, G.R. (1992) Psychophysical scaling: Judgements of attributes of objects?, *Brain and Behavioral Sciences*, 15: 543-600.
Lorenzen, P. (1994) Discussion: Konstruktivismus, *Journal of General Philosophy of Science*, 25: 125-133.
Löwdin, P.-O. (1967) Nature of quantum chemistry, *International Journal of Quantum Chemistry*, 1: 7-12.
Löwdin, P.-O. (1969) Some comments on the periodic system of elements, *International Journal of Quantum Chemistry*, Symposium 3: 331-334.
Lowe, E.J. (1989) *Kinds of Being: A Study of Individuation, Identity and the Logic of Sortal Terms*, Oxford: Basil Blackwell.
Lowe, E.J. (1991) Real selves: Persons as a substantial kind, pp. 87-107 in *Human Beings* (D. Cockburn, ed.), Cambridge: Cambridge University Press.
Lowe, E.J. (1998) *The Possibility of Metaphysics: substance, identity, and time*, Oxford: Clarendon Press.
Lowe, J.P. (1993) *Quantum Chemistry*, Boston: Academic Press.
Luikov, A.V. (1966) *Heat and Mass Transfer in Capillary-porous Bodies*, Oxford: Pergamon Press.
Macagno, E.O. (1971) Historico-critical review of dimensional analysis, *Journal of the Franklin Institute*, 292: 391-402.
Maccoll, A. (1964) *Space and Time in Chemistry*, London: Lewis.
MacDonald, D.K.C. (1960) Physics and chemistry: comments on Caldin's view of chemistry, *The British Journal for the Philosophy of Science*, 11: 222-223.
Mach, E. (1883) *The Science of Mechanics: A Critical & Historical Account of Its Development*, LaSalle: Open Court Publishing Company; 6th edition (after the 9th German edition), 1960.
Mach, E. (1919) *Die Analyse der Empfindungen und des Verhältnis des Physischen zum Psychischen*, Jena: Gustav Fischer; 8th edition; first published 1886.
Maher, P. (1993) Howson and Franklin on prediction, *Philosophy of Science* 60: 329-40.
Mainzer, K. (1997) Symmetry and complexity: Fundamental concepts of research in chemistry, *Hyle*, 3: 29-49.
Mainzer, K. (1998) Computational and mathematical models in chemistry: epistemic foundations and new perspectives of research, pp. 33-50 in P. Janich and N. Psarros (eds.), *The Autonomy of Chemistry: 3rd Erlenmeyer-Colloquy for the Philosophy of Chemistry*, Würzburg: Königshausen & Neumann.
Majer, U. (1996) Ist die Sprache der Chemie eine Begriffsschrift in Freges Sinne?, pp 91-100 in *Die Sprache der Chemie* (P. Janich and N. Psarros, eds.), Würzburg: Königshausen & Neumann.

Malisoff, W.M. (1941) Chemistry: emergence without mystification, *Philosophy of Science*, 8: 39-52.
Malt, B.C. (1994) Water is not H_2O, *Cognitive Psychology*, 27: 41-70.
Malzkorn, W. (1998) Kant über die Teilbarkeit der Materie, *Kant Studien*, 89: 385-409.
Manzelli, P. (1996) To what extent do we understand chemical reactions?, pp. 133-142 in *Philosophers in the Laboratory* (V. Mosini, ed.), Rome: Euroma.
Margalit, A. (1979) Sense and Science, pp. 17-47 in E. Saarinen, R. Hilpinen, I. Niiniluoto, and M. Provence Hintikka (eds.), *Essays in Honour of Jaakko Hintikka*, Dordrecht: Reidel Publishing Company.
Margenau, H. (1944) The exclusion principle and its philosophical importance, *Philosophy of Science*, 11: 187-208.
Marino, G. (ed.) (1994) *Storia e Fondamenti della Chimica*, Roma: Accademia Nazionale delle Scienze.
Marmur, A. and R.D. Cohen (1997) Characterization of porous media by the kinetics of liquid penetration: The vertical capillaries model, *Journal of Colloid and Interface Science*, 189: 299-304.
Massey, B.S. (1971) *Units, Dimensional Analysis and Physical Similarity*, London: Van Nostrand Reinhold.
Matteson, D.S. (1974) *Organometallic Reaction Mechanisms of the Nontransition Elements*, New York: Academic Press.
Maudlin, T. (1996) On the unification of physics, *The Journal of Philosophy*, 93: 129-144.
Mauskopf, S.H. (1988) Molecular geometry in 19th-century France: Shifts in guiding assumptions, pp. 125-144 in *Scrutinising Science* (A. Donovan, L. Laudan and R. Laudan, eds.), Dordrecht: Kluwer.
Maxwell, R.J. (1873) *A Treatise on Electricity and Magnetism*, Oxford: Clarendon Press; quoted from 1954 Dover reprint of second edition of 1891.
McAllester Jones, M. (1991) *Gaston Bachelard, Subversive Humanist: Texts and Readings*, Madison: University of Wisconsin Press.
McAllister, J.W. (1993) Scientific realism and the criteria for theory-choice, *Erkenntnis*, 38: 203-222.
McEvoy, J.G. (1989) Lavoisier, Priestley, and the *Philosophes*: Epistemic and linguistic dimensions to the chemical revolution, *Man and Nature*, 8: 91-98.
McEvoy, J.G. (2000) In search of the chemical revolution: Interpretative strategies in the history of chemistry, *Foundations of Chemistry*, 2: 47-73.
McGuire, D.P., J.J. Pearson and M.L. Dobbert (1996) Dimensional analysis in cultural anthropology: Reply to Wormer, *American Anthropologist*, 88: 703-706.
McIntyre, L. (1997) Gould on laws in biological sciences, *Biology and Philosophy*, 12: 357-367.
McIntyre, L. (1999) Davidson and social scientific laws, *Synthese*, 120: 375-394.
McKinney, W.J. [1991] Experimenting on and experimenting with: polywater and experimental realism, *British Journal for the Philosophy of Science*, 42: 295-307.
McLaughlin, D.R., C.F. Bender and H.F. Schaefer III (1972) Geometry and force constant determination from correlated wave functions for polyatomic molecules: Ground states of H_2O and CH_2, *Theoretica chimica acta (Berlin)*, 25: 352-359.
McWeeny, R. (1979) *Coulson's Valence*, Oxford: Oxford University Press.
Medin, D.L (1989), Concepts and conceptual structure, *American Psychologist*, 44: 1469-1481
Mellor, D.H. (1977) Natural kinds, *The British Journal for the Philosophy of Science* 28: 299-312.
Mendeleev, D. (1889) The periodic law of the chemical elements, *Faraday Lectures*, London: Chemical Society.
Mestrallet, G.R. (1980) *Communication, Linguistique et Sémiologie. Contribution à l'étude de la sémiologie. Études semiologique des systèmes de signes de la chimie*, Barcelona: Bellaterra.
Meyer, V. (1889) The chemical problems of to-day, *Journal of the American Chemistry Society*, 11: 101-120.
Meyerson, É. (1930) *Identity and Reality*, London: Allen & Unwin.
Meyerson, É. (1991) *Explanation in the Sciences*, Dordrecht: Kluwer.

Mill, J. S. (1843) *A System of Logic*, quoted from *Collected Works of John Stuart Mill* (J.M. Robertson, ed.), Toronto: University of Toronto Press, vol. 7.
Miller, R.B. (1990) Supervenience is a two-way street, *The Journal of Philosophy*, 87: 695-701.
Miller, R.B. (1992) A purely causal solution to one of the qua problems *Australasian Journal of Philosophy*, 70: 425-434.
Millikan, R.G. (1984) *Language, Thought, and Other Biological Categories: New Foundations for Realism*, Cambridge MA: The MIT Press.
Mittasch, A. (1948) *Von der Chemie zur Philosophie*, Ulm/Donau: Ebner.
Mittelstraß, J. and G. Stock (1992) *Chemie und Geisteswissenschaften*, Berlin: Akademie Verlag.
Monod-Herzen, G. (1976) *L'analyse dimensionnelle et l'épistémologie*, Paris: Maloine & Doin.
Montague, R. (1974) *Formal Philosophy: Selected Papers of Richard Montague*, New Haven CT: Yale University Press.
Moody, L.F. (1944) Friction factors for pipe flow, *Transactions of the American Association of Mechanical Engineers*, 66: 671-684.
Morreau, M. (1999) Other things being equal, *Philosophical Studies*, 96: 163-182.
Morrison, M. (1995) Capacities, tendencies and the problem of singular cases, *Philosophy and Phenomenological Research*, 55: 163-168.
Mosini, V. (1994) Some considerations on the reducibility of chemistry to physics, *Epistemologia*, 17: 205-223.
Mosini, V. (1995) Fundamentalism, antifundamentalism, and Gibbs's paradox, *Studies in the History and Philosophy of Modern Physics*, 26: 151-162.
Mosini, V. (ed.) (1996) *Philosophers in the Laboratory*, Rome: Euroma - Musi - Accademia Nazionale di Scienze Lettere e Arti Modena.
Mounin, G. (1981) A semiology of the sign system chemistry, *Diogenes*, Spring/Summer: 216-228.
Mulckhuyse, J.J. (1961) Molecules and models, pp. 133-51 in *The Concept and the Role of the Model in Mathematics and Natural and Social Sciences*. Dordrecht: Reidel.
Müller, A. (1998) Die inhärente Potentialität materieller (chemischer) Systeme, *Philosophia Naturalis*, 35: 333-358.
Müller, A. and H. Hörz (1996) Philosophische Aspekte der Chemie. Ihr Wesen: Universalität und Beständigkeit des Wandels, pp. 193-268 in K. Mainzer, A. Müller and W.G. Saltzer (eds.), *From Simplicity to Complexity*, Wiesbaden: Vieweg.
Mulliken, R.S. (1968) Spectroscopy, quantum chemistry and molecular physics, *Physics Today*, April: 52-57.
Mundy, B. (1990) On empirical interpretation, *Erkenntnis*, 33: 345-369.
Munowitz, M. (2000) *Principles of Chemistry*, New York: Norton & Co.
Nagel, E. (1961) *The Structure of Science*, London: Routledge and Kegan Paul.
Nagel, E. (1970) Issues in the logic of reductive explanations, pp. 117-137 in H.E. Kiefer and M.K. Munitz (eds.), *Mind, Science and History*, Albany: State University of New York Press.
Nagel, E. (1974) Issues in the logic of reducing explanations, pp. 95-113 in *Teleology Revisited*, New York: Columbia University Press.
Nayak, A.C. and E. Sotnak (1995) Kant on the impossibility of the "soft sciences", *Philosophy and Phenomenological Research*, 55: 133-151.
Needham, P. (1993) Stuff, *Australasian Journal of Philosophy*, 71: 270-290.
Needham, P. (1996a) Macroscopic objects: An exercise in Duhemian ontology, *Philosophy of Science*, 63: 205-224.
Needham, P. (1996b) Substitution: Duhem's explication of a chemical paradigm, *Perspectives on Science*, 4:408-433.
Needham, P. (1996c) Aristotelian chemistry: A prelude to Duhemian metaphysics, *Studies in History and Philosophy of Science*, 27: 251-270.
Needham, P. (1998) Duhem's physicalism, *Studies in History and Philosophy of Science*, 29: 33-62.
Needham, P. (1999a) Macroscopic processes, *Philosophy of Science*, 66: 310-331.
Needham, P. (1999b) Reduction and abduction in chemistry - A response to Scerri, *International Studies in the Philosophy of Science*, forthcoming.

Needham, P. (1999c) Alexander's critique of the stoic theory of mixture, pp. 135-160 in *Philosophical Crumbs: Essays dedicated to Ann-Mari Henschen-Dahlquist on the occasion of her seventy-fifth birthday* (R. Sliwinski, ed.), Uppsala: Department of Philosophy, Uppsala University (*Uppsala philosophical studies 49*).
Needham, P. (2000) What is water?, *Analysis*, 60: 13-21.
Needham, R. (1972) *Belief, Language, and Experience*. Oxford: Blackwell.
Nerlich, G. (1995), On the one hand: Reflections on enantiomorphy, *Australian Journal of Philosophy*, 73: 432-443.
Nersessian, N.J. (1991) Discussion: The method to "meaning": A reply to Leplin, *Philosophy of Science*, 58: 678-686.
Newman, K.E. [1989] 'Hydration of surfaces with particular attention to micro-sized particles', in F. Franks (*ed.*) *Water Science Reviews 4*, Cambridge University Press, Cambridge, pp. 127-172.
Nickles, T. (1973) Two concepts of inter-theoretic reduction, *The Journal of Philosophy*, 70: 181-201.
Niedersen, U. (1983) Chemie heute: Bemerkungen zur Klassifikation der Naturwissenschaften, *Deutsche Zeitschrift für Philosophie*, 31: 363-373.
Niedersen, U. (1994) Prozeßstrukturen: Schellings Philosophie und einige ausgewählte Theorie- und Praxisbereiche der Physikochemie, *Selbstorganisation*, 5:183-199.
Nola, R. (1980) Fixing the reference of theoretical terms, *Philosophy of Science*, 47: 505-531.
Noordhof, P. (1999) Micro-based properties and the supervenience argument: A response to Kim, *Proceedings of the Aristotelian Society*, 99: 109-114.
Nye, M.J. (1972) *Molecular Reality*, New York: Watson Academic Publications.
Nye, M.J. (1993) *From Chemical Philosophy to Theoretical Chemistry. Dynamics of Matter and Dynamics of Disciplines 1800-1950*, Berkeley: University of California Press.
O'Konski, C.T. and S. Levine (1971) Instability of water polymers, *Journal of Colloid and Interface Science*, 36: 547-553.
Ogilvie, J.F. (1990) The nature of the chemical bond - 1990, *Journal of Chemical Education*, 67: 280.
Ogilvie, J.F. (1994) The nature of the chemical bond 1993: There are no such things as orbitals, pp. 171-198 in E.S. Kryachko and J.L. Calais (eds.), *Conceptual Trends in Quantum Chemistry*, Dordrecht: Kluwer.
Oppenheim, P. and H. Putnam (1958) The unity of science as a working hypothesis, pp. 3-35 in *Minnesota Studies in the Philosophy of Science* (H. Feigl, G. Maxwell, M. Scriven, eds.), vol. 2, Minneapolis: University of Minnesota Press.
Osborne, D.K. (1978) On dimensional invariance, *Quality and Quantity*, 12: 75-89.
Ostwald, W. (1902) *Vorlesungen über Naturphilosophie*, Leipzig: Veit.
Ostwald, W. (1904) Elements and compounds, *Faraday Lectures*, London: Chemical Society (1928); *Elemente und Verbindungen, Faraday-Vorlesung*, Leipzig (1904).
Ostwald, W. (1907) *Prinzipien der Chemie: Eine Einleitung in alle chemischen Lehrbücher*; Leipzig, translated as *The Fundamental Principles of Chemistry*, London: Longmans, Green, and Co. (1909).
Owens, D. (1992) *Causes and Coincidences*, Cambridge: Cambridge University Press.
Page, T.F. and R.J. Jakobsen (1971) Some evidence for the existence of water II, *Journal of Colloid and Interface Science*, 36: 427-433.
Palacios, J. (1964) *Dimensional Analysis*, London: Macmillan; 1st ed. in Spanish, Madrid: España-Calpe (1956).
Paneth, F.A. (1962) The epistemological status of the chemical concept of element, *The British Journal for the Philosophy of Science*, 13: 1-14, 144-160; originally published in German as Über die erkenntnistheoretische Stellung des Elementebegriffs, *Schriften der Königsberger Gelehrten Gesellschaft, Naturwissenschaftliche Klasse*, 8 (4) (1931) 101-125.
Pankhurst, R.C. (1971) Alternative formulation of the Pi-theorem, *Journal of the Franklin Institute*, 292: 451-462.
Paolini, L. (1981) Quantum mechanics and the logical structure of contemporary chemistry, pp. 1-16 in *Current Aspects of Quantum Chemistry 1981*, Amsterdam: Elsevier.
Papineau, D. (1993) *Philosophical Naturalism*, Oxford: Blackwell.

Pargetter, R. (1988) Goodness and redness, *Philosophical Papers*, 17: 113-126.
Pauling, L. (1939) *The Nature of the Chemical Bond, and the Structure of Molecules and Crystals: An Introduction to Modern Structural Chemistry*, Ithaca: Cornell University Press; 3rd ed. 1960.
Pauling, L. (1950) The place of chemistry in the integration of the sciences, *Main Currents in Modern Thought*, 7: 108-111.
Pauling, L. (1954) Modern structural chemistry, pp. 429-439 in *Nobel Lectures. Chemistry 1942-1962*, Amsterdam: Elsevier (1964).
Pauling, L. (1962) Teoriia rezonansa v khimii, *Zhurnal vsesoiuznogo khimicheskogo obshchestva im. D.I. Mendeleeva*, (4) 462-467.
Pauling, L. (1964) *Leben und Tod im Atomzeitalter*, Berlin: VEB.
Pauling, L. (1992) The nature of the chemical bond - 1992, *Journal of Chemical Education*, 69: 519-521.
Pauling, L. (1995) Questions and answers, *The Chemical Intelligencer*, 1(1): 5.
Peacocke, C. (1979) *Holistic Explanation: Action, Space, Interpretation*, Oxford: Clarendon Press.
Pepper, S.C. (1926) Emergence, *The Journal of Philosophy*, 23: 241-245.
Peregrin, J. (1997) Language and its models: Is model theory a theory of semantics?, *Nordic Journal of Philosophical Logic*, 2: 1-23.
Perrin, C.E. (1988) The chemical revolution: Shifts in guiding assumptions, pp. 105-124 in *Scrutinising Science* (A. Donovan, L. Laudan and R. Laudan, eds.), Dordrecht: Kluwer.
Pessin, A. and S. Goldberg (eds.) (1996) *The Twin Earth Chronicles. Twenty Years of Reflection on Hilary Putnam's "The Meaning of 'Meaning'"*, New York: Sharpe.
Petsko, G.A. (1971) Proton magnetic resonance studies of "Polywater-Water" mixtures, *Journal of Colloid and Interface Science*, 36: 503-508.
Petsko, G.A. and W.R. Massey. (1971) X-ray diffraction investigations of anomalous water, *Journal of Colloid and Interface Science*, 36: 508-512.
Pettit, P. (1992) The nature of naturalism, *Proceedings of the Aristotelian Society*, Suppl. Vol. 66: 245-266.
Pietroski, P. and G. Rey (1995) When other things aren't equal: Saving *ceteris paribus* laws from vacuity, *British Journal of Philosophy of Science*, 46: 81-110.
Pilkington, R. (1959) *Robert Boyle: Father of Chemistry*, London: Murray.
Pirie, N.W. (1952) Concepts out of context, *The British Journal for the Philosophy of Science*, 2: 269-280.
Platts, M. (1983) Explanatory kinds, *British Journal of Philosophy of Science*, 34: 133-148.
Plesch, P. H. (1999) On the distinctness of chemistry', *Foundations of Chemistry*, 1: 6-15.
Polanyi, M. (1958) *Personal Knowledge: Towards a Post-Critical Philosophy*, Chicago: The University of Chicago Press.
Poller, S. (1966) Zur Abgrenzung und Charakteristik der chemischen Bewegungsform der Materie, *Deutsche Zeitschrift für Philosophie*, 14: 328-338
Post, E.J. (1982) Physical dimensions and covariance, *Foundations of Physics*, 12: 169-195.
Prélat, C.E. (1947) *Epistemología de la Química*, Buenos Aires and Mexico: Espasa and Calpe Argentina.
Primas, H. (1982) Chemistry and complementarity, *Chimia*, 36: 293-300.
Primas, H. (1983) *Chemistry, Quantum Mechanics and Reductionism: Perspectives in Theoretical Chemistry*, Berlin: Springer.
Primas, H. (1985a) Kann Chemie auf Physik reduziert werden? I. Das molekulare Programm, *Chemie unser Zeit*, 19: 109-119.
Primas, H. (1985b) Kann Chemie auf Physik reduziert werden? II. Die Chemie der Macrowelt, *Chemie unser Zeit*, 19: 160-166.
Primas, H. (1990) The measurement process in the individual interpretation of quantum mechanics, pp. 49-68 in M. Cini and J.-M. Lévy-Leblond (eds.), *Quantum Theory without Reduction*, Bristol/New York: Adam Hilger.
Primas, H. (1991) Reductionism: Palaver without precedent, pp. 161-172 in *The Problems of Reductionism in Science* (E. Agazzi, ed.), Dordrecht: Kluwer.

Psarros, N. (1994) Sind die 'Gesetze' der konstanten und der multiplen Proportionen empirische Naturgesetze oder Normen?, pp. 53-63 in *Philosophische Perspektiven der Chemie* (P. Janich, ed.), Mannheim: Wissenschaftsverlag.
Psarros, N. (1995a), Stoffe, Verbindungen und Elemente - Eine methodische Annäherung an die Gegenstände der Chemie, *chimica didactica*, 21: 129-148.
Psarros, N. (1995b) The constructive approach to the philosophy of chemistry, *Epistemologia*, 18: 27-38.
Psarros, N. (1996a) Die Chemie als Gegenstand philosophischer Reflexion", pp. 111-141 in *Philosophie der Chemie: Bestandsaufnahme und Ausblick* (N. Psarros, K. Ruthenberg and J. Schummer, eds.), Würzburg: Königshausen & Neumann.
Psarros, N. (1996b) The mess with mass terms, pp. 123-131 in *Philosophers in the Laboratory* (V. Mosini, ed.), Rome: Euroma.
Psarros, N. (1997) Critical realism in the test tube, *Journal for General Philosophy of Science*, 28: 297-305.
Psarros, N. (1998a) The concept of molecule in chemistry, physics and biology, pp. 91-100 in P. Janich and N. Psarros (eds.), *The Autonomy of Chemistry: 3rd Erlenmeyer-Colloquy for the Philosophy of Chemistry*, Würzburg: Königshausen & Neumann.
Psarros, N. (1998b) What has philosophy to offer to chemistry?, *Foundations of Science*, 1: 183-202.
Psarros, N. (1999) *Die Chemie und ihre Methoden: Ein philosophische Betrachtung*, Weinheim: Wiley-VCH.
Psarros, N. and K. Gavroglu (eds.) (1999) *Ars mutandi: Issues in Philosophy and History of Chemistry*, Leipzig: Leipziger Universitätsverlag.
Psarros, N., K. Ruthenberg and J. Schummer, eds. (1996) *Philosophie der Chemie: Bestandsaufnahme und Ausblick*, Würzburg: Königshausen & Neumann.
Pullman, B. (1998) *The Atom in the History of Human Thought*, New York: Oxford University Press.
Putnam, H. (1975) The meaning of meaning, pp. 215-271 in his *Mind, Language and Reality [Philosophical Papers Volume 2]*. Cambridge: Cambridge University Press.
Putnam, H. (1978) *Meaning and the Moral Sciences*. London: Routledge and Kegan Paul.
Putnam, H. (1981) *Reason, Truth and History*, Cambridge: Cambridge University Press.
Putnam, H. (1983) *Realism and Reason [Philosophical Papers Volume 3]*, Cambridge: Cambridge University Press.
Putnam, H. (1987) *The Many Faces of Realism*, La Salle: Open Court.
Putnam, H. (1990a) Is water necessarily H_2O, pp. 54-79, 325-327 in his *Realism with a Human Face*, Cambridge MA: Harvard University Press.
Putnam, H. (1990b) Meaning holism, pp 278-302, 336-337 in his *Realism with a Human Face*, Cambridge MA: Harvard University Press.
Pyle, A. (1997) *Atomism and Its Critics: From Democritus to Newton*, Bristol: Thoemes Press.
Quine, W.V. (1960) *Word and Object*, Cambridge MA: The MIT Press.
Quine, W.V. (1969) Natural kinds, in his *Ontological Relativity and Other Essays*, New York: Columbia University Press, pp. 114-138.
Quine, W.V. (1976) Whither physical objects, pp. 497-504 in *Essays in Memory of Imre Lakatos*, Dordrecht: Reidel.
Quine, W.V. (1990) *The Pursuit of Truth*, Cambridge MA: Harvard University Press.
Quine, W.V. (1992) Structure and nature, *The Journal of Philosophy*, 89: 5-9.
Quine, W.V. (1993) In praise of observation sentences, *The Journal of Philosophy*, 90: 107-116.
Quine, W.V. (1994) Assuming objects, *Theoria*, 60: 171-183.
Quine, W.V. and J.S. Ullian (1970) *The Web of Belief*, New York: Random House.
Ramberg, P. (2000) Pragmatism, belief, and reduction: Stereoformulas and atomic models in early stereochemistry, *Hyle*, 6: 35-61.
Ramsey, J.L. (1990) Beyond numerical and causal accuracy: Expanding the set of justificational criteria, *PSA 1990*, 1: 485-99.

Ramsey, J.L. (1992) On refusing to be an epistemologically black box: Instruments in chemical kinetics during the 1920s and '30s, *Studies in the History and Philosophy of Science*, 23: 283-304.
Ramsey, J.L. (1994) Ideal reaction types and the reactions of real alloys, *PSA 1994*, 1: 149-159.
Ramsey, J.L. (1997) Molecular shape, reduction, explanation and approximate concepts, *Synthese*, 111: 233-251.
Ramsey, J.L. (1998) Recent work in the history and philosophy of chemistry, *Perspectives on Science*, 6: 409-427.
Ramsey, W., S.P. Stich and D.E. Rumelhart (eds.) (1991) *Philosophy and Connectionist Theory*, Hillsdale NJ: Lawrence Erlbaum.
Rayleigh (1892) On the question of stability of the flow of fluids, *Philosophical Magazine*, 34: 59-70.
Rayleigh (1904) Fluid friction on even surfaces, *Philosophical Magazine*, 8: 66-67.
Rayleigh (1915) The principle of similitude, *Nature*, 95: 66-68, 644.
Rayleigh (1945) *The Theory of Sound* (by John William Strutt, Baron Rayleigh), New York: Dover; two volumes bound as one of the second edition (1894); first edition 1877.
Redhead, M. (1980) Models in physics, *The British Journal for the Philosophy of Science*, 31: 145-163.
Redhead, M. (1991) Review of J.D. Barrow's *Theories of Everything*, *Times Literary Supplement*, July 26, 1991.
Reichenbach, H. (1978 [1929]) The aims and methods of physical knowledge, pp. 81-225 in *Hans Reichenbach: Selected Writings 1909-1953* (M. Reichenbach and R.S. Cohen, eds.; principal translations by E.H. Schneewind), volume II, Dordrecht: Reidel.
Rheinberger, H.-J. (1992a) Experiment, difference, and writing: I. Tracing protein synthesis, *Studies in History and Philosophy of Science*, 23: 305-331.
Rheinberger, H.-J. (1992b) Experiment, difference, and writing: II. The laboratory production of transfer RNA, *Studies in History and Philosophy of Science*, 23: 389-422.
Riabouchinsky, D. (1915) Letter to the Editor, *Nature*, 95: 591.
Richter, K.-H. and H. Laitko (1962) Zur Gegenstandsbestimmung der Chemie, *Deutsche Zeitschrift für Philosophie*, 10: 1278-1293.
Ritchie, A.D. (1945) The atomic theory as metaphysics and as science, *Proceedings of the Aristotelian Society*, 45: 71-88.
Rocke, A.J. (1981) Kekulé, Butlerov, and the historiography of the theory of structure, *The British Journal for the History of Science*, 14: 27-57.
Rocke, A.J. (1988) Kekulé's benzene theory and the appraisal of scientific theories, pp. 145-161 in *Scrutinising Science* (A. Donovan, L. Laudan and R. Laudan, eds.), Dordrecht: Kluwer.
Röhler, G. (1962) Zur erkenntnistheoretischen Bedeutung von Hypothese, Modellvorstellung und Theorie in der Chemie, *Deutsche Zeitschrift für Philosophie*, 10: 1294-1307.
Rosenberg, A. (1994) *Instrumental Biology and the Disunity of Science*, Chicago: The University of Chicago Press.
Rosenthal, A. (1982) Zu einigen Aspekten des Verhältnisses von Physik und Chemie, *Deutsche Zeitschrift für Philosophie*, 30: 576-590.
Rothbart, D. (1993) Discovering natural kinds through inter-theoretic prototypes, *Methodology and Science*, 26: 171-189.
Rothbart, D. (1994) Spectrometers as analogues of nature, *Philosophy of Science Association (PSA)* 1:141-148.
Rothbart, D. (1999) On the relationship between instrument and specimen in chemical research, *Foundations of Chemistry*, 1: 255-268.
Rothbart, D. and I. Scherer (1997) Kant's *Critique of Judgement* and the scientific investigation of matter, *Hyle*, 3: 65-80.
Rothbart, D. and S.W. Slayden (1994), The epistemology of a spectrometer, *Philosophy of Science*, 61: 25-38.
Rousseau, D.L. (1971) An alternative explanation for polywater, *Journal of Colloid and Interface Science*, 36: 434-442.

Rovane, C. (1993) Self-reference: The radicalisation of Locke, *The Journal of Philosophy*, 90: 73-97.
Ruschig, U. (1987) Chemische Einsichten wider willen: Hegels Theorie der Chemie, *Hegel Studien*, 22: 17-23.
Ruschig, U. (1997) *Hegels Logik und die Chemie. Fortlaufender Kommentar zum "Realen Mass"* (Hegel Studien, Beiheft 37), Bonn: Bouvier.
Russell, B. (1948) *Human Knowledge*, London: Allen and Unwin.
Ruthenberg, K. (1994) Die Regelhaftigkeit der allgemeinen Chemie, pp. 65-82 in P. Janich (ed.), *Philosophische Perspektiven der Chemie*, Mannheim: Wissenschaftsverlag.
Ruthenberg, K. (1996) Warum ist die Chemie ein Stiefkind der Philosophie?, pp. 27-35 in *Philosophie der Chemie: Bestandsaufnahme und Ausblick* (N. Psarros, K. Ruthenberg and J. Schummer, eds.), Würzburg: Königshausen & Neumann.
Ruthenberg, K. (1997) Friedrich Adolf Paneth, *Hyle*, 3: 103-106.
Ruthenberg, K. and Psarros, N. (1994) Frantisek Wald und die phänomenologische Chemie, *Mitteilungen der Fachgruppe Geschichte der Chemie der GDCh*, 10: 17-30.
Salmon, N.U. (1982) *Reference and Essence*, Oxford: Blackwell.
Salmon, W.C. (1999) The spirit of logical empiricism: Carl G. Hempel's role in twentieth century philosophy of science, *Philosophy of Science*, 66: 333-350.
Sanchez Perez, E.A. and Sanchez Marin, J. (1997) Sobre algunas propiedades formales de los sistemas de representación en química (On some formal properties of the chemical representation systems), *Theoria*, 12: 567-588.
Sankey, H. (1991) Translation failure between theories, *Studies in the History and Philosophy of Science* 22: 223-236.
Sarkar, S. (1992) Models of reduction and categories of reductionism, *Synthese*, 91: 167-194.
Satheesh, V.K., R.P Chhabra, V. Eswaran (1999) Steady incompressible fluid flow over a bundle of cylinders at moderate Reynolds numbers, *The Canadian Journal of Chemical Engineering*, 77: 978-989.
Saunders, B.A.C. and J. van Brakel (1997a) Are there non-trivial constraints on colour categorisation?, *Behavioral and Brain Sciences*, 20: 167-232.
Saunders, B.A.C. and J. van Brakel (1997b) The phantom objectivity of colour: With reference to the works of Franz Boas on the Kwakiutl, pp. 87-95 in *Translation of Sensitive Texts* (K. Simms, ed.), Amsterdam: Rodopi.
Savellos, E. (1992) Criteria of identity and the individuation of natural-kind events, *Philosophy and Phenomenological Research*, 52: 807-831.
Scerri, E.R. (1991a) The electronic configuration model, quantum mechanics and reduction, *The British Journal for the Philosophy of Science*, 42: 309-25.
Scerri, E.R. (1991b) Chemistry, spectroscopy, and the question of reduction, *Journal of Chemical Education*, 68: 122-126.
Scerri, E. R. (1992a) Quantum chemistry truth, *Chemistry in Britain*, 28: 326.
Scerri, E. R. (1992b) Quantum extrapolation, *Chemistry in Britain*, 28: 781.
Scerri, E.R. (1993) Correspondence and reduction in chemistry, pp. 45-64 in *Correspondence, Invariance and Heuristics* (S. French and H. Kamminga, eds.), Dordrecht: Kluwer.
Scerri, E.R. (1994) Has chemistry been at least approximately reduced to quantum mechanics, *PSA 1994*, 1: 160-170.
Scerri, E.R. (1995) The exclusion principle, chemistry and hidden variables, *Synthese*, 102: 165-169.
Scerri, E.R. (1997a) The periodic table and the electron, *American Scientist*, 85: 546-553.
Scerri, E.R. (1997b) Has the periodic table been successfully axiomatised?, *Erkenntnis*, 47:229-243.
Scerri, E.R. (1998a) Popper's naturalised approach to the reduction of chemistry, *International Studies in the Philosophy of Science*, 12: 33-44.
Scerri, E. R. (1998b) How good is the quantum mechanical explanation of the periodic system?, *Journal of Chemical Education*, 75: 1384-1385.
Scerri, E.R. (1999a) Response to Needham, *International Studies in the Philosophy of Science*, 13: 185-192;

Scerri, E.R. (1999b) A critique of Atkins' periodic kingdom and some writings on electronic structure, *Foundations of Chemistry*, 1: 297-305.
Scerri, E.R. and L. McIntyre (1997) The case for the philosophy of chemistry, *Synthese*, 111: 213-232.
Schaefer III, H.F. (1986) Methylene: A paradigm for computational quantum chemistry, *Science*, 231: 1100-1107.
Schaefer, H.F. (1992) Quantum dispute, *Chemistry in Britain*, 28: 604.
Scheibe, E. (1991) Substances, physical systems, and quantum mechanics, in *Advances in Scientific Philosophy* (G. Schurz and G.J.W. Dorn, eds.), Amsterdam: Rodopi.
Scheibe, E. (1999) *Die Reduktion physikalischer Theorien: Ein Beitrag zur Einheit der Physik, vol. 2.*, Berlin: Springer.
Schelling, E.W.J. von (1988) *Ideas for a Philosophy of Nature*, translated by E.E. Harris and P. Heach from the first German edition (1797), Cambridge: Cambridge University Press.
Schicketanz, W. (1974) Eine Anmerkung zur kapillaren Flüssigkeitsbewegung in Pulvern, *Powder Technology*, 9: 49-52.
Schilpp, P.A. (1959) *The Philosophy of C.D. Broad* (P.A. Schilpp, ed.), New York: Tudor.
Schlesinger, G. (1961) The prejudice of micro-reduction, *The British Journal for the Philosophy of Science*, 12: 215-224.
Schröder, J. (1998) Emergence: Non-reducibility or downwards causation?, *The Philosophical Quarterly*, 48: 433-452.
Schulze, S. (1994) *Kants Verteidigung der Metaphysik: eine Untersuchung zur Problemgeschichte des Opus Postumum*, Marburg: Tectum.
Schummer, J. (1994) Die Rolle des Experiments in der Chemie, pp. 27-52 in P. Janich (ed.), *Philosophische Perspektiven der Chemie*, Mannheim: Wissenschaftsverlag.
Schummer, J. (1995) Ist die Chemie eine schöne Kunst? Ein Beitrag zum Verhältnis von Kunst und Wissenschaft, *Zeitschrift für Ästhetik und allgemeine Kunst-Wissenschaft*, 40: 145-178.
Schummer, J. (1996a) *Realismus und Chemie: Philosophische Untersuchungen der Wissenschaft von den Stoffen*, Würzburg: Königshausen & Neumann.
Schummer, J. (1996b) Bibliographie chemiephilosophischer Literatur der DDR, *Hyle*, 2: 3-11.
Schummer, J. (1996c) Zur Semiotik der chemischen Zeichensprache: Die Repräsentation dynamischer Verhältnisse mit statischen Mitteln", pp. 113-126 in *Die Sprache der Chemie* (P. Janich and N. Psarros, eds.), Würzburg: Königshausen & Neumann.
Schummer, J. (1997a) Towards a philosophy of chemistry, *Journal for General Philosophy of Science*, 28: 307-335.
Schummer, J. (1997b) Scientometric studies on chemistry I: The exponential growth of chemical substances, 1800-1995", *Scientometrics*, 39: 107-123.
Schummer, J. (1997c) Scientometric studies on chemistry II: Aims and methods of producing new chemical substances, *Scientometrics*, 39: 125-140.
Schummer, J. (1997d) Challenging standard distinctions between science and technology: The case of preparative chemistry, *Hyle*, 3: 81-94.
Schummer, J. (1998a) Physical chemistry: neither fish nor fowl?, pp. 135-148 in P. Janich and N. Psarros (eds.), *The Autonomy of Chemistry: 3rd Erlenmeyer-Colloquy for the Philosophy of Chemistry*, Würzburg: Königshausen & Neumann.
Schummer, J. (1998b) The chemical core of chemistry I: a conceptual approach, *Hyle*, 4: 129-162.
Schwab, G.-M. (1959) *Die Erkenntniskrise der Chemie und ihre Überwindung*, München: Verlag der Bayerischen Akademie der Wissenschaften.
Schweber, S.S. (1997) The metaphysics of science at the end of a heroic age, pp. 171-198 in *Experimental Metaphysics* (R.S. Cohen et al., eds.), Dordrecht: Kluwer
Scott, M. (1992) The nature of the chemical bond - once more, *Journal of Chemical Education*, 69: 600-601.
Searle, J.R. (1983) *Intentionality*, Cambridge: Cambridge University Press.
Searle, J.R. and D. Vanderveken (1985) *Foundations of Illocutionary Logic*, Cambridge: Cambridge University Press.

Sedov, L.I. (1959) *Similarity and Dimensional Methods in Mechanics*, New York: Academic Press.
Seibt, J. (1990) *Properties as Processes: A Synoptic Study of Wilfrid Sellars' Nominalism*, Atascadero CA: Ridgeview.
Sellars, W. (1963) *Science, Perception, and Reality*, London: Routledge and Kegan Paul.
Sellars, W. (1967) *Science, and Metaphysics: Variations on Kantian Themes*, London: Routledge and Kegan Paul.
Semjonov, N.N. (1959) On the relation of chemistry and biology [in Russian], *Voprosy filosofii*, (10) 95-102.
Shain, R. (1993) Mill, Quine and natural kinds, *Metaphilosophy*, 24(3): 275-292.
Shakhparanov, M.I. (1957) *Outlines of the Philosophical Problems of Chemistry* [in Russian], Moscow: Moscow University Press.
Shakhparanov, M.I. (1962) *Philosophical Chemistry* [in Russian], Moscow: Gos. Ind. Polit. Lit.; published as *Chemie und Philosophie* (mimeographed), by the Institute for Marxism-Leninism at the Technischen Hochschule für Chemie in Leuna-Merseburg (1963).
Shapere, D. (1984) *Reason and the Search for Knowledge: Investigations in the Philosophy of Science*, Boston: Reidel.
Shapere, D. (1991) Discussion: Leplin on essentialism, *Philosophy of Science*, 58: 655-677.
Shapiro, L. (1999) Toward 'perfect collections of properties: Locke on the constitution of substantial sorts, *Canadian Journal of Philosophy*, 29: 551-592.
Shelton, J. (1995) Seeing and paradigms in the chemical revolution, *Philosophy of Science (Tucson)*, 6: 129-141.
Shrader-Frechette, K.S. (1993) *Burying Uncertainty: Risk and the Case Against Geological Disposal of Nuclear Waste*, Berkeley: University of California Press.
Shrader-Frechette, K.S. (1997) Hydrogeology and framing questions having policy consequences, *Philosophy of Science*, 64 (Proceedings): S149-S160.
Sicha, J. (1988) Sellarsian realism, *Philosophical Studies*, 54: 229-256.
Siebold, A., A. Walliser, M. Nardin, M. Oppliger, and J. Schultz (1997) Capillary rise for thermodynamic characterisation of solid particle surface, *Journal of Colloid and Interface Science*, 186: 60-70.
Siegfried, R. and B. J. Dobbs (1968) Composition: A neglected aspect of the chemical revolution, *Annals of Science*, 24: 275-293.
Simoes, A. and Gavroglu, K. (1999) Quantum chemistry *qua* applied mathematics: The contributions of Charles Alfred Coulson (1910-1974), *Historical Studies in the Physical Sciences*, 29(2): 363-406.
Simon, R. (1975) Dialektik und Chemie, *Deutsche Zeitschrift für Philosophie*, 23: 980-984.
Simon, R., U. Niedersen and G. Kertscher (1982) *Philosophische Probleme der Chemie*, Berlin: VEB.
Simons, J. (1991) There are no such things as orbitals - act two!, *Journal of Chemical Education*, 68: 131-132.
Sisakjan, N.M. (1959a) Some philosophical questions concerning biochemistry [in Russian], *Voprosy filosofii*, (2) 89-104.
Sisakjan, N.M. (1959b) The ideas of Lenin are confirmed in the results of modern biochemistry [in Russian], *Voprosy filosofii*, (5) 123-129.
Sklar, L. (1990) How free are initial conditions?, *PSA 1990*, 2: 551-564.
Sklar, L. (1993) *Physics and Chance: Philosophical Issues in the Foundations of Statistical Mechanics*, Cambridge: Cambridge University Press.
Smart, J.J.C. (1964) *Philosophy and Scientific Realism*, London: Routledge.
Snelders, H.A.M. (1993) The significance of Hegel's treatment of chemical affinity, in *Hegel and Newtonianism* (M.J. Petry, ed.), Dordrecht: Kluwer. in Hegel and
Sober, E. (1997) Two outbreaks of lawlessness in recent philosophy of biology, *Philosophy of Science*, 64 (Proceedings): S458-S467.
Sober, E. (1999) The multiple reliability argument against reductionism, *Philosophy of Science*, 66: 542-564.
Spencer-Smith, R. (1995) Reductionism and emergent properties, *Proceedings of the Aristotelian Society*, 95: 113-129.

Splitter, L.J. (1988) Species and identity, *Philosophy of Science*, 55: 323-348.
Stalker, D. (ed.) (1994) *Grue! The New Riddle of Induction*, La Salle: Open Court.
Stanford, P.K. (1998) Reference and natural kind terms: the real essence of Locke's view, *Pacific Philosophical Quarterly*, 79: 78-97.
Stanford, P.K. and Kitcher, P. (2000) Refining the causal theory of reference for natural kind terms, *Philosophical Studies*, 97: 99-129.
Stengers, I. (1995) Ambiguous affinity: The Newtonian dream of chemistry in the eighteenth century, pp. 372-400 in *A History of Scientific Thought: Elements of a History of Science* (M. Serres, ed.), Oxford: Blackwell.
Stöckler, M. (1991) A short history of emergence and reductionism, pp. 71-90 in *The Problem of Reductionism in Science* (E. Agazzi, ed.), Dordrecht/Boston/London: Kluwer.
Stoneham, T. (1999) Boghossian on empty natural kind concepts, *Aristotelian Society Proceedings*, 99:119-122.
Stork, H. (1963) Über die Diskussion und Ablehnung der Resonanz-Mesomerie-Theorie in der Sowjetunion, *Chemiker-Zeitung/ Chemische Apparatur*, 87: 608-619.
Ströker, E. (1967) *Denkwege der Chemie - Elemente Ihrer Wissenschaftstheorie*, Freiburg: Alber.
Stroll, A. (1988) *Surfaces*, Minneapolis: University of Minnesota Press.
Stroll, A. (1991) Observation and the hidden, *Dialectica*, 45(2-3): 165-179.
Stuart, M. (1999) Locke on natural kinds, *History of Philosophy Quarterly*, 16(3): 277-296.
Sutcliffe, B.T. (1996) The development of the idea of a chemical bond, *International Journal of Quantum Chemistry*, 58: 645-656.
Syrkin, Y.K. and Dyatkina, M.E. (1952) Remarks on the theory of resonance or mesomerism, *Bulletin of the Academy of Sciences of the USSR / Division of chemical science*, 973-979.
Tarski, A. (1956) The concept of truth in formalised languages, in his *Logic, Semantics, Metamathematics*, Oxford: Clarendon Press, pp. 152-278.
Tatevskii, V.M. and M.I. Shakhparanov (1949), Ob odnoi makhistskoi teorii v khimii i ee propagandistakh, *Voprosy filosofii*, (3) 176-192; excerpts translated as "About a Machistic theory in chemistry and its propagandists" in *Journal of Chemical Education*, 29: 14-15.
Taylor, E.H. (1971) A kinetic argument against the existence of anomalous water, *Journal of Colloid and Interface Science*, 36: 543-546.
Thagard, P. (1989) Explanatory coherence, *Behavioral and Brain Sciences* (12: 435-502.
Thagard, P. (1990) The conceptual structure of the chemical revolution, *Philosophy of Science* 57: 183-209.
Theobald, D.W. (1976) Some considerations on the philosophy of chemistry, *Chemical Society Reviews*, 5: 203-213.
Theobald, D.W. (1982) Gaston Bachelard et la philosophie de la chimie, *Archives de philosophie*, 45: 63-83.
Thornton, M. (1981) Sellars' scientific realism: A reply to van Fraassen, *Dialogue*, 20: 79-83.
Tijiattas, M. (1991) Bachelard and scientific realism, *Philosophical Forum*, (Spring) 203-210.
Tiles, M. (1985) *Bachelard: Science and Objectivity*, Cambridge: Cambridge University Press.
Tiles, M. (1987) Epistemological history: The legacy of Bachelard and Canguilhem, *Philosophy*, 21 (Suppl.): 141-156.
Timmermans, J. (1963) *The Concept of Species in Chemistry*, New York: Chemical Publishing Company; revised translation of *La notion d'espèce chimie* (Paris 1928).
Tolman, R.C. (1917) The measurable quantities of physics, *The Physical Review*, 9: 237-253.
Tomasi, J. (1999) Towards 'chemical congruence' of the models in theoretical chemistry, *Hyle*, 5: 79-115.
Tontini, A. (1999) Developmental aspects of contemporary chemistry, *Hyle*, 5: 57-76.

Toulmin, S.E. (1957) Crucial experiments: Priestley and Lavoisier, *Journal of the History of Ideas*, 18: 205-220.
Trindle, C. (1984) The hierarchy of models in chemistry, *Croatica Chemica Acta*, 57: 1231-1245.
Trindle, C. (1999) Entering modeling space. An apprenticeship in molecular modeling, *Hyle*, 5: 145-160.
Tuomela, R. (1988) The myth of the given and realism, *Erkenntnis*, 29: 181-200.
Tursman, R. (1989) Phanerochemistry and semiotic, *Transactions of the Charles S. Peirce Society*, 25: 453-468.
Tye, M. (1983) On the possibility of disembodied existence, *Australasian Journal of Philosophy*, 61: 275-282.
Unwin, N. (1984) Substance, essence and conceptualism, *Ratio*, 26: 41-53.
Vallentyne, P. (1998) The nomic role account of carving reality at the joints, *Synthese*, 115:171-198.
van 't Hoff, J.H. (1905) The relation of physical chemistry to physics and chemistry, *Journal of Physical Chemistry*, 9: 81-89.
van Brakel, J. (1975a) Pore space models for transport phenomena in porous media: review and evaluation with special emphasis on capillary liquid transport, *Powder Technology*, 11: 205-236.
van Brakel, J. (1975b) Comments on "Eine Anmerkung zur kapillaren Flüssigkeitsbewegung in Pulver", *Powder Technology*, 11: 91-91.
van Brakel, J. (1984) Norms and facts in measurements, *Measurement*, 2 (1) 45-51.
van Brakel, J. (1986) The chemistry of substances and the philosophy of natural kinds, *Synthese*, 69: 291-324.
van Brakel, J. (1990) Units of measurement: some Kripkean considerations, *Erkenntnis*, 33: 297-317.
van Brakel, J. (1991a) Meaning, prototypes and the future of cognitive science, *Minds and Machines*, 1: 233-257.
van Brakel, J. (1991b) The limited belief in chance, *Studies in History and Philosophy of Science*, 22: 499-513.
van Brakel, J. (1991c) Chemistry, *Handbook of Metaphysics and Ontology*, München: Philosophia, 1: 146-147.
van Brakel, J. (1992a) Natural kinds and manifest forms of life, *Dialectica*, 46: 243-263.
van Brakel, J. (1992b) The complete description of the frame problem, *Psycoloquy*, 1992.3.60.frame-problem.2; also in *Vivek (Bombay)*, 5 (3) 11-16.
van Brakel, J. (1993a) Polywater and experimental realism, *The British Journal for the Philosophy of Science*, 44: 775-784.
van Brakel, J. (1993b) Peirce's limited belief in chance, in E.S. Moore and R.S. Robin (eds.), *From Time to Chance to Consciousness: Studies in the Metaphysics of Charles Peirce*, Providence RI: Berg, pp. 128-148.
van Brakel, J. (1993c) The analysis of sensations as the foundation of all sciences, *Behavioral and Brain Sciences*, 16: 163-164.
van Brakel, J. (1994) Emotions as the fabric of forms of life: A cross-cultural perspective, in W.M. Wentworth and J. Ryan (eds.), *Social Perspectives on Emotion*, Vol. II, Greenwich USA: JAI Press, pp. 179-237.
van Brakel, J. (1996a) Interdiscourse or supervenience relations: the priority of the manifest image, *Synthese*, 106 (1996) 253-297.
van Brakel, J. (1996b) Empiricism and the manifest image, pp. 147-164 in *Realism in the Sciences* (I. Douven en L. Horsten, eds.), Leuven: Leuven University Press.
van Brakel, J. (1997) Chemistry as the science of the transformation of substances, *Synthese*, 111: 253-282.
van Brakel, J. (1998a) *Interculturele Communicatie en Multiculturalisme: Enige Filosofische Voorbemerkingen*, Assen: van Gorcum.
van Brakel, J. (1998b) Epistemische deugden en hun verantwoording, *Tijdschrift voor Filosofie*, 60: 243-268.
van Brakel, J. (1998c) Peirce's natural kinds, pp. 31-45 in *C.S. Peirce: Cosmology to Constantinople* (J. van Brakel and M. van Heerden, eds.), Leuven University Press.

van Brakel, J. (1999a) On the neglect of the philosophy of chemistry, *Foundations of Chemistry*, 1: 111-174.
van Brakel, J. (1999b) Supervenience and anomalous monism, *Dialectica*, 53: 3-25.
van Brakel, J. (1999c) We, *Ethical Perspectives*, 6: 224-235.
van Brakel, J. (2000a) The nature of chemical substances, forthcoming in a book to be published by Oxford University Press, edited by N. Bhushan and S. Rosenfeld.
van Brakel, J. (2000b) Quine and innate similarity spaces, pp. 81-99 in *Poznan Studies in the Philosophy of the Sciences and the Humanities*, vol. 70, Amsterdam: Rodopi.
van Brakel, J. (2000c) Modelling in chemical engineering, *Hyle*, 6: 101-116.
van Brakel, J. and A.A. van der Peut (1979) Alternative physical concatenation procedures in length measurements, *Abstracts 6th International Congress of Logic, Methodology and Philosophy of Science*, Hannover 1979, section 8, pp. 36-40.
van Brakel, J. and J.P.M. Geurts (1988) Pragmatic identity of meaning and metaphor, *International Studies in the Philosophy of Science*, 2: 205-226.
van Brakel, J. and P.M. Heertjes (1975) Capillary rise in porous media, *Nature*, 254: 585-586.
van Brakel, J. and P.M. Heertjes (1975/76) Capillary rise and the contact-angle glasshexane), *Proceedings International Conference Colloid and Surface Science*, E. Wolfram (ed.), Budapest: Akademiai Kiado, 1 (1975) 141-147, 2 (1976) 75-78.
van Brakel, J. and P.M. Heertjes (1977) Contact angle and film jumps, *Nature*, 268: 44-45.
van Brakel, J. and P.M. Heertjes (1978) Geometry and contact-angle in capillarity-induced motion, pp. 119-132 in *Physico Chemical Hydrodynamics: V.G. Levich Festschrift* (D.B. Spalding, ed.), vol. I, St Peter Point (Guernsey): Advance Publications.
Van Cleve, J. and R.E. Frederick (1991) *The Philosophy of Right and Left: Incongruent Counterparts and the Nature of Space*, Dordrecht: Kluwer.
van den Brink, J.T. and J. van Brakel (1988) Internal realism, causal reference and the private language argument, *Reports of the 13th International Wittgenstein-Symposium*, Vienna: Hölder-Pichler-Tempsky.
van der Vet, P.E. (1979a) The debate between F.A. Paneth, G. von Heresy, and K. Fajans on the concept of chemical identity, *Janis*, 66: 285-303.
van der Vet, P.E. (1979b) Overdetermined problems and anomalies, *Studies in History and Philosophy of Science*, 10(3): 259-261.
van der Vet, P.E. (1987) *The Aborted Take-over of Chemistry by Physics*, Ph.D. Thesis University of Amsterdam.
van der Vet, P.E. (1989) *Filosofie van de Scheikunde*, Leiden: Nijhoff.
van der Waals, J.D. (1927) *Lehrbuch der Themostatik: das heißt des thermischen Gleichgewichtes materieller Systeme*, Leipzig: Ambrosius Barth.
van Fraassen, B.C. (1975) Wilfrid Sellars on scientific realism, *Dialogue*, 14: 606-616.
van Fraassen, B.C. (1980) *The Scientific Image*, Oxford: Clarendon Press.
van Fraassen, B.C. (1986) The world we speak of, and the language we live in, pp. 213-221 in *Philosophy and Culture: Proceedings of the XVIIth World Congress of Philosophy*, Montreal: Editions du Beffroi.
van Fraassen, B.C. (1989) *Laws and Symmetry*, Oxford: Clarendon Press.
van Fraassen, B.C. (1992) From vicious circle to infinite regress, and back again, *PSA 1992*, 2: 6-29.
van Fraassen, B.C. (1995) 'World' is not a count noun?, *Noûs*, 29: 139-157.
van Fraassen, B.C. (1998) The manifest image, in *Einstein Meets Magritte* (D. Aerts, ed.), New York: Kluwer Academic Publishers.
Van Gulick, R. (1992) Three bad arguments for intentional property epiphenomenalism, *Erkenntnis*, 36: 311-332.
van Vleck, J. and A. Sherman (1935) The quantum theory of valence, *Reviews of Modern Physics*, 7: 167-228.
Vancik, H. (1999) Opus magnum: an outline for the philosophy of chemistry, *Foundations of Chemistry*, 1: 239-254.
Vasconi, P. (1996), Kant and Lavoisier's chemistry, pp. 155-162 in *Philosophers in the Laboratory* (V. Mosini, ed.), Rome: Euroma.

Vemulapalli, G. K. and H. Byerly (1999) Remnants of reductionism, *Foundations of Chemistry*, 1: 17-41.
Vermeeren, H.P. (1986) Controversies and existence claims in chemistry: the theory of resonance, *Synthese*, 69: 273-90.
Vihalemm, R. (1974) Elaboration of philosophical problems in chemistry [in Russian], *Voprosy filosofii*, 4 (6) 90-95, 187-188.
Villani, G. (1993) Sostanze e reazioni chimiche: Concetti di chimica teorica di interesse generale, *Epistemologia*, 16: 191-212.
Villani, G. (1996) Specificità della chimica, pp. 163-180 in *Philosophers in the Laboratory* (V. Mosini, ed.), Rome: Euroma.
Vinti, C. (1996) Bachelard: Ragione e realtà nella chimica contemporanea, pp. 181-217 in *Philosophers in the Laboratory* (V. Mosini, ed.), Rome: Euroma.
von Engelhardt, D. (1984) The chemical system of substances, forces and processes in Hegel's philosophy, pp. 41-54 in *Hegel and the Sciences* (R.S. Cohen, ed.), Dordrecht: Reidel.
von Engelhardt, D. (1986) Philosophie und Theorie der Chemie um 1800, *Philosophia Naturalis*, 23: 223-237.
von Engelhardt, D. (1993) Hegel on chemistry and the organic sciences" in *Hegel and Newtonianism* (M.J. Petry, ed.), Dordrecht: Kluwer.
Wald, F. (1897) Die chemischen Proportionen, *Zeitschrift für physikalische Chemie*, 22: 253-267.
Wang, C.-C., Y-J Du, Y-J Chang and W-H Tao (1999) Airside performance of herringbone fin-and-tube heat exchangers in wet conditions, *The Canadian Journal of Chemical Engineering*, 77: 1225-1230.
Wang, Z., J. Feyen and D.E. Elrick (1998) Prediction of fingering in porous media, *Water Resources Research*, 34: 2183-2190.
Watkins, E. (1998) The argumentative structure of Kant's metaphysical foundations of natural science, *Journal of the History of Philosophy*, 36(4): 567-593.
Weatherall, P. (1993) Tarski's theory of truth and field's solution to the problem of intentionality, *Australasian Journal of Philosophy* 71: 291-303.
Weinert, F. (1991) Introducing events, successful reference and reference-fixing, *Journal for General Philosophy of Science*, 22: 155-167.
Weingartner, P. (1997) Can the laws of nature (physics) be complete?, pp 429-446 in *Logic and Scientific Method* (M.L. Dalla Chiara *et al.*), Dordrecht: Kluwer.
Weininger, S.J. (1984) The molecular structure conundrum: Can classical chemistry be reduced to quantum chemistry, *Journal of Chemical Education*, 61: 939-944.
Weininger, S.J. (1998) Contemplating the finger: visuality and the semiotics of chemistry, *Hyle*, 4: 3-27.
Westphal, J. (1987) *Colour: Some Philosophical Problems from Wittgenstein*, Oxford: Basil Blackwell.
Wheland, G. (1944) *The Theory of Resonance and its Applications to Organic Chemistry*, New York: John Wiley.
Wheland, G.W. (1955) *Resonance in Organic Chemistry*, New York: Wiley.
Whitaker, A. (1996) *Einstein, Bohr and the Quantum Dilemma*, Cambridge: Cambridge University Press.
Whitney, J. (1968) The mathematics of physical quantities, part II: Quantity structures and dimensional analysis, *American Mathematical Monthly*, 75: 227-256.
Whyte, L.L. (1954) A Dimensionless physics?, *The British Journal for the Philosophy of Science*, 5: 1-17.
Wiggins, D. (1980) *Sameness and Substance*, Oxford: Blackwell.
Wilkerson, T. (1988) Natural kinds, *Philosophy*, 63: 29-42.
Wilkerson, T. (1995) *Natural Kinds*, Aldershot: Avebury.
Wilkerson, T.E. (1998) Recent work on natural kinds, *Philosophical Books*, 31: 225-233.
Williams, B. (1978) *Descartes: The Project of Pure Enquiry*, Harmondsworth: Penguin Books.
Williams, B. (1985) *Ethics and the Limits of Philosophy*, Cambridge MA: Harvard University Press.
Williams, M.B. (1985) Species are individuals: theoretic foundations for the claim, *Philosophy of Science*, 52: 578-591.

Williams, W. (1892) On the relation of the dimensions of physical quantities to directions in space, *Philosophical Magazine*, 34: 234-271.
Wilson, M. (1982) Predicate meets property, *The Philosophical Review*, 91: 549-589.
Wilson, M. (1990) Law along the frontier: Differential equations and their boundary conditions, *PSA 1990*, 2: 565-575.
Wilson, R.A. (ed.) (1999) *Species. New Interdisciplinary Essays*, Cambridge MA: MIT Press.
Wilson, S. (1984) *Electron Correlation in Molecules*, Oxford: Pergamon.
Wimsatt, W.C. (1996) Aggregativity: reductive heuristics for finding emergence, *Philosophy of Science*, 64 (Proceedings) S371-384.
Witmer, D.G. and Sarnecki, J. (1998) Is natural kindness a natural kind?, *Philosophical Studies*, 90: 245-264.
Witt, C. (1989a) *Substance and Essence in Aristotle: An Interpretation of* Metaphysics *VII-IX*, Ithaca: Cornell University Press.
Witt, C. (1989b) Aristotelian essentialism revisited, *Journal of the History of Philosophy*, 27: 285-298.
Wittgenstein, L. (1953) *Philosophical Investigations*, Oxford: Basil Blackwell.
Wittgenstein, L. (1969) *On Certainty*, G.E.M. Anscombe and G.H. von Wright, eds., Oxford: Basil Blackwell.
Woodward, W.R. and R.S. Cohen (1991) *World Views and Scientific Discipline Formation*, Dordrecht: Kluwer.
Woolley, R.G. (1978) Must a molecule have a shape?, *Journal of the American Chemical Society*, 100: 1073-1078.
Woolley, R.G. (1986) Molecular shapes and molecular structures, *Chemical Physics Letters*, 125: 200-205.
Woolley, R.G. (1991) Quantum chemistry beyond the Born-Oppenheimer approximation, *Journal of Molecular Structure (Theochem)*, 230: 17-46.
Wormer, A.J. (1986) Dimensional analysis: a critique and cautionary note, *American Anthropologist*, 88: 448-452.
Yaminsky, V.V. (1997) Long range attraction in water vapour: Capillary forces relevant to 'polywater', *Langmuir*, 13: 2-7.
Zandvoort, H. (1985) *Models of Scientific Development and the Case of Nuclear Magnetic Resonance*, Ph.D. Thesis University of Groningen.
Zandvoort, H. (1988) Macromolecules, dogmatism, and scientific change, *Studies in the History and Philosophy of Science*, 19: 489-515.
Zbarsky, I.B. (1963) Molecular regulatory mechanism of protein biosynthesis and the problem of development [in Russian], *Voprosy filosofii*, (9) 51-63.
Zeidler, P. (2000) The epistemological status of theoretical models of molecular structure, *Hyle*, 6: 17-34.
Zeidler, P. and D. Sobczynska (1995/96) The idea of realism in the new experimentalism and the problem of the existence of theoretical entities in chemistry, *Foundations of Science*, 4: 517-535.
Zemach, E.M. (1976) Putnam's theory of reference of substance terms, *Journal of Philosophy*, 73: 116-127.
Zemansky, M.W. (1957) *Heat and Thermodynamics. An Intermediate Textbook*, New York: McGraw-Hill.
Zemansky, M.W. and R.H. Dittman (1981) *Heat and Thermodynamics: An Intermediate Textbook*, London: McGraw-Hill.
Zhdanov, Y.A. (1960) *Icherki metotologii organicheskoi khimii* [*Outline of the Methodology of Organic Chemistry*], Moscow.
Zhdanov, Y.A. (1963) The question of modelling in organic chemistry [in Russian], *Voprosy filosofii*, (6) 63-74.
Zhdanov, Y.A. (1965) The problem of the general definition of chemistry, *Journal of General Chemistry of the USSR*, 2643-2644.
Zhdanov, Y.A. (ed.) (1972) *Philosophical Problems of Chemistry* [in Russian], Rostov-on-Don: Rostov University Press.
Zollinger, H. (1997) Logic and psychology of scientific discoveries: a case study in contemporary chemistry, *Perspectives on Science*, 5(4): 516-532.
Zucker, A. (1988) Davy refuted Lavoisier not Lakatos, *The British Journal for the Philosophy of Science*, 39: 537-540.

index of names

Abbott, 99
Abbri, 2, 39
Abramova, 32
Achinstein, 154
Ackerman, 105
Adams, 48
Adickes, 11
Akeroyd, 36, 39
Aldrich, 98
Alexander, 3
Allchin, 38-9, 141
Allen, 87, 89, 95, 98, 138
Amann, 131-2, 143-5
Arabatzis, 67-8, 122
Arbuzov, 29
Archimedes, 10, 174
Aristotle, vi, 1, 17, 43, 71, 76-8, 118, 126, 197
Armstrong, 54-5, 60, 61, 64, 79
Aronson, 154
Arrhenius, 97, 174
Arteca, 145, 150
Aspect, 142
Au, 43
Aune, 45
Austin, 43
Auyang, 143
Averill, 108-9
Averroës, 1
Avogadro, 4-5, 19, 34, 36, 74
Ayer, 103, 196
Ayers, 1
Bachelard, 13, 15-6, 21, 24
Bader, 120, 148-9
Baierlein, 128
Baird, 39
Bakhtin, 43
Balashov, 157
Balzer, 35-6
Bantz, 36
Baracca, 127
Barbier, 39
Barnet, 101
Barzakovsky, 25
Bealer, 200
Bear, 163, 168, 189
Beattie, 103, 105
Beatty, 151
Bechtel, 36, 57
Beckermann, 56-7
Beeckmans, 5
Bell, 64, 143, 201
Bellu, 22, 24, 38
Belusov, 126
Bender, 138
Benfey, 2, 21
Bensaude, 13, 16, 18, 39, 72
Bernal, 89
Bernstein, 44, 46, 201

Berzelius, 14, 35
Bhargava, 114
Bhushan, 154
Bigelow, 59, 62, 64
Bingham, 174
Biot, 174
Bird, 178, 181, 186
Bittrich, 23
Black, 154
Blackburn, 81
Bochénski, 24
Bodenstein, 174
Bogaard, 38, 120, 129, 139, 152
Bogdan, 64
Boghossian, 115
Böhme, 1, 161
Bohr, 119, 125, 129, 139, 201
Boltzmann, 4, 126, 128
Born, 129-31, 133, 144, 147
Bourdieu, 43
Boyd, 58-9, 62, 64, 67, 105, 107, 191
Boyle, 1, 3, 34-5, 158
Bradley, 6, 19, 22
Brandon, 151
Brandt, 11
Bridgman, 19, 160, 176-7, 181-2, 186
Brinkman, 174
Broad, 19, 21-2, 56
Brønsted, 27, 35, 154
Brown, 183, 185
Brownian motion, 5, 127
Browning, 58, 62
Brummer, 90, 98
Bruner, 58
Brush, 17, 28, 39
Buchanan, 20
Buchwald, 59
Buck, 121
Buckingham, 164-5, 168, 177, 182
Budreiko, 22, 31
Bumstead, 128
Bunge, 13, 18, 21-2, 82, 145
Burbidge, 13-4
Burger, 154
Burnstead, 128
Butlerov, 29-32, 34
Buttker, 23
Byerly, 52, 124-6
Cadenhead, 98
Caldin, 19-20, 22, 151-2
Callaway, 69
Callender, 126
Campbell, 116-7, 155, 176-7, 182-3
Cannizzaro, 19
Cao, 143
Carman, 165, 181
Carmeliet, 164, 168
Carnap, 45, 46, 197
Carpenter, 154
Carrier, 7, 11, 36, 64, 68, 88, 153

Carroll, 128, 137, 140
Cartwright, 88, 132, 142-3, 154-5, 157-9, 161-2, 188, 192, 194, 197
Carvallo, 177
Cassam, 105
Cassirer, 2, 15, 57, 151-2
Cauchy, 184
Causey, 18, 182
Cerruti, 39
Chang, 142-3
Chelintsev, 27, 29, 31, 34
Chhabra, 188
Child, 196
Chilton, 179
Chomsky, 58, 99
Chow, 98
Christie, 151
Churaev, 95
Churayev, 94, 98
Churchland, 43, 58-9, 63-4, 125
Churchman, 20
Clark, 45, 58
Claverie, 129, 131, 145-7
Clifford, 87
Cohen, 23, 164
Colburn, 173-4, 179
Copernicus, 34
Cornforth, 39
Coulomb, 68, 130, 132
Coulson, 30, 139-41
Couper, 29
Crane, 115, 155, 201
Crick, 34
Crosland, 39
Currie, 58, 65
Cushing, 129
D'Agostino, 175
d'Alembert, 175
D'Amico, 58
Dalibard, 142
Dalton, 1, 3, 5-6, 14-5, 19, 23, 34-6, 151
Daly, 58
Darcy, 164-6, 181
Davidson, 54, 69, 192, 194-5
Davies, 102
Davy, 35-6
Day, 154, 185
de Certeau, 43
de Regt, 2, 55
de Sousa, 101, 107
de Waele, 181
Deborah, 174
Del Re, 18, 39, 129, 131, 144-5, 154
Delhez, 16
Denbigh, 126, 128
Dennett, 195
Derjaguin, 89-95, 98, 174
Derrida, 50
Descamps, 164, 168
Descartes, 1
DeVidi, 45
Devitt, 67-8
Diamond, 36-7, 82, 99
Diderot, 175
Dieks, 2, 55
Diesel, 34
Diner, 129, 131, 145-7

Dingle, 18, 122, 160, 180
Dirac, vii, 64, 119-22, 129, 139-41
Dittman, 126, 128
Dobbert, 177
Dobbs, 3
Dobrotin, 25
Donahoe, 89
Donovan, 34, 36
Douven, 64, 68
Downes, 155
Drews, 8
Duhem, vi, 1, 6, 11, 13-5, 17, 123, 126-7, 151-3, 155
Dumas, 61
Dummett, 100, 103
Duncan, 39, 182
Duncanson, 183
Dupré, 27, 58, 61-3, 155, 192
Düring, 1
Dussort, 7, 11
Dyatkina, 30
Earley, 37-8
Early, 20-1, 39, 52, 83
Earman, 155, 162
Eddington, 41, 183
Edmiston, 137
Ehrenfest-Afanassjewa, 177
Einstein, 34, 129, 142, 144
Eisvogel, 18, 38
Elder, 62
Elders, 104
Elgin, 69, 118
Ellis, 19, 59, 62, 64, 176, 182, 185
Elrick, 164
Endicott, 52
Engels, 24-7, 31, 34
Ereshefsky, 58, 60
Esfeld, 143
Eswaran, 188
Eu, 128
Euler, 174
Everett, 87, 89-91, 93, 96, 98, 138
Eyring, 120, 148
Fajans, 22
Falvey, 115
Fanning, 173, 184
Faraday, 1, 4, 19, 36, 89
Farmer, 37
Farrell, 104
Faustov, 25
Fedyakin, 89, 91
Feibleman, 19
Feyen, 164
Feyerabend, 157
Feynman, 120-1, 129-30, 201
Fichte, 13
Fick, 122-3, 158
Field, 37
Flanagan, 58
Fleischer, 23
Flohr, 56
Focken, 176, 177
Fodor, 8, 155
Forbes, 74, 78, 101, 105, 108
Förster, 9, 11
Foucault, 15, 36, 50
Fourier, 122-3, 171, 174-5

NAME INDEX

Fowkes, 98
Francoeur, 154
Frank, 17
Frankland, 14
Franklin, 35, 49
Franks, 87, 89, 92
Fredrick, 37
Frege, 37, 67, 103
French, 4, 13, 15-6, 22, 36, 132, 141
Frenking, 120, 122
Freundlich, 95
Frické, 36
Friedman, 7, 9, 11
Froude, 173-4, 180, 184
Fryer, 173
Fuchs, 23, 26
Gabay, 175
Galilei, 8, 124
Gallis, 174
Gans, 132
García-Sucre, 145
Garfinkel, 43
Garkovenko, 22-3, 25, 32
Gascoigne, 20
Gasper, 64
Gassendi, 1, 5
Gavroglu, 7, 30, 39, 132
Gay, 4, 19, 36, 158
Gay-Lussac, 4, 19, 36, 158
Gee, 94
Gell-Mann, 120-1, 129
Geoffroy, 4
Geurts, 69
Gey, 32
Ghiselin, 58
Gibbins, 133
Gibbs, 85, 126-8, 152, 177
Giere, 35, 155
Gingold, 87
Glas, 39
Gleick, 201
Goodman, 15, 42, 46, 63, 118, 160, 196-7
Graham, 27, 30-1
Grashof, 174
Guggenheim, 183
Gukhman, 172, 174
Gutman, 160
Gutting, 15
Haack, 62, 64
Haberditzl, 26
Habermas, 34, 43
Hacking, 35, 58-9, 61, 68, 88, 96, 159
Hadorn, 154
Hagen, 165, 185
Haldane, 25, 31, 193
Hall, 40, 133
Han, 18
Hanekamp, 121, 160
Hanfling, 105
Hare, 54, 73-8
Harré, 101, 154
Hartmann, 18, 129, 141, 154
Hartree, 130, 133-7
Haugeland, 197
Haynes, 87, 89, 91-3, 95, 96, 138
Healey, 142
Healy, 94
Heber, 33
Heertjes, 95, 166
Hegel, 13-5, 22, 24, 50
Heidegger, 43
Heinig, 7, 13
Heisenberg, 107, 120-1, 129, 201
Hempel, 17, 19-20, 74, 155
Hendry, 38, 130, 132, 154, 159
Hersey, 177
Hesse, 154
Hettema, 35-6, 152
Hiebert, 121
Hirsch, 62-3, 65, 106
Hoffman, 71
Hoffmann, 18, 21, 36, 39, 72, 82, 119, 132, 141, 154
Hofmann, 38, 154, 166
Hooker, 52, 124
Hoppe, 9
Horgan, 65, 155
Hornsby, 194-5
Hörz, 14, 73
Houvenaghel, 164, 168
Howson, 34-5
Hulswit, 59
Humberstone, 102
Hume, 13
Hundt, 107, 119, 152
Hunsberger, 27
Huntley, 183, 185
Husserl, 43
Hüttemann, 132, 154
Ingold, 28, 32
Ionidi, 22
Isaacson, 176
Jackson, 111, 115, 117
Jakobsen, 90
Jakobson, 98
James, 43, 45, 196
Janich, 18, 38-9, 121, 160
Johnson, 62, 64, 68
Johnston, 37
Jones, 143
Jordan, 120
Kamlah, 35-6
Kant, vi, 1, 3, 7-13, 17-8, 33, 37, 42-3, 54, 71, 122, 198
Karpinskaya, 23
Karplus, 75
Kedrov, 2, 22-7, 31
Keith, 148-9
Kekulé, 1, 2, 4, 7, 28-9, 31-2, 35-6, 83
Kemeny, 51, 119-20
Kent, 19
Kern, 75
Kertscher, 22, 24
Khalidi, 59
Kiesewetter, 9, 11
Kim, 54-6, 64, 75, 78, 192
Kimball, 120, 148
Kirpichev, 172, 174
Kitaygorodsky, 26
Kitcher, 36, 62, 68, 107, 153, 193
Klee, 155, 161
Klein, 3
Kluge, 58
Klüver, 34

Knudsen, 174
Kollman, 98, 138
Körner, 8
Kossovich, 174
Kozeny, 164-6, 181
Kragh, 107
Kramers, 178
Krantz, 177, 182
Kripke, 1, 58, 67, 99-102, 104-11, 116, 118
Kroon, 68
Krypton, 109-10
Kubbinga, 5
Kuhn, 3, 17, 34, 45-6, 59
Kuipers, 35-6, 152
Kultgen, 2, 19
Kuznetsov, 23-5, 31-3
Ladenburg, 34
Ladyman, 132, 141
Lagrange, 128, 174
Laitko, 22-3, 25-7, 32-3, 153
Lakatos, 20, 36
Lakoff, 42, 154
Landau, 95
Lange, 155
Langhaar, 176
Laplace, 165
LaPorte, 99, 105
Laszlo, 21, 39, 154
Latour, 7, 34-5, 160, 163
Laudan, 34, 36
Lauth, 35-6
Lavoisier, 1, 5, 7, 9-12, 17, 19, 22, 34-6, 84
Laymon, 177
Le Poidevin, 37
Lee, 181
Lei, 38
Leibniz, 1, 80, 103, 113
Lenin, 27, 34
Lennard-Jones, 126
Leplin, 67
LePore, 194
Lequan, 9
Levere, 13
Levine, 98, 120, 135
Levinson, 62
Lévy, 20-1, 39, 51-2, 129
Lewis, 1, 35, 54, 149, 154, 174
Li, 63, 116
Liegener, 18, 39, 129
Lierse, 59, 62, 64
Lightfoot, 178, 181, 186
Lippincott, 89-90, 93
Liu, 173
Locke, 1, 22, 79-80, 84, 99, 103, 107, 118
Lockhead, 197
Loewer, 194
Lomonossov, 13
London, 1, 7, 36, 38, 89, 132
Lorentz, 34
Lorenzen, 160
Lovejoy, 98
Löwdin, 119, 128-9, 141, 145
Lowe, 1, 58, 62, 84, 134-5, 197
Lowry, 154
Luce, 177
Luikov, 168-9, 172
Macagno, 177

Maccoll, 146
MacDonald, 20, 22, 151
Mach, 1, 4, 6-7, 27, 86, 182, 184, 196-7
MacKinnon, 107
Madden, 101
Maher, 35
Mainzer, 73, 122, 149, 154
Majer, 37
Malisoff, 19
Malt, 74
Malzkorn, 1
Manzelli, 82, 126
Margalit, 58, 105, 111
Margenau, 133
Marino, 39
Markovnikov, 31
Marmur, 164
Marx, 24, 26-7, 34, 50
Massey, 90, 182
Matteson, 137
Maudlin, 201
Mauskopf, 36
Maxey Flats, 188
Maxwell, 7, 175
Mayer, 14
McAllester Jones, 16
McAllister, 36
McElroy, 87, 89, 91, 93, 95-6, 98, 138
McEvoy, 36, 39
McGuire, 177
McIntyre, 18, 151-2, 191, 194
McKinney, 89-92, 94, 96
McLaughlin, 138
McWeeny, 140
Medin, 74
Mellor, 105, 107, 111, 118, 155, 201
Mendeleev, 2-3, 14, 19, 31, 34-5, 37, 82, 133, 152, 154
Mestrallet, 21
Metzner, 181
Meyer, 13, 123
Meyerson, 13, 21-2
Mezey, 145, 150
Mill, 15, 62
Miller, 65
Millikan, 62
Minkin, 154
Mittasch, 14, 24
Mittelstraß, 38
Monod-Herzen, 177
Moody, 184
Morreau, 155
Morris, 159
Morrison, 159
Mosini, 38, 39, 126
Moulines, 35
Mounin, 21
Moutier, 152
Mulckhuyse, 18, 154
Müller, 14, 34, 73
Mulliken, 119
Mundy, 155
Munowitz, 139
Nagel, 17, 19, 41, 51-3, 56, 120-1, 125
Navier, 185, 188
Nayak, 8

NAME INDEX

Needham, 6, 58, 101, 124, 126-7, 139, 146, 193
Nerlich, 37
Nersessian, 69
Newman, 95
Newton, 1, 4, 10, 12, 34, 158, 175
Nguyen-Dang, 120, 149
Nickles, 51
Niedersen, 14, 22-6, 33
Nietzsche, 14, 50
Nola, 36
Noordhof, 56
Nusselt, 174, 179
Nye, 5, 13, 18, 127
O'Konski, 98
Ogilvie, 136-7, 150
Oppenheim, 20, 51-2, 119-20
Oppenheimer, 130-1, 133, 144, 147
Osborne, 182
Ostwald, vi, 1-2, 4-7, 15, 17, 19, 24, 84, 86, 123, 126, 152, 181
Overbeek, 95
Owens, 115, 196
Page, 7, 90, 98
Pain, 173
Palacios, 180, 182, 184
Paley, 13
Paneth, 2, 13, 22, 107, 152
Pankhurst, 184
Paolini, 39, 129
Papineau, 116
Paracelsus, 35
Pargetter, 65
Parthey, 32
Pasteur, 34, 35
Pauli, 107, 119, 132-4
Pauling, 1, 3, 7, 27-8, 30-2, 137, 139, 141, 153
Peacocke, 196
Pearson, 177
Pechenkin, 23, 32
Péclet, 174
Peirce, 2, 14, 45, 59, 62, 199
Pepper, 56
Peregrin, 155
Perrin, 5, 36, 127
Petsko, 90, 98
Pettit, 194
Pietroski, 155
Pilkington, 3
Pirie, 19
Planck, 176
Platts, 107
Plesch, 18, 72
Podolsky, 142, 144
Poiseuille, 165, 185
Polanyi, 3, 18, 22, 151
Poller, 25-7
Popelier, 148-9
Popper, 34, 119
Post, 39, 89, 183, 197
Prandtl, 174, 179
Prélat, 86
Priestley, 1, 34, 36
Prigogine, 126
Primas, 119, 123-5, 127, 129, 132, 143-5, 147

Proust, 5, 151
Prout, 35
Psarros, 6, 18, 38-9, 84, 86, 121, 160-1
Pullman, 5
Putnam, 1, 36-7, 41, 51-2, 58-9, 61, 64, 67, 78, 88, 99-114, 116, 120, 155, 197
Pyle, 5
Quine, 15, 43, 46, 58, 60-1, 63, 69, 71, 106, 192, 195, 197, 200
Ramberg, 154
Ramsey, 35, 38, 58, 145-7, 154, 161, 198
Rayleigh, 175, 180
Redhead, 112, 126, 154, 201
Regnault, 61
Reichenbach, 120-1
Rey, 155
Reynolds, 34, 171, 173-4, 176, 179-81, 184-6
Rheinberger, 36
Riabouchinsky, 177
Richardson, 57
Richter, 5, 13, 25, 151
Rip, 39
Ritchie, 5
Roberts, 155, 162
Rocke, 31, 36
Roger, 142
Röhler, 154
Rollins, 43
Rosen, 142, 144
Rosenberg, 151
Rosenfeld, 154
Rosenkrantz, 71
Rosenthal, 25
Rothbart, 9, 36, 38-9, 83
Rousseau, 92, 98
Rovane, 58
Rumelhart, 58
Ruschig, 14
Ruse, 60
Russell, 59, 62
Ruthenberg, 5, 18, 22, 38-9, 86
Rutherford, 7, 69
Salmon, 17, 103, 106
Sanchez Marin, 39
Sanchez Perez, 39
Sankey, 36
Sarkar, 52
Sarnecki, 58, 60
Satheesh, 188
Saunders, 63, 66, 127, 197
Savellos, 59
Scerri, 18, 35-6, 38, 107, 132-3, 135-9, 152-4, 193
Schaefer, 63, 136, 138, 148
Scheibe, 143, 155
Schelling, 13, 14, 22
Scherer, 9, 39
Schicketanz, 164
Schilpp, 22
Schlesinger, 52
Schmidt, 174, 179
Schopenhauer, 13-4
Schorlemmer, 23, 25
Schröder, 56
Schrödinger, v, 41, 120, 129-31, 133, 138-9, 141, 144, 161

Schulze, 9, 11
Schummer, 1, 16, 18, 21-3, 33, 38-9, 71-2, 83, 85, 121, 123, 146
Schweber, 120-1
Scott, 120
Searle, 37, 43, 58, 67
Sedov, 176, 183
Seibt, 45
Sellars, v-vi, 41-6, 50, 61, 198-9, 202
Semjonov, 23
Shain, 58, 62, 64
Shakhparanov, 22, 26-8, 31-3
Shamin, 23, 31, 33
Shapere, 36, 69
Sharpe, 115
Shelton, 39
Sherman, 120, 132, 134, 140
Sherwood, 174, 179
Shrader-Frechette, 188-9
Sicha, 45, 202
Sidle, 43
Siebold, 164
Siegfried, 3
Simoes, 7, 30
Simon, 22-5, 27
Simons, 137
Sisakjan, 23
Sklar, 124-7, 157, 162
Smart, 41
Sneed, 35
Snelders, 13
Sobczynska, 88, 147
Sober, 151, 195
Solomon, 45
Solonov, 32
Sotnak, 8
Spencer-Smith, 56
Spinoza, 1, 192
Splitter, 58
Sprung, 22-3, 25, 33
Stahl, 8, 10-2, 36, 88
Stalin, 31
Stalker, 63
Stanford, 68, 80, 107
Stanton, 174
Stengers, 13, 16, 18
Sterelny, 68
Stewart, 178, 181, 186
Stich, 58
Stock, 38
Stöckler, 56
Stokes, 185, 188
Stoneham, 115
Stork, 27, 31-2
Ströker, 18, 20
Stroll, 48-9, 104, 109
Strube, 23
Suppes, 177
Suratman, 174
Sutcliffe, 33, 141
Syrkin, 30
Tal, 120, 149
Tarski, 46
Tatevskii, 27-8
Taylor, 98
Thagard, 35-6
Thales, 1, 71

Theobald, 16, 19-20
Thornton, 43
Tienson, 155
Tijiattas, 15
Tiles, 15
Timmermans, 85-6
Tolman, 182-3
Tomasi, 154
Tontini, 39, 72
Toricelli, 8
Toulmin, 36
Trindle, 154
Trout, 64
Tuomela, 162
Tursman, 2, 15
Tversky, 177
Ullian, 46
Unwin, 77
van 't Hoff, 6, 83, 123
van Brakel, 1, 36, 38, 40-3, 47, 50, 54, 58-60, 63-4, 66, 68-9, 73, 87, 95, 104, 109, 119, 127, 155, 163-4, 166, 168, 171, 182, 192, 194, 196-7, 199, 201
Van Cleve, 37
van den Brink, 68
van der Peut, 182
van der Vet, 22, 39, 152
van der Waals, 83, 86-7, 126
van Fraassen, 20, 43, 45-7, 89, 154-5, 157, 201
Van Gulick, 155
van 't Hoff, 18, 123, 145, 152
van Vleck, 120, 132, 134, 140
Vancik, 18, 140
Vanderveken, 58
Vaschy, 177
Vasconi, 7, 9
Vemulapalli, 52, 124-6
Vermeeren, 27, 33, 38-40
Verwey, 95
Vihalemm, 26, 33, 38
Villani, 18, 39
Vincent, 18, 39, 72
Vinti, 16
von Bertalanffy, 19
von Engelhardt, 13-4
von Kármán, 186
von Liebig, 35, 84
von Neumann, 129
Wald, 6, 86
Walter, 120, 148
Wang, 164, 173, 188
Washburn, 164-6
Watkins, 8
Way, 154
Weatherall, 37
Weber, 174, 184
Weinert, 68
Weingartner, 195
Weininger, 39, 145, 148
Weissenberg, 174
Welsch, 23
Westphal, 66
Wheland, 27-8, 30, 32
Whitaker, 3
White, 94
Whitehead, 192, 196

NAME INDEX

Whiteheadian, 197
Whitney, 182
Whyte, 183
Wiggins, 76, 103, 107, 197
Wilkerson, 58, 62, 65
Williams, 58, 181, 183
Williamson, 2
Wilson, 60, 63, 135, 157
Wimsatt, 56
Witmer, 58, 60
Witt, 1, 71, 118
Wittgenstein, 43, 68, 105, 117, 169, 185-6
Woodward, 23
Woody, 38
Woolley, 113, 129-31, 138, 144-8
Wormer, 177
Yaminsky, 95
Young, 34
Zandvoort, 36, 39
Zbarsky, 23
Zeidler, 88, 145, 147, 154
Zemach, 107, 111
Zemansky, 126, 128
Zhdanov, 22-3, 26, 31, 154
Zielonacka-Lis, 38
Zollinger, 39
Zucker, 36

index of substances

alcohol, 82
anthracene, 83
$ArCl_2$, 83
benzene, 7, 27-8, 31, 36, 83, 86, 130
buckminsterfullerine, 82, 193
bullvalene, 83
$Ca_{0.75}Nb_3O_6$, 82
carbon, 31, 36, 82, 132, 136
cellulose, 82
charcoal, 99
chromium, 119
coal, 95, 99
Coca-Cola, 84
corundum, 99
crystals, 20, 22, 25, 82, 86, 126, 141, 146, 151
cyclopropenylindene, 63
D_2O, see heavy water
dendrimers, 72
enzymes, 72, 82, 141, 154
ethanol, 82
glass, 48-9, 85, 91, 93, 101, 116, 146
glue, viii, 16, 48, 193
glycol, 164-5, 168
gold, 58-60, 71, 103, 108, 113, 115, 117
grook, 113
H_2O, vii, 37, 53, 55, 57, 73-81, 84, 94, 97, 99-113, 115-18, 122-3, 138, 155, 193
heavy water, 76, 99, 102, 106-7, 155
helium, 82, 119
hydrochloric acid, 116
hydrogen, 31, 76, 79-81, 84, 94, 97, 104, 106-8, 110, 113, 115, 118, 131, 133, 136, 139, 146, 154
iodine, 152
jade, 2, 99
Kronenether, 72
Li_xWO_3, 82
magnesium, 119
mercury, 6, 48-9, 85

metals, 10, 59, 82, 126, 136
Na_xWO_3, 82
nephrite, 99
n-heptane, 164
nickel, 119
nylon, 82
oil, 48-9, 91
oxygen, 6, 34, 36, 68, 76, 79, 80-1, 88, 104, 106-7, 110, 113, 115, 118, 139, 146
paint, 47, 63, 85
palladium, 119
phlogiston, 8, 11, 34, 36, 68, 88
platinum, 119
polystyrene, 164, 168
polywater, vii, viii, 36, 73, 87-97, 102, 124, 138, 154, 163, 168-9, 191
porcelain, 86
proteins, 72, 82, 154
quartz, 91-4, 96, 101
ruby, 82, 99
sand, 48, 95, 101, 154, 166, 188
scandium, 120
silica, 92-4, 96
silicic acid sol, 2, 93-4, 96
silk, 100
silver, 22, 159
soap, 48-9
sodium chloride, 22, 57, 92
soot, 37, 82
sulphuric acid, 91, 116
tellurium, 152
topaz, 99
tourmaline, 82
water, v, vii, 2, 5, 10, 16, 37, 41-3, 48-9, 53, 55, 57-61, 63, 71, 73-82, 84-5, 87-118, 122-3, 139, 146-7, 154-5, 166, 183, 191, 193, 197
XYZ, 111-4, 116-8
zinc, 64

index of subjects

'ab initio' methods, 83, 96, 133-4, 136, 138-9, 144, 148, 187-8
acid, 2, 15, 27, 86, 92-4, 96, 115-6, 126, 154, 191
affinity, 4, 5, 13, 49, 71
anomalous monism, viii, 51, 192, 194-8
aromaticity, 27-8, 132
atomism, 1-7, 34, 51, 126-7
autocatalytic, 37, 82
autonomy of chemistry, 2, 121-2, 128
baptism event, 100, 104
basis set, 134-6, 138, 150
beryllium, 119
bootstrapping, 200
bridge laws, vi, 50, 52, 53, 100, 192, 193, 196
caloric particles, 68
capillary rise, 154, 164-6, 168, 174
catalyst, 14, 21, 48, 72, 82, 97, 153
causal theory of reference, 64, 66-7, 69, 87, 100, 107, 111, 113
ceteris paribus, vii, 50, 63, 91, 94, 101, 112, 151, 153, 155, 157-9, 161-3, 169, 171-2, 175, 177, 180, 184-9, 193, 196-8
chemical bonding, v, 17, 21, 28, 30, 33, 36, 81, 83, 96, 122, 129, 131-2, 134, 136-41, 145, 147, 149, 174
chemical engineering, vii, ix, 85, 122, 124, 154, 163, 171-2, 177-8, 186-8
chemical laws, 25, 39, 121, 151, 152-3, 161, 201
chemical reaction, xiii, 3, 5-6, 14, 16, 21, 35, 48, 55, 71-2, 83, 86, 94, 114, 123-4, 126, 132, 146, 152-3, 161, 168, 172, 174, 220
chemical revolution, 35-6
chemical space, 21, 39, 57, 71
chirality, 132, 146
common sense, 8, 41-4, 48, 58, 74-6, 79, 82, 102, 111, 162, 169, 199, 201
contact angle, 91-2, 164, 166
Copenhagen interpretation, 26
determinism, 14, 129, 142, 192
dialectical materialism, 24-5, 27, 29, 31-4
dimensional 'constants', 176
dimensional analysis, viii, ix, 154, 164, 166, 172, 175-80, 182-8
dimensional variables, 176-7
dimensionless numbers, viii, 163, 165-6, 171-6, 178-89
drag coefficient, 180
electron configurations, 109-10, 119-20, 131, 133-8, 146, 152
electronegativity, 55, 84, 132
electrons, 5, 7, 20-1, 28, 35, 49, 54, 57, 59, 64, 67-9, 76, 83, 109-10, 119, 122, 130-7, 139-40, 142-6, 148-50, 154, 186, 193, 210, 226
eliminativism, 51, 56, 66, 124, 193, 201

emergence, v-vi, 1, 19, 21-2, 34, 50, 53, 56-7, 122-3, 125, 192-3
enantiomers, 86, 146
entity realism, 35, 67, 87-8, 96
epistemic virtues, 47, 199
EPR-correlations, 143-5
essences, vii, 58, 65-7, 73, 78-81, 87, 99-102, 105-6, 108-10, 113, 116, 118
essentialistic realism, vii, 67, 99, 106, 113
filtration, 85, 124, 165, 181
folk psychology, 43, 51, 194
form(s) of life, 42-3, 47, 51, 98
frame problem, 43, 155
friction factor, 173-4, 184
functional groups, 61, 148-9
fundamental magnitudes, 176, 183-4
fundamental units, 176, 182, 187
Gibbs' paradox, 126
Hamiltonian, 7, 131-4, 140, 154
Hartree-Fock method, 134-5
hybridisation, 60, 136-7, 139
hydrodynamics, 49, 161
hyle, vi, 17, 39, 160
hysteresis, 91-2, 164, 189
IDEAL PHYSICS, 200
interdiscourse relations, vi, viii, 50, 57, 74-5, 122, 125, 191-3
ions, 20, 25, 80, 109
isomers, 28, 63, 83, 86, 131, 146
isotopes, 6, 76, 81, 86, 97, 106-10, 118
Kelvin equation, 89, 93, 95
laminar flow, 181, 185
language of chemistry, 21, 37, 39-40
law of definite proportions, 5, 151-2, 161
law of multiple proportions, 5, 15, 151-2, 161
magnetic resonance, 36
manifest image, vi, 41-7, 50, 69, 75, 78, 81, 87, 101, 194, 198-202
mass transfer, viii, 122, 164, 171, 178-9, 186
mesomer, 28-9, 33
metamers, 83
metaphysical realism, 64, 113
microstructure, 75, 80, 99, 111-2, 115, 127, 148
model theory, 156
models, vi-viii, xii, xiv, 4, 17, 19-20, 26, 28, 31, 35-6, 49- 50, 72-4, 82-3, 87, 90, 96-8, 123, 128, 130, 132, 134, 136-41, 146, 149, 151, 153-9, 161-6, 168-9, 171-3, 175, 177, 181, 184-9, 193, 198, 203, 206, 210, 212, 214-5, 218-9, 221, 223-4, 229, 230, 233
molecular diffusion, 21, 122, 168, 175, 178
molecular structure, 4, 20, 26, 28, 36, 50, 57, 73, 76, 81, 83-4, 92, 96, 106, 108-9, 112-3, 115-8, 129, 132, 137, 141, 144-7, 149, 154, 199
Myth of the Given, 161-2

natural kinds, vi-vii, 42, 52, 58-69, 75, 80, 87, 99, 101, 104, 107, 117-8, 123, 154, 156, 191-2, 202
natural laws, vii, 61-2, 64, 84, 121, 159, 161, 202
naturalism, 29, 34, 41, 51, 60-1, 63, 69, 116
neutron scattering, 83
Newtonian fluids, 161, 181, 188
ontology, vi, 1-3, 16, 46, 63, 92, 143, 192
orbitals, 6, 21, 28, 33, 81, 96, 119, 131-6, 137, 145, 148
particle shape, 173
periodic table, 6, 35-6, 105, 119, 133, 152, 215
Π-theorem, 176-7, 179, 182-4, 187
phase, vii, 48-9, 61, 71, 76, 83-5, 87, 97, 123, 127, 152, 163, 173, 177
physical chemistry, vii-viii, 11, 21, 47-8, 93, 123, 126, 152, 154, 193
physicalism, 41, 56, 66, 196
porous media, viii, 154, 163-8, 188
possible worlds, vii, 54, 64, 100, 102, 107, 109-14, 118
primary quantities, 183-4
projectible predicates, 59, 61-3, 65-6, 191
protochemistry, 73, 86, 160-1, 199
pure substance, vi, 4, 71, 74, 77, 84-7, 110
quantum chemistry, vii, viii, 1, 7, 21, 26-7, 32-3, 39, 49, 119-20, 123, 128-33, 136, 140, 144-5, 147, 149, 195
quantum field theory, 120, 143, 157
quantum mechanics, v-viii, 3, 5, 7, 26-8, 31, 33, 47, 57, 75-6, 84, 87, 110, 119-20, 122, 123-4, 128-9, 131-3, 136-7, 139-49, 157-9, 193, 197, 199, 201
quarks, v, 35, 41, 54-5, 59, 63, 66, 81, 87, 89, 193
radicals, 25, 36, 109
realism, 20, 35-6, 50, 68, 87-9, 113, 124, 147, 159, 160, 191-2, 198
reduction, vi-viii, 16, 18, 26, 40, 50-3, 55-6, 69, 71, 107, 110, 119, 121-5, 127, 129, 131-2, 141, 145-50, 157, 162, 192-3, 196, 200
resonance structures, 6-7, 27-34, 83, 96, 117, 131-2, 136
rigid designators, 100-2, 105, 110, 118, 198
sameness relation, 78, 100, 116
scale-model, 171

scientific image, v-vi, viii-ix, 41-7, 50, 55, 69, 75, 78, 87, 99, 161, 194, 197-8, 199, 201-2
shape factor, 166, 173, 186-7
similarity, 2, 37, 58-61, 114, 156, 169, 171-2, 175, 176, 179, 182, 187
spectroscopic, 73, 83-4, 86, 89, 96-7 120, 136, 138, 147, 154
spin, 81, 122, 132-4, 159
standard meter, 104, 185
statistical mechanics, vii, 47, 53, 122-8, 162, 193
stereotype, 66, 76, 102-3, 110-1, 117
Stern-Gerlach experiment, 158-9
stoichiometry, 4, 36, 151
strict laws, vi, 112, 152-3, 155, 157, 159, 162, 191, 195-6, 198
stuff, vi, 17, 51, 71, 73-4, 77, 79, 100, 111-3, 117, 160, 169
substances, v, vi, vii, 1, 3, 5-7, 16-7, 19-22, 39-40, 51, 57, 63, 66, 71-4, 76, 78-87, 90, 92, 96-7, 99, 100-2, 104-6, 108-10, 113-4, 116-8, 123-4, 126-7, 137, 141, 146-7, 152-3, 160-1, 168, 172, 175, 177, 191, 193, 199
superposition, 28, 41, 76, 81, 131, 135, 143
supervenience, vi, ix, 26, 34, 50-1, 53-7, 65-66, 73, 82, 99, 123, 125, 143, 192-4, 200-1
surface tension, 49, 102, 165, 171, 187
symmetry breaking, 112, 122, 203
tautomers, 28, 83, 117
temperature, vii, 50, 53-4, 73, 75, 78-81, 83-7, 91, 93, 97, 101, 111-2, 114, 124-5, 127-8, 152-3, 171-2, 175, 182-3, 186, 193, 211
Theory of Everything, 200-1
thermodynamics, vi, 17, 21, 34, 39, 47, 49, 53, 85, 87, 120, 122-8, 147, 162, 168-9, 177, 193
transport phenomena, 163-4, 168, 177-8, 180-1
Twin Earth, vii, ix, 103, 105, 108, 110-7
unification, 2, 18, 26, 36, 47, 51, 60, 153, 193, 196, 198-201
unit operations, 124, 181
unity of science, 35, 50, 52
universal constants, 176, 186
universals, 60-1, 67
valence, v, 1-2, 5, 13-4, 21, 26, 28-30, 37, 71, 83, 132, 134, 140
wave equation, 57, 122, 128-31, 133-6, 139-41
X-ray diffraction, 83, 96